NEUROMETHODS

Series Editor
Wolfgang Walz
University of Saskatchewan
Saskatoon, SK, Canada

For further volumes:
http://www.springer.com/series/7657

The Cytoskeleton

Imaging, Isolation, and Interaction

Edited by

Rolf Dermietzel

Faculty of Medicine, Neuroanatomy and Molecular Brain Research, Ruhr University Bochum, Bochum, Germany

Humana Press

Editors
Rolf Dermietzel
Faculty of Medicine, Neuroanatomy
and Molecular Brain Research
Ruhr University Bochum
Bochum, Germany

ISSN 0893-2336 ISSN 1940-6045 (electronic)
ISBN 978-1-4939-5937-2 ISBN 978-1-62703-266-7 (eBook)
DOI 10.1007/978-1-62703-266-7
Springer New York Heidelberg Dordrecht London

Cover Illustration: This image appears, in full, as Figure 3 of Chapter 10.

Printed on acid-free paper

Humana Press is a brand of Springer
Springer is part of Springer Science+Business Media (www.springer.com)

Dedication

This book is dedicated to Eva for her patience and love.

Preface to the Series

Under the guidance of its founders Alan Boulton and Glen Baker, the Neuromethods series by Humana Press has been very successful since the first volume appeared in 1985. In about 17 years, 37 volumes have been published. In 2006, Springer Science+Business Media made a renewed commitment to this series. The new program will focus on methods that are either unique to the nervous system and excitable cells or which need special consideration to be applied to the neurosciences. The program will strike a balance between recent and exciting developments such as those concerning new animal models of disease, imaging, in vivo methods, and more established techniques. These include immunocytochemistry and electrophysiological technologies. New trainees in neurosciences still need a sound footing in these older methods in order to apply a critical approach to their results. The careful application of methods is probably the most important step in the process of scientific inquiry. In the past, new methodologies led the way in developing new disciplines in the biological and medical sciences. For example, Physiology emerged out of Anatomy in the 19th century by harnessing new methods based on the newly discovered phenomenon of electricity. Nowadays, the relationships between disciplines and methods are more complex. Methods are now widely shared between disciplines and research areas. New developments in electronic publishing also make it possible for scientists to download chapters or protocols selectively within a very short time of encountering them. This new approach has been taken into account in the design of individual volumes and chapters in this series.

Saskatoon, SK, Canada *Wolfgang Walz*

Preface

This volume of *Neuromethods* provides a collection of reviews and hands-on protocols on diverse aspects of the cytoskeleton. Conceptually, the book was designed as an interdisciplinary approach, assembling chapters on general aspects of the cytoskeleton in the form of reviews, which are thought to be helpful for the curious reader to polish his/her knowledge on the cytoskeleton in a brief and informative way. These reviews are flanked by protocols on diverse emerging techniques including in vivo and in vitro imaging of the cytoskeleton at work (Chaps. 1, 2, and 10). Since the book is aimed to present state-of-the art techniques, the editor had to acknowledge that the most favorable trend for unearthing the molecular machinery of the cytoskeleton today is the proteomic approach. It was therefore inevitable to include a general article on proteomic techniques (Chap. 5), which is supplemented by additional, more specific protocols addressing preparative strategies for the isolation of cytoskeletal components of the central nervous system (Chaps. 7, 11, and 15). Although the predominance of proteomic techniques in this collection of papers seems to appear a little biased from an editor's point of view, it pays credit to the fact that for future research, this arsenal of high-end molecular biological techniques will substantially foster our understanding of the cerebral cytoskeleton, the working horse in brain tissues. This notion becomes particularly important when degenerative diseases of the nervous system are considered. As such, it is not surprising that most of the chapters, in particular the extended review on the neuronal intermediate filaments (Chap. 9), contain a profound part on the pathology of the brain in which components of the cytoskeleton are involved. A limping horse, to stay in the frame of the picture, is not worth a penny. The editor therefore hopes that this collection of papers written by a team of experts may strengthen our efforts to overcome at least some of the fatal outcomes of the diseased brain cytoskeleton.

Bochum, Germany ***Rolf Dermietzel***

Contents

Part IV Astroglia and the Cytoskeleton

Part V Oligodendroglia and Myelin

Contributors

CANTAS ALEV • *Department of Neuroanatomy and Molecular Brain Research, Faculty of Medicine, Ruhr-University Bochum, Bochum, Germany*
NICOLE BARTSCH • *Max Planck Institute of Experimental Medicine, Göttingen, Germany*
ROLF DERMIETZEL • *Department of Neuroanatomy and Molecular Brain Research, Faculty of Medicine, Ruhr-University Bochum, Bochum, Germany*
ENO E. EBONG • *Department of Neuroscience, Albert Einstein College of Medicine, Bronx, NY, USA; Department of Biomedical Engineering, The City College of New York, New York, NY, USA*
HADAS EREZ • *Department of Neurobiology, Institute of Life Science, The Hebrew University of Jerusalem, Jerusalem, Israel*
JOEL EYER • *Laboratoire de Neurobiologie & Transgenese, UPRES-EA3143, Institut de Biologie en Sante—IRIS, CHU, Angers, France*
ILIANA GOZES • *The Elton Laboratory for Neuroendocrinology, Department of Human Molecular Genetics and Biochemistry, Sackler Faculty of Medicine, The Adams Super Center for Brain Studies, The Lily and Avraham Gildor Chair for the Investigation of Growth Factors, Sagol School of Neuroscience, Tel Aviv University, Tel Aviv, Israel*
ELENA E. GRINTSEVICH • *Department of Chemistry & Biochemistry, University of California Los Angeles, Los Angeles, CA, USA*
DUMITRU ANDREI IACOBAS • *D.P. Purpura Department of Neuroscience, Albert Einstein College of Medicine, Bronx, NY, USA*
SANDA IACOBAS • *D.P. Purpura Department of Neuroscience, Albert Einstein College of Medicine, Bronx, NY, USA*
OLAF JAHN • *Max Planck Institute of Experimental Medicine, Göttingen, Germany*
PATRICIA KLEMMER • *Leibniz Institute for Neurobiology, Magdeburg, Germany*
CHRISTINA LOOSSE • *Department of Functional Proteomics, Medizinisches Proteom-Center, Ruhr-Universitaet Bochum, Bochum, Germany*
KATRIN MARCUS • *Department of Functional Proteomics, Medizinisches Proteom-Center, Ruhr-Universitaet Bochum, Bochum, Germany*
KARL MELLER • *Department of Cytology, Ruhr University of Bochum, Bochum, Germany*
JOHN W. MURRAY • *Department of Anatomy and Structural Biology, Marion Bessin Liver Research Center, Albert Einstein College of Medicine, Bronx, NY, USA; Division of Gastroenterology and Liver Disease, Albert Einstein College of Medicine, Bronx, NY, USA*
GENEVIEVE NEAL-PERRY • *D.P. Purpura Department of Neuroscience, Albert Einstein College of Medicine, Bronx, NY, USA*
STEFAN OLK • *Department of Neuroanatomy and Molecular Brain Research, Ruhr University of Bochum, Bochum, Germany*
SAAR OZ • *The Elton Laboratory for Neuroendocrinology, Department of Human Molecular Genetics and Biochemistry, Sackler Faculty of Medicine, The Adams Super Center for Brain Studies, The Lily and Avraham Gildor Chair for the Investigation of Growth Factors, Sagol School of Neuroscience, Tel Aviv University, Tel Aviv, Israel*

YOLANDA DE PABLO • *Center for Brain Repair and Rehabilitation, Department of Clinical Neuroscience and Rehabilitation, Institute of Neuroscience and Physiology, Sahlgrenska Academy at the University of Gothenburg, Gothenburg, Sweden*
JULIA PATZIG • *Max Planck Institute of Experimental Medicine, Göttingen, Germany*
MARCELA PEKNA • *Center for Brain Repair and Rehabilitation, Department of Clinical Neuroscience and Rehabilitation, Institute of Neuroscience and Physiology, Sahlgrenska Academy at the University of Gothenburg, Gothenburg, Sweden*
MILOS PEKNY • *Center for Brain Repair and Rehabilitation, Department of Clinical Neuroscience and Rehabilitation, Institute of Neuroscience and Physiology, Sahlgrenska Academy at the University of Gothenburg, Gothenburg, Sweden*
RODOLPHE PERROT • *Service Commun d'Imageries et d'Analyses Microscopiques, Université d'Angers, Institut de Biologie en Sante—IRIS, CHU, Angers, France*
MICHELLE PORRITT • *Center for Brain Repair and Rehabilitation, Department of Clinical Neuroscience and Rehabilitation, Institute of Neuroscience and Physiology, Sahlgrenska Academy at the University of Gothenburg, Gothenburg, Sweden*
EMIL REISLER • *Department of Chemistry & Biochemistry, University of California Los Angeles, Los Angeles, CA, USA*
CHRISTIANE RICHTER-LANDSBERG • *Department of Biology, Molecular Neurobiology, University of Oldenburg, Oldenburg, Germany*
BODO SCHOENEBECK • *Department of Neuroanatomy and Molecular Brain Research, Institute of Anatomy, Ruhr-Universitaet Bochum, Bochum, Germany*
KARL-HEINZ SMALLA • *Leibniz Institute for Neurobiology, Magdeburg, Germany*
MICHA E. SPIRA • *Department of Neurobiology, Institute of Life Science, The Hebrew University of Jerusalem, Jerusalem, Israel*
DAVID C. SPRAY • *Department of Neuroscience, Albert Einstein College of Medicine, Bronx, NY, USA*
SYLVIA O. SUADICANI • *Department of Neuroscience, Albert Einstein College of Medicine, Bronx, NY, USA; Department of Urology, Albert Einstein College of Medicine, Bronx, NY, USA*
STEFAN TENZER • *University Medical Center, Johannes Gutenberg University Mainz, Mainz, Germany*
CARSTEN THEISS • *Department of Cytology, Ruhr University of Bochum, Bochum, Germany*
MIA M. THI • *Department of Orthopedic Surgery, Albert Einstein College of Medicine and Montefiore Medical Center, Bronx, NY, USA; Department of Neuroscience, Albert Einstein College of Medicine, Bronx, NY, USA*
ANDREY TUCHINOWITZ • *Department of Neuroanatomy and Molecular Brain Research, Ruhr University of Bochum, Bochum, Germany*
HAUKE B. WERNER • *Max Planck Institute of Experimental Medicine, Göttingen, Germany*
ULRIKA WILHELMSSON • *Center for Brain Repair and Rehabilitation, Department of Clinical Neuroscience and Rehabilitation, Institute of Neuroscience and Physiology, Sahlgrenska Academy at the University of Gothenburg, Gothenburg, Sweden*
URSULA WYNEKEN • *Universidad de los Andes, San Carlos de Apoquindo, Santiago, Chile*
GEORG ZOIDL • *Department of Neuroanatomy and Molecular Brain Research, Faculty of Medicine, Ruhr-University Bochum, Bochum, Germany; Department of Psychology, York University, Toronto, ON, Canada*

Part I

Imaging

Chapter 1

Fluorescence Proteins and Time-Lapse Imaging of the Cytoskeleton

Carsten Theiss and Karl Meller

Abstract

Visualization of cytoskeletal dynamics in real time is of paramount interest in cell biological research. With the aid of fluorescent-cytoskeletal fusion proteins and enhancement of confocal laser scanning microscopes with high-end objectives and cell-incubation chambers, high-resolution time-lapse imaging is nowadays possible for long time periods. However, most of the cytoskeletal dynamics can be detected during short observation periods.

In this chapter, we provide a detailed description for time-lapse imaging of microtubules, neurofilaments, and microfilaments within primary neurons. We use two primary neuronal cell culture systems for the analysis of different aspects of cytoskeletal motion and organization: (1) dissociated dorsal root ganglia, which are highly practical to study cytoskeletal dynamics along their neurites or within the growth cone, and (2) cerebellar slice cultures, which are characterized by their organotypic morphology even after 30 days in vitro. In particular in these slice cultures Purkinje cells exhibit highly dynamic dendritic spines within a functional neuronal network.

Key words: Actin, Cerebellar slice cultures, Dorsal root ganglia, Growth cone, Microinjection, Microtubules, Neurofilaments, Purkinje cells, Spines, Time-lapse imaging, Confocal laser scanning microscopy

1. Introduction

Live-cell imaging of the cytoskeletal network is fundamental to understand its role in several cellular processes. Based on the rapid technological advances in live-cell imaging, our knowledge about molecular and functional cytoskeletal organization and dynamics has been significantly increased during the last years.

The cytoskeleton consisting of microfilaments, intermediate filaments, and microtubules is involved in many critical functions in a cell's life span. To generally outline: the cytoskeleton is involved

Rolf Dermietzel (ed.), *The Cytoskeleton: Imaging, Isolation, and Interaction*, Neuromethods, vol. 79,
DOI 10.1007/978-1-62703-266-7_1, © Springer Science+Business Media, LLC 2013

in the assembly of the mitotic spindle during cell division and forms the cell architecture for cell migration and for intracellular transport mechanisms. It is therefore an essential requirement for the cytoskeleton and its constituting parts to be highly dynamic. This dynamic is achieved through a balanced polymerization and depolymerization in cooperation with accessory anterograde and retrograde transports of cytoskeletal subunits. For instance, microtubules remain attached to the microtubule-organizing center by their less dynamic minus-end, while the dynamic plus-end is unstable and grows outwards to the cell membrane through polymerization of tubulin alpha/beta dimers. This growth and shortening of microtubules at the free ends is called dynamic instability. Another subtype of the highly dynamic cytoskeleton are microfilaments, built of actin, which are involved in cargo transport as well as in cell motility. In fibroblasts and lens epithelial cells microfilaments are directly implicated in changes of cell morphology (1), whereas in neurons short transport of proteins is performed along actin filaments (2, 3). In glial cells microfilaments are distributed along the cell membrane and here they have a direct impact on gap junctional organization (4, 5 and Chapter 1 this volume). At this juncture the performance of microfilaments within different cell types is probably correlated to their function. Bundles of actin filaments called stress fibers are acto-myosin structures found in many types of non-muscle cells, where they are involved in adhesion, motility, and morphogenesis (6). In contrast actin filaments are organized in form of dynamic short filaments in neurons, for example in the axonal growth cone (7) as well as in dendritic spines (8), where these proteins are highly dynamic (9). The third class of cytoskeletal proteins are the intermediate filaments see Chapter 9 this volume, which also show a tissue-specific expression. These are solid and ropy fibers with several micrometers length (10) that seem to have a direct and indirect effect on transport mechanisms in nervous cells (11, 12), glial cells (13), and lens epithelial cells (5, 14).

As the cytoskeleton is involved in numerous dynamic tasks, it is highly attractive for various aspects of biological and biomedical research. Currently different experimental approaches using fluorescent microscopy are used to visualize and investigate the organization and dynamics of the cytoskeleton. Where as the organization can be studied with specific antibodies and indirect immunofluorescence staining in fixed cells, direct analysis of cytoskeletal dynamics is possible by microinjection of fluorescently labeled tubulin subunits (e.g., rhodamine tubulin) into the cytoplasm of living cells. With this method the first data of microtubules motility were obtained (15, 16). Even though this technique is still in use, today visualization of dynamic processes of the cytoskeleton is mostly done by fluorescent-cytoskeletal fusion proteins. Since the introduction of the original green fluorescent protein (GFP) (17, 18) and its almost unlimited possibility to be fused to other proteins, fluorescent protein (FP) constructs can be exploited as a reporter of protein expression and distribution to study intracellular trafficking in living cells, tissues, and organisms.

In the last years over 40,730 reports have been published using FP techniques, including approximately 8704 investigations dealing with cytoskeleton research (based on a PubMed research). The advantage of this method is that proteins can be tagged with FP with little or no disturbance of their normal distribution or function (19). To broaden the application spectrum many new variants of GFP have been engineered based on different mutations, including color mutants (blue, cyan, green, yellow, orange, red), ph-sensitive mutants, and many more.

Studying dynamic mechanisms of the cytoskeletal organization or intracellular transport along the cytoskeleton needs some sophisticated techniques like microinjection and live-recording. Moreover the choice of cell lines or primary cells within an organotypic environment is an important criterion for studying cytoskeletal dynamics. Therefore the objective of this chapter is to give an overview of different techniques to observe and analyze different cytoskeletal dynamics in dissociated as well as organotypic tissues focusing on the nervous system.

Neurons typically are composed of a perikaryon, multiple branched dendrites, and one axon, with different functional properties served by these cellular substructures. In primary cell cultures neurons develop axons and dendrites that are comparably organized and structured like in vivo. Therefore they are highly suitable to study different aspects of cytoskeletal rearrangement within subdomains like the axonal growth (see Chapter 10 this volume) cone or dendritic spines under various experimental conditions.

2. Methods

2.1. Dorsal Root Ganglion Cultures

The cell culture model of trypsinated dorsal root ganglia (DRG) neurons has been previously reported (20). Typically 30 chicken eggs are incubated for 10 days; thereafter approximately 250 DRGs are collected and trimmed of residual connective tissue. Before dissociation procedures start prepare a 5× Gibco A stocking solution (0.4 g/l KCl; 2.2 g/l $NaHCO_3$; 6.8 g/l NaCl; 1.0 g/l glucose) and the 30 ml dissociation solution (6 ml trypsin (0.25% trypsin, Gibco 25050, USA); 4.8 ml 5× Gibco stocking solution; 19.2 ml A dest.).

2.1.1. Dissociation Procedures

1. Add 5 ml dissociation solution to the isolated DRGs in a 10 ml beaker. After 5 min of agitation, the supernatant containing the dissociated cells is transferred into a centrifuge tube containing 20 ml cold minimal essential medium (MEM, M2279,

Sigma, Germany), supplemented with 10% horse serum (HS), 1% L-glutamine, and 1% penicillin (A321-42, Biochrom AG, Germany).

2. Following the addition of fresh trypsin solution (37°C), the DRG material is gently stirred for another 5 min. This procedure of stirring and removing the supernatant is repeated four times.
3. The suspension with the collected cells is centrifuged for 10 min at 1,200 × *g* to obtain a pellet, which is resuspended in 4 ml fresh nutrient medium, consisting of MEM, supplemented with 10% HS, 6% glucose, 1% chicken embryo extract, 1% L-glutamine, 1% nerve growth factor-7 S (N0513, Sigma), and 1% penicillin.

2.1.2. Culture Preparation

Aliquots of 350 µl (~1×10^5 cells/ml) of the cell suspension are cultivated on custom-made glass coverslips (Ø 32 mm, Kindler, Germany) covered with rat-tail collagen (C7661, Sigma) and placed in culture dishes (60 × 15 mm, No. 353004 BD Falcon) in a CO_2 incubator (5% CO_2, 37°C, 90% humidity). After 1 day cells are attached to the coverslips, clearly visible by short prolongations growing from the rounded neurons. At this date coverslips are carefully covered with 2 ml medium and can be incubated for up to 3 weeks. Medium has to be changed twice a week.

2.1.3. Cerebellar Slice Cultures

Slice cultures of rat cerebellum are obtained from 10-day-old animals (P10) according to the roller-tube technique (21, 22). This organotypic cell culture system is ideal to study spine dynamics of Purkinje cells within a functional neuronal network.

1. Decapitate the head of the pup, and collect the cerebella in ice-cold HANKS.
2. Brains must be trimmed of residual connective tissue by means of binocular inspection. Thereafter the cerebellum has to be detached and collected in fresh ice-cold HANKS. Do not bruise the tissue.
3. Cerebella of neonatal rats are cut sagittally into approximately 250 µm thick slices with aid of a McIlwain tissue chopper (Mickle Laboratory Engineering, Gomshall, Surrey, UK) and collected in ice-cold Gey-glucose (0.37 g/l KCl; 0.22 g/l $CaCl_2 \cdot 2H_2O$; 0.21 g/l $MgCl_2 \cdot 6H_2O$; 0.07 g/l $MgSO_4 \cdot 7H_2O$; 7.0 g/l NaCl; 0.03 g/l KH_2PO_4; 0.12 g/l Na_2HPO_4; 0.03 g/l $NaH_2PO_4 \cdot H_2O$; 1.0 g/l glucose).
4. Under visual control with the binocular microscope approximately 30 cerebellar slices are elected without any damage of the pial surface and attached to collagen-faced cover glasses (12 × 24 mm, Kindler, Germany) with a plasma clot (10 µl; P3266, Sigma, Germany) coagulated with thrombin (60 U/10 µl; 605157, Calbiochem, La Jolla, USA).

5. Incubation is done in 1.3 ml medium in closed tubes (Nunclon TM Surface, Nunc, Denmark) at 37°C in a roller-drum incubator (TC-7, New Brunswick Scientific, USA). The nutrient "Basal Medium Eagle" (BME, B1522, Sigma) is supplemented with 25% Gey's-salts, 25% horse serum (S9135, Biochrom, Germany), 1% glutamine (G7513, Germany), 6.5 mg/ml glucose, and 25 ng/ml nerve growth factor (N0513, Sigma) which has to be renewed twice weekly.
6. During cultivation slice cultures flatten to approximately 100 μm in thickness, but the characteristic features, a three-layered organization, of the cytoarchitecture of the cerebellum is still retained. In P10 cerebellar slice cultures dendritic arborization arises from a stem-dendrite at the apical pole of the cell body with extensive branching in the molecular layer. Along the dendritic tree numerous spiny branches with electrical and chemical synapses are detectable.
7. This organotypic tissue can be cultured for up to 6 weeks; however, the proliferation of glial cell and fibroblasts can raise problems for microinjection later than day 25 in vitro. To study cytoskeletal dynamics, time period between days 12 and 20 in vitro is most practical.

2.2. Pressure Microinjection

2.2.1. Prearrangement

The microinjection is achieved under visual control on an inverted microscope with aid of long-distance phase-contrast optics (10× and 40×) and a pressure injection device. For optimal results the microscope should be equipped with a heating stage and a vibration-free table.

For the injection of DRG neurons we use sterile glass capillaries (0.2–0.5 μm; Femtotips, Eppendorf), whereas microinjection of Purkinje cells in roller-tube slice cultures is performed with capillaries (Hilgenberg, Nr. 1403512, Germany; borosilicate glass: 1.5 mm/0.2 mm), pulled with the aid of an injection puller (Sutter Instruments P97, USA) with a diameter of 0.5–0.6 mm according to Graessmann and Graessmann (1986). Several pulled micro-needles can be stored in petri dishes, with stripes of modelling clay on the bottom, to evade breakage. To avoid dust petri dishes are packed with parafilm until use.

2.2.2. Microinjection Procedure

Right before microinjection starts the vector solution (pEGFP-NF-M, 0.2 μg/μl in A dest. (12); RFP-actin, pLifeAct-TagRFP 0.2 μg/μl in A dest. (LifeAct, Ibidi, Germany); pEYFP-actin 0.3 μg/μl in A dest. (Clontech, Germany); pYFP-tubulin 0.4 μg/μl in A dest. (Invitrogen, USA)) is centrifuged at 5,000 rpm for 10 min at room temperature (Eppendorf mini spin, Germany). Capillaries are filled with 2 μl of the supernatant using microloaders (Eppendorf, No. 930001007) immediately before being used. In our lab we inject a variety of GFP-tagged cytoskeletal single vectors or in various combinations with varying concentrations of RNA depending on the probe and the cell system being used. RNA concentration in the range of 0.1–1.0 mg/ml is appropriate for most applications; however the lowest concentration that shows sufficient expression levels should be used.

We use the InjectMan N12 microinjector (Eppendorf) and FemtoJet (Eppendorf), though other microinjection devices should work as well. The pressure microinjector maintains a constant pressure of 5–10 hPa in DRG and of 80–100 hPa in Purkinje cells on the tip, to avoid clogging at the tip of the needle. Normally the injection of neuronal cells can be performed by increasing the pressure up to 15–20 hPa (DRG) and 100–120 hPa (Purkinje cells) for about 0.1–0.6 s. The success of injection is visible as a brief swelling of the neuronal soma. Injection periods longer than 30 min should be avoided. Usually 50–70 neurons in a single cell culture dish could be injected in a single period. Subsequent to the injection procedure, cell cultures have to be rinsed thoroughly in fresh medium before postincubation for 1 up to 7 days in the incubator. These intervals are sufficient to demonstrate a bright expression of GFP-tagged cytoskeletal proteins over several hundred micrometers of length in neurites of DRG or axons and dendrites in Purkinje cells.

2.3. Live-Cell Imaging and Image Analysis

The microscope system including the excitation lamp, the detection camera, as well as a heated microscope stage and a vibration-free table is essential for the quality and the reproducibility of live-cell imaging.

2.3.1. Microscopical Setup

For imaging cytoskeletal motions we use a confocal laser scanning microscope (Zeiss LSM 510, Germany) equipped with a laser module containing a diode laser (405 nm), an Ar laser (458, 477, 488, 514 nm), a HeNe laser (543 nm), and an additional HeNe laser (633 nm). To maintain the incubation settings at 37°C and 5% CO_2 on the microscope stage, a CTI controller 3700 digital, O_2 controller, Tempcontrol 37-2 digital, and the Incubator S_{oxygen} together with the heating insert P (Zeiss) are used. Additionally the immersion oil objective is heated with a tempcontrol mini system (Zeiss).

2.3.2. Live-Cell Imaging

Typically, live-cell imaging of cytoskeletal protein dynamics in primary neurons starts 24 h after transfection. Therefore coverslips are placed in a Rose chamber and covered with 2 ml phenol red-free nutrient medium to get the best results in terms of brightness, contrast, acuity, and resolution. DRG neurons (Fig. 1) or cerebellar slice cultures (Figs. 2 and 3) maintained in Rose chambers retain active motile growth cones or dendritic spines and apparently normal organelle movements for many hours. To discover the whole thickness of the region of interest optical sections (z-stacks) are typically collected in several steps at each time point and are reconstructed with an appropriate software.

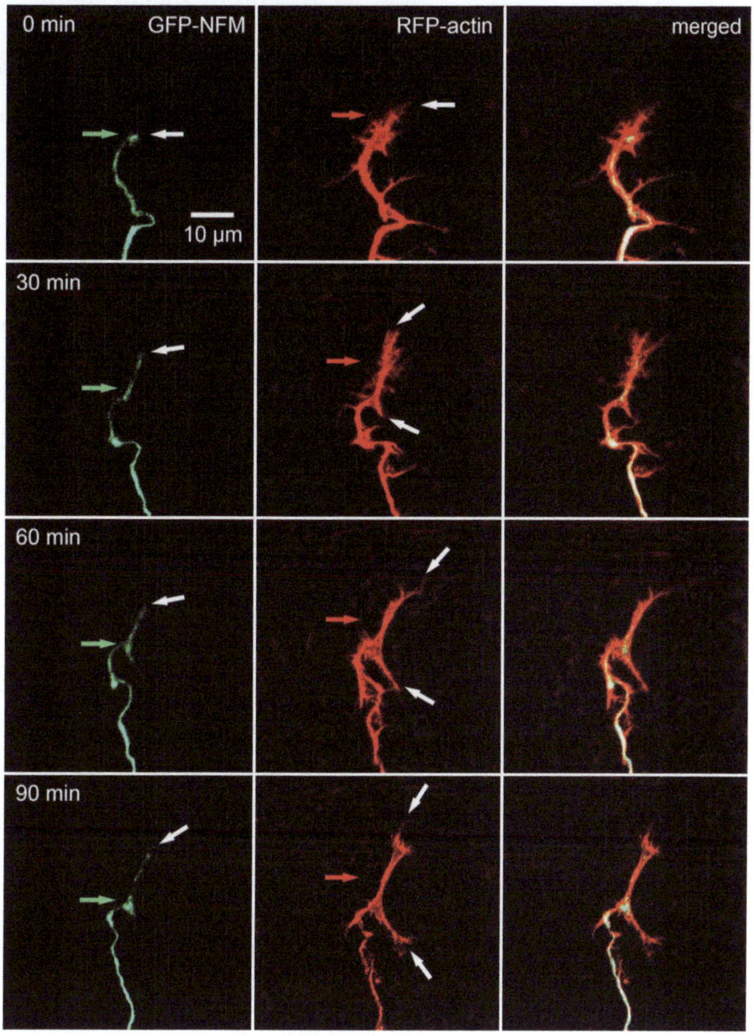

Fig. 1. Turnover of actin and neurofilaments in the neuronal growth cone. In trypsinated dorsal root ganglia microinjection was performed using a mixture of GFP-neurofilament and RFP-actin after 1 day in vitro. After 1 day time-lapse imaging of the growth cone is done by confocal laser scanning microscopy. Time-lapse intervals are set at 60 s, having a total imaging time of approximately 2 h. Apparently neurofilaments are located centrally and seem to be almost stationary; in contrast actin is found in the periphery and highly variable in its distribution, being the equivalent of the growth cone's motility. As the *arrows* demonstrate, the red stained actin continuously moves forward, leading to an outspread and increase of the growth cone's extent, while the neurofilament follows slowly, apparently stabilizing the new alignments (Laura Olbrich, Institut für Anatomie, Ruhr-Universität Bochum).

To reduce phototoxicity of the cells and photobleaching of the FP-signal the excitation intensity should be reduced to a minimum and the number of z-slices and time-intervals should be as low as possible. Additionally, the parameters for the scanning time, the

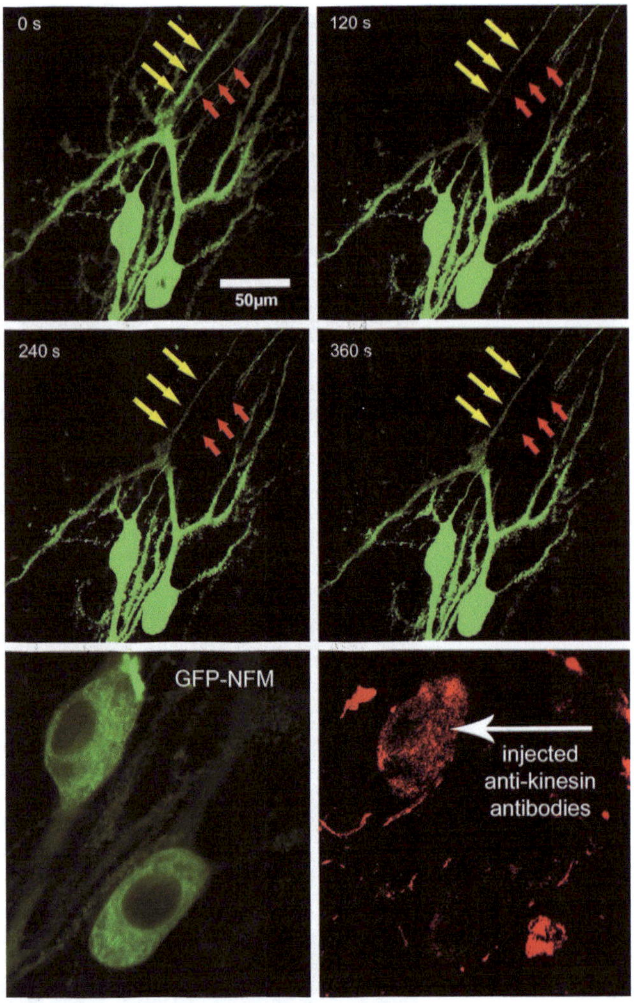

Fig. 2. Axonal transport of neurofilaments is blocked by anti-kinesin microinjection in Purkinje cells. In Purkinjce cells the axonal transport of GFP-NFM is impaired by microinjected anti-kinesin antibodies (*red arrows*), whereas dendritic transport of GFP-NFM in a control Purkinje cell is not affected (*yellow arrows*). Time-lapse images of GFP-NF-M transport are recorded at 30-s intervals subsequent to a 30-min postincubation after microinjection of anti-kinesin antibodies. In the montage of representative pictures particular points of time of the GFP-NFM transport study are shown (pre-bleaching, bleaching, and recovery). In higher magnifications lower two pictures the filamentous distribution of GFP-tagged NF and kinesin-antibodies in the perikaryon is obvious.

detector gain, and amplifier offset of the detection system should be increased to levels obtaining bright FP-signal even with low excitation power.

2.3.3. Image Analysis

To analyze data of cytoskeletal dynamics photobleaching techniques can be combined with time-lapse imaging. Therefore the FP can be photo-bleached using the Zeiss LSM510 in combination with Zeiss 40× (Plan-Neofluar, NA 1.30) or Zeiss 63× (Plan-Neofluar, NA 1.40) oil immersion lenses (Figs. 2 and 3). To determine

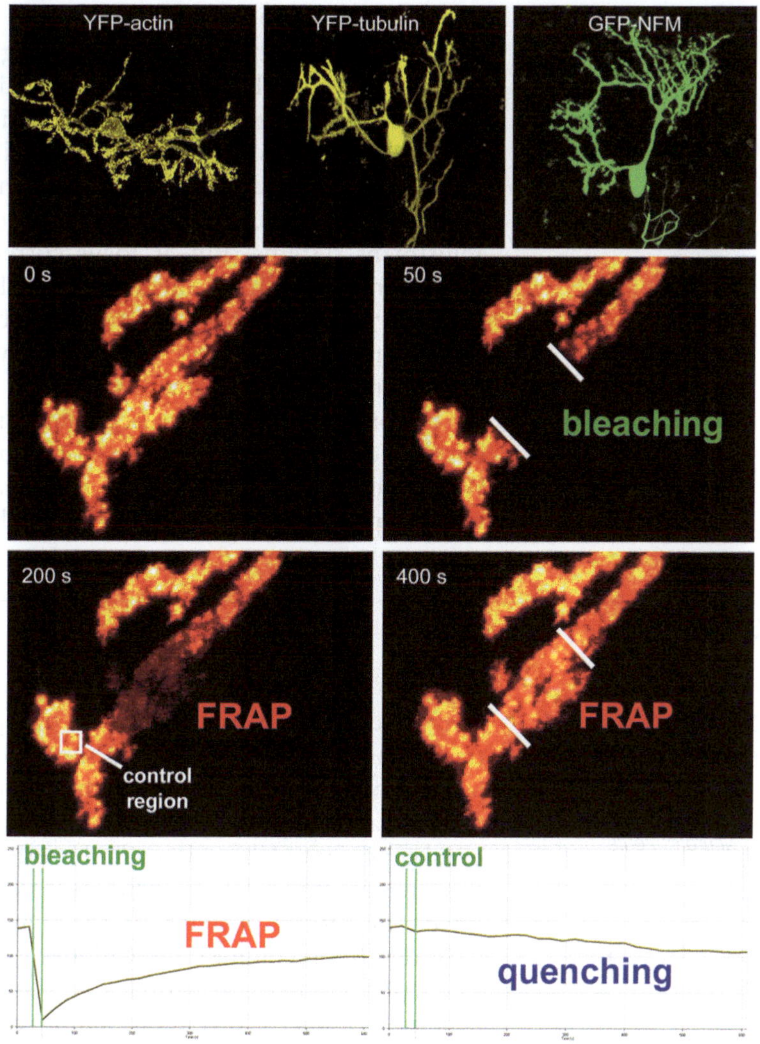

Fig. 3. Cytoskeletal distribution and actin motility in Purkinje cells PC of organ-specific slice cultures. The cytoskeletal distribution of PC can be discovered by microinjection of YFP-actin, YFP-tubulin, and GFP-neurofilament in single PC. After a postincubation period of 24 h, the transfected PC can be analyzed by the CLSM. The real-time analysis revealed a high density of YFP-labeled tubulin and GFP-labeled neurofilament in the Purkinje cell soma and the dendritic tree, while the YFP-actin signal accumulates especially in the dendritic spines. The actin turnover in dendrites and dendritic spines can be investigated with the aid of time-lapse imaging in combination with fluorescence recovery after photobleaching (FRAP) in YFP-actin transfected Pukinje cells 24 h after microinjection. The local fluorescence recovery is quantified with the aid of the Zeiss physiology kit (Lisa Wessel, Institut für Anatomie, Ruhr-Universität Bochum).

qualitative fluorescence recovery after photobleaching (FRAP), we first generate a prebleach image with a 40×/63× objective (zoom level 1, 1024_1024 pixels) with a slight laser power (e.g., 0.2–5%). Thereafter, the laser is zoomed to a region of interest, e.g., on one cell prolongation or dendritic spine, and this area is scanned approximately 100 times with a laser power of 100%, until fluorescence is completely removed through the entire depth of this region of interest. The laser is then zoomed out again, and its power is reduced to the initial conditions, before time-lapse imaging starts.

One of the major difficulties during time-lapse imaging is that the region of interest has to be within the focal plane for hours. In the last years several manufacturers offered automatic focus adjustment systems. But as focus correction is usually due to thermal instability in the environment of the microscope system, in particular by the thermal emission of the lasers, we aim to set a constant room temperature of 27°C. However, we set the first and the last image of the z-stack far outside from the focal plane of the region of interest at the beginning of the experiment so that up- and downturns within one live-recording are still in focus. When following these advices in vivo analysis of cytoskeletal "at work" is possible over a period of several hours.

3. Notes

3.1. Specifications of Cytoskeletal Dynamics

3.1.1. Axonal Transport of Neurofilaments

Over a period of 30 years neuronal transport along microtubules tracks was divided into fast and slow transport. In detail, it was well established that transport of membrane-bound vesicles occurs at a fast rate (23–26). The anterograde transport was known to be usually promoted by microtubule-activated motor proteins of the kinesin family (27–29), whereas retrograde transport is mostly driven by motor proteins of the dynein family (30, 31 and Chapter 10 this volume). But with the aid of time-lapse imaging of FP-tagged cytoskeletal proteins it was possible to specify the definition of neuronal transports more accurately. Transfection of neuronal cells with neurofilament-GFP demonstrated that the slow rate of anterograde axonal transport is the result of rapid movements interrupted by prolonged pauses (32, 33). Additionally, these studies showed that neurofilament transport occurs in both directions, anterograde and retrograde, but with an anterograde bias. In our laboratory we could demonstrate that neurofilaments are transported along the microtubules network by the use of the motor proteins kinesin in anterograde direction and dynein in retrograde direction (12). The form in which neurofilaments undergo transport (subunits or oligomers) is still controversially discussed; however, on the light-microscopic level the dimension of a fluorescent dot is likely to represent filamentous or oligomeric neurofilaments, and varies during axonal initiation, elongation, and maturation (34). Simple diffusion of FP can be excluded as disruption of the microtubules by nocodazole treatment in neuroblastoma cells (35) or colchicine incubation of chicken DRG (12) completely blocks neurofilament transport. Besides this, other intermediate filaments like vimentin and keratin are also transported as nonfilamentous punctuate structures during cell remodeling, probably representing oligomers (36–38). In our primary chicken DRGs, dots of neurofilament-FP moved through a photobleached region. It is well established that photobleaching does not impair neurofilament movement (39). Therefore FRAP can be used as an instrumental tool to observe translocation of FP in closely fluorescent regions of interest. Studying neurofilament organization and transport is of paramount interest, as a variety of neuropathological disorders are characterized by accumulation of neurofilament proteins (40–42 and Chapter 9 this volume). Unfortunately, the mechanisms that lead to neurofilament bundling are not known to date. There is evidence that aluminum exposure leads to bundling of neurofilaments in neurons (11, 43, 44). Additionally phosphorylation of neurofilament subunits seems to impair their transport into the neurites (11, 45–47). Indeed it was demonstrated that phosphorylation of neurofilaments slowed their transport by increased pausing in neurofilament movement (48). Nonetheless in some neurodegenerative diseases abnormal phosphorylation of neurofilaments can promote their detachment from the motor proteins and as a consequence results in irregular transport properties (49).

3.1.2. Cytoskeletal Dynamics Within the Axonal Growth Cone

Ramon y Cajal was the first author to describe the neuronal growth cone as the leading edge of an axon (50). Nowadays numerous studies have classified the growth cone to be important in axonal pathfinding during neurite outgrowth (51). At this juncture a number of guidance cues navigate the growth cone and directly modulate the assembly of cytoskeletal proteins to mediate growth cone dynamics (ref. 52 and Chapter 10 this volume).

Within the growth cone different domains can be distinguished based on the morphology and distribution of cytoskeletal proteins. Whereas the central region consists of microtubules and neurofilaments, the surrounding peripheral region is characterized by receptive fingerlike filopodia and flat veil-like lamellipodia that both show a dense actin network with isolated microtubules (53, 54). At the junction between the central and the peripheral zone is the transition zone. Microtubule–actin filament interactions occur in the transition zone and central domain of the growth cone, where actin exerts compressive forces to corral microtubules into the core of the growth cone and thereby facilitate microtubule bundling (55, 56).

This morphology of three growth cone domains with distinct distribution of cytoskeletal elements forms the basis of its motility and allows the growth cones to extend, retract, and collapse in order to lead the axon towards a determined direction (51, 52). In particular actin filaments are directly involved in growth cone dynamics by actin-treadmilling and actin retrograde flow (57). Actin-treadmilling means actin polymerization at the leading edge in the peripheral zone and concurrent actin severing in the transition zone with transport of the recycled subunits to the leading edge. In contrast retrograde flow of actin goes from the leading edge to the transition zone. But also microtubules in growth cones exhibit dynamic instability, alternating between relatively slow growth from their plus-end and rapid plus-end disassembly, which might be followed by recovery of plus-end assembly (56 for detailed description see Chapter 10 this volume).

In the last 20 years numerous receptors were identified at the growth cone surface, which integrate stimulating or repelling guidance cues as well as neurotrophic factors to the cytoskeleton via multiple signaling pathways. In this context we recently demonstrated an unknown effect of the vascular endothelial growth factor (VEGF) in neurons. VEGF directly stimulates the growth cone by increasing the growth cone size and additionally act as an attractant for the growth cone in cultured DRG (9). This effect is mediated by actin reorganization downstream of VEGF-receptor 2, through phosphorylation of tyr1214 (9, 58). Although the pathway of VEGFR-2 in the growth cone downstream of Cdc42 leading to actin dynamics and cell motility is not yet fully understood, there is evidence that several messenger molecules and actin-binding proteins like heat-shock protein27 (Hsp27), WASP, Arp2/3, or ADF/cofilin are involved (59–66).

3.1.3. Cytoskeletal Dynamics in Dendrites and Spines

To study cytoskeletal dynamics in dendrites and spines we use organotypic cerebellar slice cultures. There are several advantages of these slice cultures, which flatten to approximately 100 μm thickness without losing the characteristic feature of a three-layered cerebellar cortex (67). In particular, the dendritic arborization in cerebellar slice cultures arises from a stem-dendrite at the apical pole of the cell body with extensive branching in the molecular layer. Along the dendrites numerous prominent spines with functional chemical and electrical synapses are obvious, clearly indicating the organotypic morphology (67, 68). It is well described that they serve as sites of synaptic input and constitute postsynaptic signaling micro-compartments (69).

The cytoskeletal backbone of dendrites is built of microtubules, neurofilaments, and microfilaments, whereas dendritic spines of Purkinje cells mainly contain actin filaments (70, 71). Recently, it was shown that in hippocampal neurons also microtubules are able to transiently enter spines (72, 73). Typical dendritic spines consist of a bulbous head and a narrow neck; however a variety of shapes and sizes is described, which is likely to reflect distinct roles in Purkinje cell function (74). As summarized by Hayashi and coworkers (75), it is quite established that actin and actin-binding proteins regulate dendritic spine morphology (76–78). Besides this, the transient invasion of microtubules into dendritic spines seems to have an impact in mature spine morphology (72, 73), as inhibition of microtubule dynamics by a microtubule-disrupting agent or

EB3 knockdown (KD) has been shown to transform dendritic spines into filopodia-like protrusions (73). But besides this there are numerous open questions as to how extrinsic and intrinsic signals affect cytoskeletal dynamics that in consequence regulate the dendritic morphology including spine appearance during development and in the adult. Time-lapse imaging of FP-tagged cytoskeletal proteins in organotypic slice cultures can serve as a very handy tool to advance our understanding of the cytoskeletal dynamics.

References

1. Lascola CD, Nelson DJ, Kraig RP (1998) Cytoskeletal actin gates an C1-channel in neocortical astrocytes. J Neurosci 18:1679–1692
2. Kuznetsov SA, Langford GM, Weiss DG (1992) Actin-dependent organelle movement in squid axoplasm. Nature 356:722–725
3. Meller K (1992) Axoplasmic transport of horseradish peroxidase in single neurons of the dorsal root ganglion studied in vitro by microinjection. Cell Tissue Res 270:139–148
4. Theiss C, Meller K (2002) Microinjected anti-actin antibodies impaired cell-communication in astrocytes in vitro. Exp Cell Res 281: 197–204
5. Giessmann D, Theiss C, Breipohl W et al (2003) Microinjection of anti-actin antibodies impaired gap junctional intercellular communication in lens epithelial cells in vitro. Curr Eye Res 27(3):157–164
6. Naumanen P, Lappalainen P, Hotulainen P (2008) Mechanisms of actin stress fibre assembly. J Microsc 231(3):446–454
7. Gallo G, Letourneau PC (2004) Regulation of growth cone actin filaments by guidance cues. J Neurobiol 58:92–102
8. Matus A (2000) Actin-based plasticity in dendritic spines. Science 290:754–758
9. Foehring D, Brand-Saberi B, Theiss C (2012) VEGF induced growth cone enhancement is diminished by inhibiting tyrosine-residue 1214 of VEGFR-2. Cells Tissues Organs. 196(3):195–205
10. Herrmann H, Aebi U (2004) Intermediate filaments: molecular structure, assembly mechanism, and integration into functionally distinct intracellular Scaffolds. Annu Rev Biochem 73:749–789
11. Theiss C, Meller K (2001) Inhibitory effect of aluminum on the axonal transport of microinjected HRP on DRG neurons in vitro. J Neurocytol 30:59–71
12. Theiss C, Napirei M, Meller K (2005) Impairment of anterograde and retrograde neurofilament transport after anti-kinesin antibody microinjection in chicken dorsal root ganglia. Eur J Cell Biol 84(1):29–43
13. Theiss C, Meller K (2002) Aluminum impairs gap junctional intercellular communication between astroglial cells in vitro. Cell Tissue Res 310:143–154
14. Giessmann D, Theiss C, Meller K (2005) Decreased gap junctional communication in microinjected lens epithelial cells after taxol treatment. Anat Embryol 209:391–400
15. Sabry JH, O'Connor TP, Evans L et al (1991) Mircrotubule behavior during guidance of pioneer neuron growth cones in situ. J Cell Biol 115:381–395
16. Miller KE, Joshi HC (1996) Tubulin transport in neurons. J Cell Biol 133:1355–1366
17. Chalfie M, Tu Y, Euskirchen G et al (1994) Green fluorescent protein as a marker for gene expression. Science 263:802–805
18. Chalfie M (1995) Green fluorescent protein. Photochem Photobiol 62:651–656
19. Haseloff J (1999) GFP variants for multispectral imaging of living cells. Meth Cell Biol 58: 139–151
20. Meller K (1974) The reaggregation of neurons and their satellite cells in cultures of trypsin-dissociated spinal ganglia. Cell Tissue Res 152:175–183
21. Gähwiler BH (1981) Morphological differentiation of nerve cells in thin organotypic cultures derived from rat hippocampus and cerebellum. Proc R Soc Biol Sci 2111: 287–290
22. Krah K, Meller K (1999) Axonal and dendritic transport in Purkinje cells of cerebellar slice cultures studied by microinjection of horseradish peroxidase. Cell Tissue Res 295:55–64
23. Lasek R (1968) Axoplasmic transport in cat dorsal root ganglion cells: as studied with ^{3}H-L-leucine. Brain Res 7:360–377
24. Lasek R (1970) Axonal transport of proteins in dorsal root ganglion cells of the growing cat:

a comparison of growing and mature neurons. Brain Res 20:121–126
25. McClure WO (1972) Effect of drugs upon axoplasmic transport. Adv Pharmacol Chemother 10:185–220
26. Theiss C, Meller K (2000) Taxol impairs anterograde axonal transport of microinjected horseradish peroxidase in dorsal root ganglia neurons in vitro. Cell Tissue Res 299:213–224
27. Brady ST (1985) A novel brain ATPase with properties expected for the fast axonal transport motor. Nature 317:73–75
28. Vale RD, Reese TS, Sheetz MP (1985) Identification of a novel force-generating protein, kinesin, involved in microtubule-based motility. Cell 42:39–50
29. Kuznetsov SA, Gelfand VI (1986) Bovine brain kinesin is a microtubule-activated ATPase. Proc Natl Acad Sci U S A 83:8530–8534
30. Susalka SJ, Pfister KK (2000) Cytoplasmic dynein subunit heterogeneity: implications for axonal transport. J Neurocytol 29:819–829
31. Shea TB, Flanagan LA (2001) Kinesin, dynein and neurofilament transport. Trends Neurosci 24:644–648
32. Roy S, Coffee P, Smith G et al (2000) Neurofilaments are transported rapidly but intermittently in axons: implications for slow axonal transport. J Neurosci 20:6849–6861
33. Wang L, Ho CI, Sun D et al (2000) Rapid movement of axonal neurofilaments interrupted by prolonged pauses. Nat Cell Biol 2:137–141
34. Yabe JT, Chan WKH, Chylinski TM et al (2001) The predominant form in which neurofilament subunits undergo axonal transport varies during axonal initiation, elongation, und maturation. Cell Motil Cytoskeleton 48:61–83
35. Yabe JT, Chylinski T, Wang FS et al (2001) Neurofilaments consist of distinct populations that can be distinguished by C-terminal phosphorylation, bundling, and axonal transport rate in growing axonal neurites. J Neurosci 21:2195–2205
36. Prahlad V, Yoon M, Moir RD et al (1998) Rapid movements of vimentin on microtubule tracks: kinesin-dependent assembly of intermediate filament networks. J Cell Biol 143: 159–170
37. Martys JL, Ho CL, Liem RK et al (1999) Intermediate filaments in motion: observations of intermediate filaments in cells using green fluorescent protein-vimentin. Mol Biol Cell 10:1289–1295
38. Windoffer R, Leube RE (1999) Detection of cytokeratin dynamics by time-lapse fluorescence microscopy in living cells. J Cell Sci 112:4521–4534
39. Wang L, Brown A (2001) Rapid intermittent movement of axonal neurofilaments observed by flourescene photobleaching. Mol Biol Cell 12:3257–3267
40. Hirano A (1991) Cytopathology of amyotrophic lateral sclerosis. In: Rowland LP (ed) Advances in neurology, amyotrophic lateral sclerosis and other motor neuron diseases, vol 56. Raven Press, New York, NY, pp 91–101
41. Trojanowski JY, Schmidt ML, Shin RW et al (1993) Altered tau and neurofilament proteins in neurodegenerative diseases: diagnostic implications for Alzheimer's disease and Lewy body dementias. Brain Pathol 3:45–54
42. Schmidt ML, Martin JA, Lee VMY et al (1996) Convergence of Lewy bodies and neurofibrillary tangles in amygdale neurons of Alzheimer's disease and Lewy body disorders. Acta Neuropathologica 91:475–481
43. Katsetos CD, Savory J, Herman MM et al (1990) Neuronal cytoskeletal lesions induced in the CNS by intraventricular and intravenous aluminium maltol in rabbits. Neuropathol Appl Neurobiol 16:511–528
44. Gilbert MR, Harding BL, Hoffman PN et al (1992) Aluminum-induced neurofilamentous changes in cultured rat dorsal root ganglia explants. J Neurosci 12:1763–1771
45. Bizzi A, Gambetti P (1986) Phosphorylation of neurofilaments is altered in aluminium intoxication. Acta Neuropathologica 71:154–158
46. Troncoso JC, Hoffman PN, Griffin JW et al (1985) Aluminum intoxication: a disorder of neurofilament transport in motor neurons. Brain Res 342:172–175
47. Shea TB, Wheeler E, Jung C (1997) Aluminum inhibits neurofilament assembly, cytoskeletal incorporation, and axonal transport. Dynamic nature of aluminum-induced perikaryal neurofilament accumulations as revealed by subunit turnover. Mol Chem Neuropathol 32:17–39
48. Ackerley S, Thornhill P, Grierson AJ et al (2003) Neurofilament heavy chain side arm phosphorylation regulates axonal transport of neurofilaments. J Cell Biol 161:489–495
49. Miller CC, Ackerley S, Brownless J (2002) Axonal transport of neurofilaments in normal and disease states. Cell Mol Life Sci 59:323–330
50. Cajal SRY (1890) À quelle époque apparaissent les expansions des cellules nerveuses de la moëlle épinière du poulet? Anat Anz 21–22:609–639
51. Dent EW, Gertler FB (2003) Cytoskeletal dynamics and transport in growth cone motility and axon guidance. Neuron 40:209–227

52. Dent EW, Gupton SL, Gertler FB (2011) The growth cone cytoskeleton in axon outgrowth and guidance. Cold Spring Harb Perspect Biol 3(3):pii: a001800. doi:10.1101/cshperspect.a001800
53. Huber AB, Kolodkin AL, Ginty DD et al (2003) Signaling at the growth cone: ligand–receptor complexes and the control of axon growth and guidance. Annu Rev Neurosci 26:509–563
54. Maskery S, Shinbrot T (2005) Deterministic and stochastic elements of axonal guidance. Annu Rev Biomed Eng 7:187–221
55. Schaefer AW, Kabir N, Forscher P (2002) Filopodia and actin arcs guide the assembly and transport of two populations of microtubules with unique dynamic parameters in neuronal growth cones. J Cell Biol 158(1):139–152
56. Geraldo S, Gordon-Weeks PR (2009) Cytoskeletal dynamics in growth-cone steering. J Cell Sci 122:3595–3604
57. Lowery LA, Van Vactor D (2009) The trip of the tip: understanding the growth cone machinery. Nat Rev Mol Cell Biol 10:332–343
58. Lamalice L, Houle F, Jourdan G et al (2004) Phosphorylation of tyrosine 1214 on VEGFR2 is required for VEGF-induced activation of Cdc42 upstream of SAPK2/p38. Oncogene 23:434–445
59. Dickson BJ (2001) Rho GTPases in growth cone guidance. Curr Opin Neurobiol 11(1):103–110
60. Meyer G, Feldman EL (2002) Signaling mechanisms that regulate actin-based motility processes in the nervous system. J Neurochem 83(3):490–503
61. Matsumoto T, Mugishima H (2006) Signal transduction via vascular endothelial growth factor (VEGF) receptors and their roles in atherogenesis. J Atheroscler Thromb 13:130–135
62. Rousseau S, Houle F, Landry J, Huot J (1997) p38 MAP kinase activation by vascular endothelial growth factor mediates actin reorganization and cell migration in human endothelial cells. Oncogene 15:2169–2177
63. Rousseau S, Houle F, Hout J (2000) Integrating the VEGF signals leading to actin-based motility in vascular endothelial cells. Trends Cardiovasc Med 10:321–327
64. Rousseau S, Houle F, Kotanides H, Witte L, Waltenberger J, Landry J, Huot J (2000) Vascular endothelial growth factor (VEGF)-driven actin-based motility is mediated by VEGFR2 and requires concerted activation of stress-activated protein kinase 2 (SAPK2/p38) and geldanamycin-sensitive phosphorylation of focal adhesion kinase. J Biol Chem 275:10661–10672
65. Bromberg KD, Iyengar R, He JC (2008) Regulation of neurite outgrowth by $G_{i/o}$ signalling pathways. Front Biosci 13:4544–4557
66. Van Troys M, Huyck L, Leyman S, Dhaese S, Vandekerkhove J, Ampe C (2008) Ins and outs of ADF/cofilin activity and regulation. Eur J Cell Biol 87:649–667
67. Meller K, Krah K, Theiss C (2005) Dye coupling in Purkinje cells of organotypic slice cultures. Dev Brain Res 160(1):101–105
68. Theiss C, Meller K (2010) Microscopic techniques to study cytoskeletal dynamics in neurons. In: Méndez-Vilas A, Díaz J (eds) Microscopy: science, technology, applications and education. Formatex Research Center, Badajoz
69. Bourne JN, Harris KM (2008) Balancing structure and function at hippocampal dendritic spines. Annu Rev Neurosci 31:47–67
70. Fifková E, Morales M (1992) Actin matrix of dendritic spines, synaptic plasticity, and long-term potentiation. Int Rev Cytol 139:267–307
71. Harris KM, Kater SB (1994) Dendritic spines: cellular specializations imparing both stability and flexibility to synaptic function. Annu Rev Neurosci 17:341–371
72. Hu X, Viesselmann C, Nam S et al (2008) Activity-dependent dynamic microtubule invasion of dendritic spines. J Neurosci 28:13094–13105
73. Jaworski J, Kapitein LC, Gouveia SM (2009) Dynamic microtubules regulate dendritic spine morphology and synaptic plasticity. Neuron 61:85–100
74. Lee KJ, Kim H, Rhyu IJ (2005) The roles of dendritic spine shapes in Purkinje cells. Cerebellum 4:97–104
75. Hayashi K, Suzuki A, Hirai S, Kurihara Y, Hoogenraad CC, Ohno S (2011) Maintenance of dendritic spine morphology by partitioning-defective 1b through regulation of microtubule growth. J Neurosci 31(34):12094–12103
76. Hering H, Sheng M (2003) Activity-dependent redistribution and essential role of cortactin in dendritic spine morphogenesis. J Neurosci 23:11759–11769
77. Li Z, Okamoto K, Hayashi Y, Sheng M (2004) The importance of dendritic mitochondria in the morphogenesis and plasticity of spines and synapses. Cell 119:873–887
78. Newey SE, Velamoor V, Govek EE, Van Aelst L (2005) Rho GTPases, dendritic structure, and mental retardation. J Neurobiol 64:58–74

Chapter 2

Visualizing In Vitro Trafficking

John W. Murray

Abstract

Here we present a detailed guide for performing in vitro trafficking assays. These are high-resolution light microscopy assays designed to look at the cytoskeletal filament-based trafficking of cellular organelles. The strategy is to partially purify organelles from lysed mammalian cells and freeze them as single-use aliquots. The organelles are then thawed and allowed to bind microtubule and actin filaments that have been coated onto handmade optical microchambers. Time lapse multichannel fluorescence microscopy is then performed to identify specific vesicles and associated proteins and to observe and quantify how the material is transported. These protocols were initially developed to study rodent liver endosomes but are adapted here for the study of cultured cells and include commentary on their use with other types of organelles.

Key words: Microtubules, Actin, In vitro trafficking, Cellular organelles, Time lapse imaging, Endosomes

1. Introduction

Vesicular trafficking involves the movement of membrane and protein under nonintuitive topological and thermodynamic constraints. In vitro trafficking assays have been designed to investigate this partially purified world that is neither a fully functioning cell nor a precise assembly of known components. The assays described here follow this general scheme:

1. Biochemically isolate intracellular vesicles.
2. Incubate vesicles with cytoskeletal filaments within disposable microscope chambers, and wash.
3. Begin live cell fluorescence microscopy and add ATP to activate the motor proteins.

Vesicle trafficking requires active motor proteins and ATP-releasable attachment to the filaments. For endocytic vesicles isolated

Rolf Dermietzel (ed.), *The Cytoskeleton: Imaging, Isolation, and Interaction*, Neuromethods, vol. 79,
DOI 10.1007/978-1-62703-266-7_2, © Springer Science+Business Media, LLC 2013

from rat liver, ATP induces motility in less than a second and in our hands lasts for 1–2 min. In living cells, motility is continuous, oscillatory, 3-dimensional, and very complex, all of which can hinder quantification. The trafficking assays provide a 2-dimensional, optically clean matrix, and an experimental starting point for quantitation. They are ideal for dissecting the direct chemical or protein contribution to motility since the responsible machinery is solution accessible. For instance, in vitro assays were crucial in identifying the cytosolic microtubule motor proteins, kinesin and dynein (1, 2). In contrast, in whole cell knockdown studies cells have time to adapt and changes in activities may stem from secondary effects. This Chapter attempts to provide an easy-to-understand step-by-step description of all that is needed for performing in vitro vesicle trafficking assays. Significant commentary is also included with observations that have helped our own exploration of these assays.

2. Special Materials

The following is a list of required materials that may not be present in a bioscience laboratory: large coverslips, DEAE-Dextran, polylysine, phalloidin, paclitaxel (Taxol), double-sided tape, clear nail polish, tubulin, rhodamine-labeled tubulin, GTP, ATP, Airfuge, kinesin motor domain, rhodamine-labeled actin, orosomucoid, Texas Red sulfonyl chloride, fluorescent transferrin, fluorescent EGF, protease inhibitors, tabletop ultracentrifuge, casein, primary antibodies, fluorescent secondary antibodies, multi-fluorescence channel heated microscope system, and image analysis software.

2.1. Microscope Chambers

The chamber described here is designed for an inverted microscope where the objective lens is beneath the chamber and out of the way. It is easily customized (e.g., (3)). A treated coverslip forms the optical, bottom surface, supports a 5 mm wide channel formed from two pieces of double-sided tape, and is topped with a piece of glass cut from a slide and held in place by the tape (Fig. 1). The chamber is inexpensive, easy to construct, contains an internal volume of approximately 5 μL, and allows rapid exchange of contents by pipetting at one end and wicking from the other. Microtubules and actin filaments do not readily adhere to glass and therefore coverslips are first coated with a binding agent such as polylysine, sigmacote, or DEAE-dextran. In some cases adequate microtubule attachment may be observed in the absence of a binding agent (4). We have found that DEAE-dextran is easy to handle, does not form aggregates, and allows microtubules to bind to the surface as single filaments. The binding agents can bundle microtubules and this can inhibit or alter motility, so the coverslip should be washed thoroughly following treatment. We have not found it necessary to additionally clean the coverslips.

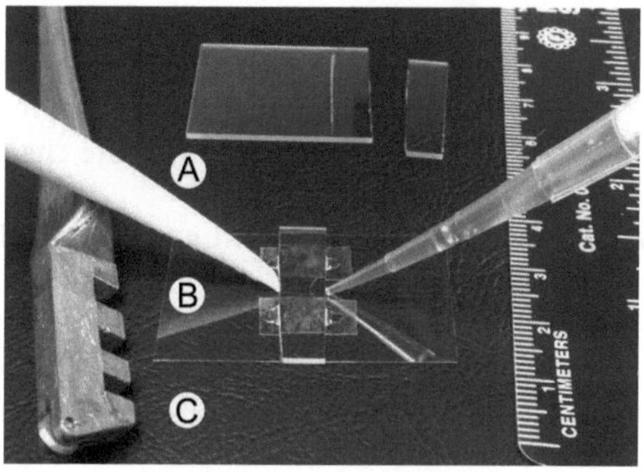

Fig. 1. Construction of disposable glass microchambers. (A) A standard glass slide is scored with a wheeled glass cutter and broken into small rectangles that are used for the tops of chambers. (B) The assembled chamber with pipette tip dispensing 5–10 μL into the chamber where they are wicked at the other end with a folded Kimwipe. The chamber should not be allowed to dry once material has been added. (C) A wheeled glass cutter is shown. These can be found at hardware stores or from companies specializing in stained glass work.

2.2. Construction of Chambers

1. Sequentially submerge large rectangular coverslips (22 × 40 mm, Corning #2940 224) into deionized water containing 30 μg/mL DEAE-dextran (e.g., Sigma-Aldrich #D9885) for 1 min and wash extensively. Allow coverslips to air-dry, or use compressed air to dry. For attachment of F-actin which requires stronger adhesive, substitute 50 μg/mL polylysine (Sigma-Aldrich P8954) for DEAE-dextran.
2. Using a razor blade and appropriate cutting surface, cut strips of double-sided tape (3M Scotch #655, 12 mm wide, 0.076 mm thick) into rectangles of approximately 15 mm × 6 mm and attach these to the DEAE-coated coverslips leaving a 5 mm wide channel.
3. Construct glass tops of the chambers by scoring standard slides (e.g., 25 × 75 × 1 mm, Fisher Brand #12-550D) with a wheeled glass cutter and breaking at the score, creating multiple 25 × 6 mm pieces. Affix the tops to the double-sided tape channel as shown in Fig. 1.
4. Add small drop of clear nail polish at the cut glass–tape junction to improve the chamber integrity. A piece of folded Scotch or lab tape at the corner can serve as a handle and provide a label.

The optical surface of the chamber should remain clean and the entire bottom surface should be free of tape or debris to avoid affecting the focal plane. 10–20 chambers can be made for a day's experiments and mounted to a black platform for handling. The black surface helps to visualize the chambers and their contents. The DEAE-coating will last for a day and possibly longer. Once solutions are added to the chamber, care must be taken to prevent drying or formation of air bubbles as these can destroy the contents. If perfusions are completed but the contents require further imaging, a drop of glycerol can be added to both ends to semi-permanently seal the chambers.

2.3. Polymerization of Fluorescent Microtubules

Chemical labeling of purified tubulin with a fluorophore has provided a convenient means to view microtubules. Unlabeled microtubules can also be detected by bright field or DIC microscopy (2) but it can be difficult to distinguish whether microtubules are bundled. On the other hand, fluorescent modification of tubulin may affect the binding of proteins to microtubules. This, however, can be addressed by diluting labeled tubulin with unlabeled tubulin to the limit of detection, or by comparing motor activities with unlabeled tubulin under DIC microscopy.

Our recommended protocol is the following:

1. Purchase unlabeled (cat #TL238-A, 250 μg lyophilized aliquots) and rhodamine-labeled tubulin (cat #TL590M, 20 μg lyophilized aliquots) from Cytoskeleton Inc. (Denver, CO). We have found that these tubulins preserve best at –80°C, dehydrated.
2. Add 25 μL of cold MT buffer (80 mM K_2-PIPES, 1 mM ethyleneglycoltetraacectic acid (EGTA), 1 mM $MgCl_2$, 3% glycerol, 1 mM GTP, pH 7.0) to lyophilized unlabeled tubulin to create "U-tub" stock (concentration, 10 μg/μL), and store on ice.
3. Add 12 μL of U-tub to a 20 μg aliquot of rhodamine tubulin to create "Rh-tub" stock (concentration, approximately 11.7 μg/μL) at 1:6 molar ratio of labeled to unlabeled tubulin, and store on ice.
4. Clarify (i.e., remove aggregates and debris from) both tubes by centrifugation for 5 min at 14,000 rpm (20,000×*g*), 4°C. Collect the supernatant, carefully avoiding any potential pellet.
5. Optionally, dilute Rh-Tub with U-tub to obtain optimal brightness for your imaging conditions.
6. Polymerize batches of Rh-tub by warming 6 μL to 37°C in a water bath for 15 min. Polymerize a second batch for 25 min.
7. Stabilize and terminate the polymerization reaction by addition of 190 μL of pre-warmed MT buffer containing 20 μM Taxol (Sigma-Aldrich #T1912).
8. Pellet the microtubules in a Beckman Airfuge (Beckman Coulter, Brea, CA) at 15 psi for 5 min to remove unpolymerized tubulin. The pink microtubules should be just visible as a pellet.
9. Remove supernatant and resuspend the microtubules to approximately 0.7 μg/μL with 100 μL MT buffer plus 20 μM

Taxol by repeated pipetting, being careful not to create air bubbles.

10. Store the microtubules at room temperature in the dark. They should be usable for 2 weeks. Dilute 50-fold (or to desired density) to use in motility assays.

Microtubule quality should be examined by dilution onto standard glass slides or chambers in MT buffer plus Taxol. Thick filaments indicate MT bundling and that attachment reagent should be washed more thoroughly or buffers should be remade from fresh stocks. The 15-min or 25-min polymerization point may show longer filaments or better polymerization. Asters of microtubules generally indicate Taxol-induced polymerization while puncta can indicate aggregates that should have been removed in the clarification step. Some differences may be observed between different lots of tubulin. Longer microtubules can be obtained by polymerizing at lower concentration but polymerization may then be less robust. The microtubules will typically anneal and lengthen over 1–2 days. Note that some loss occurs with pipetting and centrifugation of small volumes; minor adjustments may be required.

2.4. Polarity-Marked Microtubules

Microtubules have a fast-growing "plus" end and slow-growing "minus" end, and this polarity is critical to their cellular function. Our preferred strategy for visualizing the polarity is to polymerize dimly labeled tubulin and shear these into short "seeds" from which to grow bright fluorescent tubulin. The plus end is then identified as the long bright end of the filament.

1. Prepare and clarify Rh-tub and U-tub as for "polymerization of fluorescent microtubules," steps 1–4 above (Section 2.3), and store on ice.
2. Make dim "Seed-tub" stock by adding 1 μL of Rh-tub to the 13 μL of the U-tub stock, giving 14 μL of approximately 10 μg/mL stock at molar ratio of approximately 84:1 unlabeled to labeled tubulin (or adjust ratio as desired).
3. Polymerize 4 μL of Seed-tub by warming to 37°C for 5 min. Warm up 8 μL of MT buffer.
4. Shear the polymerized seeds by rapidly pipetting up and down 15 times with pipette set to 2 μL, press pipette tip against tube, and avoid air bubbles.
5. Add 2 μL of the seeds to the 8 μL of warm MT buffer and quickly add 3 μL of Rh-tub using a cut tip, and mix by up-and-down pipetting, avoiding shearing.
6. Allow to polymerize at 37°C for 6 min. This polymerizes 2.7 μg/μL tubulin from 1.5 μg/μL seeds, and the Rh-tub is approximately 14× brighter than the seeds.

7. Stabilize and terminate the polymerization by addition of 190 μL of pre-warmed MT buffer containing 20 μM Taxol, final tubulin concentration, approximately 0.4 μg/μL.
8. Pellet the microtubules in a Beckman Airfuge at 15 psi for 5 min to remove unpolymerized tubulin.
9. Remove supernatant and resuspend the polarity-marked microtubules to approximately 0.1 μg/μL with 500 μL MT buffer plus 20 μM Taxol by gentle, repeated pipetting. This lower concentration will reduce the amount of microtubule annealing.
10. Store microtubules at room temperature and use within 3 h. The accuracy will degrade over time due to annealing and breaking.

The accuracy of the polarity marks can be assessed with gliding assays that utilize a single species of motor protein. In these, motor protein is added directly to untreated chambers followed by washing, and addition of microtubules and then ATP (5). Commercially available kinesin motor domain (KR01, Cytoskeleton Inc.) can be used for this purpose, although motility may be quite slow and light sensitive. This construct has recently changed but is indicated to function in motility assays. Some microtubules will contain ambiguous marks and must be eliminated from scoring. Since marked microtubules can only be used for a single day, and because the microtubules must be carefully scored by hand, these assays require considerable effort, and there is certainly room for improvement. Axonemes from tetrahymena or sea urchin have been used to nucleate microtubules and indicate the microtubule minus ends (6, 7). The GTP analogue, GMPCPP, will stabilize microtubule seeds and improve reproducibility of microtubule marks (8), but this can be difficult to obtain commercially. Treatment of tubulin with N-ethylmaleimide inhibits polymerization of tubulin more strongly at the minus end, and tubulin can be diluted with NEM-treated tubulin to preferentially inhibit minus end polymerization and increase polarity accuracy (6). Soppina and colleagues have used magnetic avidin beads to bind biotin-labeled, nonfluorescent tubulin seeds to mark microtubule minus ends (9). Other groups have shown that MTs will align with the direction of force from either an electric field (10) or fluid flow (11) during kinesin-based gliding assays. The microtubules could then be chemically fixed in place and potentially used for vesicle trafficking assays.

2.5. Polymerization of Fluorescent Actin

Actin filaments can also be imaged directly by fluorescence microscopy in vitro (12, 13), and we have developed actin trafficking assays in the style of the microtubule assays. Because of their different diameters and effective brightness, it is possible to distinguish actin and microtubule filaments in the same fluorescent

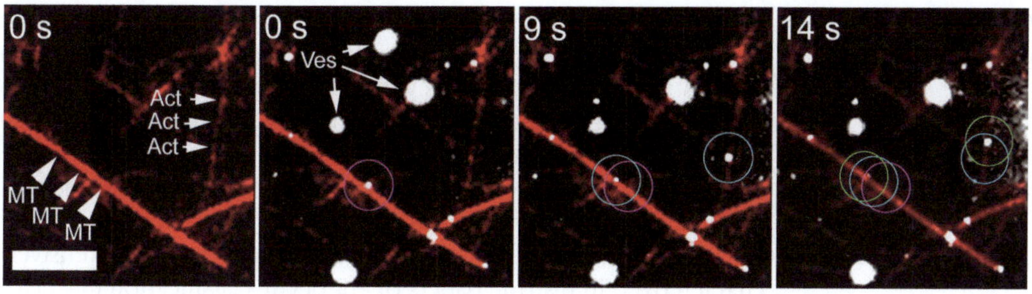

Fig. 2. Motility of rat liver vesicles along microtubule and actin filaments. Following the procedures described in Sect. 2.9, rat liver vesicles were bound to actin and microtubules within microchambers and 1 mM ATP was added to initiate vesicle trafficking. In this example 2 μM bodipy ceramide was used to stain the vesicles. Imaging data was captured in the bright field (not shown), FITC (vesicles), and rhodamine (microtubules and actin) fluorescence channels. Microtubules (MTs) are seen as thick bright filaments while actin (Act) filaments are dimmer. A vesicle (*circled* at time 0 s) moves to the left on the MT, while another vesicle at 9 s (*circled*) binds to an Act filament and moves upwards. This demonstrates that both MT and Act motors are active in the vesicle preparation. Bar = 5 μm.

channel (Fig. 2), and many options for imaging actin, microtubules, and vesicles are possible. Our recommended protocol is the following:

1. Obtain rhodamine-labeled actin (Cat #AR05, lyophilized aliquots, Cytoskeleton Inc.), and store at −80°C.
2. To a single 20 μg aliquot add 50 μL of cold "G Buffer" (5 mM Tris, 0.2 mM $CaCl_2$, 0.2 mM ATP, 0.5 mM DTT pH 8.3), incubate on ice for 1 h to allow depolymerization of actin oligomers, and clarify by centrifugation for 5 min at 14,000 rpm (20,000 × *g*), 4°C. Collect the supernatant, containing 0.4 μg/μL (9.3 μM) globular "G-actin."
3. Polymerize overnight at approximately 6.2 μM (0.27 mg/mL) by adding 10 μL G-actin, 3 μL of 5× polymerization buffer (the 5× polymerization buffer contains 250 mM KCl, 10 mM $MgCl_2$, 5 mM ATP, made fresh from single species stocks of KCl and $MgCl_2$ and frozen stocks of ATP), and 2 μL deionized H_2O.
4. Warm to room temperature and stabilize actin filaments (F-actin) by addition of 0.07 μM phalloidin (1.7 μL from a 0.7 μM phallodin stock).
5. Store both G- and F-actin on ice in the dark. Both are usable for approximately 1 week.

Actin, unlike tubulin, does not depolymerize at low temperature and both F- and G-actin are stored cold to prevent degradation. Polymerization is induced by raising the KCl and magnesium concentration.

2.6. Isolation of Vesicles

We have developed protocols for isolating endocytic vesicles from rodent liver that contain active, endogenous microtubule and actin motors. In these, hepatocytes within the liver are allowed to endocytose fluorescent ligand, and vesicles are isolated by floatation through a sucrose gradient.

Further purification is not necessary since the endosomes are defined by their fluorescent ligand. We have traditionally utilized the ligand, asialoorosomucoid (ASOR), which has been extensively characterized in historic papers of endocytic classification (14, 15), and which is well behaved in solution and released upon lysis of endosomes. Fluorescent ASOR is prepared by removing terminal sialic acid residues from orosomucoid (alpha-1 acid glycoprotein, Sigma Cat. #G-9885) and labeling with an amine-reactive fluorescent dye (e.g., Texas Red sulfonyl chloride, cat #T1905, Life Technologies (5)). Numerous alternate ligands (e.g., Transferrin, EGF, or LDL, GFP—your favorite protein) can be utilized; although if endocytic uptake is not performed, the identity of the vesicles should be confirmed by other methods. ASOR binds specifically to hepatocytes via the highly expressed asialoglycoprotein receptor (ASGPR) (16, 17), and a single rat liver contains approximately a billion hepatocytes (18), providing an abundance of concentrated endosomes that are required for the microscopy assays. Although in some cases freezing can reduce motor protein activity (19), it has been essential to freeze vesicles as single-use aliquots so that a reliable stock of vesicles is available for the quantitative assays. For mice, we use between 2 and 5 livers and maintain these on ice prior to homogenization. For cultured cells, attaining enough concentrated vesicles can be difficult and therefore a gel filtration step is omitted and screening the cells and ligands for high binding capacity is recommended (see below). Many related strategies have been used to isolate populations of vesicles that are active for in vitro assays and these can be combined with live cell imaging studies to bolster results (4, 20, 21).

2.7. Protocol for Isolating Motile Fluorescent Early Endocytic Vesicles from Cultured Cells

1. Culture Huh7 or other cell line to confluence in sixteen 15-cm dishes.
2. Wash cells in Hepes-buffered saline (HBS, 135 mM NaCl, 1.2 mM $MgCl_2$, 0.8 mM $MgSO_4$, 28 mM dextrose, 2.5 mM $CaCl_2$, 25 mM Hepes, pH 7.4) and incubate in HBS for 30 min to remove cell surface-bound proteins that may occupy the receptor.
3. Cool cells to 4°C by addition of cold HBS, place cells on ice, and add approximately (see below) 5 μg/mL fluorescent ligand (i.e., ~0.1 μM ASOR) for 30 min. Fluorescent ASOR, transferrin (Life Technologies Cat #T23364), EGF (Life Technologies Cat #E-35350), dextran, and other ligands have been used in our studies. At 4°C, cell surface binding will occur without

internalization (note that ASOR binding requires calcium). After 30 min, wash cells in cold HBS.

4. Initiate endocytosis by addition of 37°C HBS, and allow endocytosis to proceed for 5 min.
5. Wash cells in cold MEPS buffer (5 mM $MgCl_2$, 5 mM EGTA, 35 mM K_2–PIPES, 0.25 M sucrose, pH 7.4). This low-calcium buffer will remove surface-bound ASOR. Subsequent steps are performed at 4°C.
6. Scrape cells from dishes with rubber policeman and pellet whole cells at 240 × *g* (1,500 rpm) for 5 min.
7. Resuspend the cells in 600 μL MEPS buffer plus 4 mM DTT, 2 mM phenylmethanesulfonyl fluoride (PMSF) and a 1:50 dilution of protease inhibitor cocktail (Sigma Cat. #P-8340).
8. Lyse the cells by 20–30 passages through a 25-gauge syringe needle.
9. Remove nuclei and cell debris by centrifugation at 950 × *g* (3,000 rpm) for 10 min; collect supernatant (PNS), which should be approximately 600 μL; and re-add 4 mM DTT, 2 mM PMSF, and 1:50 dilution of protease inhibitor cocktail. An opaque supernatant will indicate effective lysis; if clear, re-lyse the pellet.
10. Bring PNS to 1.4 M sucrose by the addition of 2.5 M sucrose MEPS buffer (i.e., MEPS buffer that contains 2.5 M sucrose instead of 0.25 M). A handy calculation is that the volume of added 2.5 M sucrose should equal 1.045× the volume of the PNS (total volume equals 2.045× the PNS).
11. Add the 1.4 M sucrose PNS to the bottom of tube fitted for a tabletop ultracentrifuge swinging bucket rotor (e.g., Beckman Coulter TL-100 centrifuge, TLS-55 rotor, and 347357 tube). On top of this load approximately 1 mL of 1.2 M MEPS buffer, followed by 0.1 mL of 0.25 MEPS buffer, creating a 1.4 M/1.2 M/0.25 M sucrose step gradient.
12. Centrifuge for 2 h at 101,000 × *g* (39,000 rpm) during which time the low-density membranes, including the endocytic vesicles, will float up through the 1.2 M sucrose.
13. Collect approximately 200 μL of the opaque, membrane-containing material from the 1.2 M/0.25 M interface and store frozen in 12 μL aliquots at −80°C. It is critical that the vesicle density be very high so that 20–60 vesicles can be seen in a single microscope field during the trafficking assays.

The binding of ligand to the surface of the cells should be examined prior to performing vesicle preparations to ensure adequate uptake and to determine optimal ligand concentration. This can be done by culturing cells on coverslip-bottomed chambers

(MatTek, Ashland, MA) and treating cells with ligand for 5 min at 37°C followed by washing and imaging. The receptor for ASOR is expressed in hepatocyte cell lines but often at low levels. Similarly, transferrin receptor expression can be low, but this may be increased by addition of deferoxamine to the culture medium 1–2 days prior to the experiment (22). Cell density, passage number, or culture conditions may affect cell surface receptor expression. Investigators may wish to purify other types of organelles, and for this a survey of the vesicle trafficking literature is recommended (e.g., (4, 7, 21, 23)). Liver, brain, and different kinds of cultured cells may require different strategies and our recommendation is to plan for many isolations with various permutations (buffers, lysis conditions, etc.), and to perform fast, crude isolations in the initial stages, and test these for trafficking activity as isolation conditions become more complex.

2.8. Vesicle Trafficking Assay

Here we describe how to combine the vesicles with cytoskeletal filaments to produce reliable trafficking assays. Trial experiments should be performed to assess proper dilutions of microtubules, vesicles, antibodies, and imaging exposure times and lamp intensity. Once these are established, a series of DEAE-dextran-treated chambers are assembled, coated with microtubules, washed, incubated with vesicles, washed, and optionally incubated with antibodies, washed, and placed on ice in humid container to prevent dehydration. The chambers are brought to the microscope and one by one placed on the microscope stage and imaged during addition of ATP. Vesicles bound to microtubules and washed free of soluble material in this manner should retain their motile capacity for many hours, whereas solutions of thawed vesicles can rapidly lose activity.

1. Prepare 2.5 mL of "10×" PMEE (350 mM K_2-PIPES, 50 mM $MgCl_2$, 10 mM EGTA, 5 mM EDTA, 20 mg/mL bovine serum albumin (BSA) pH 7.4, made daily from single chemical liquid stocks and powdered BSA), 25 µL of 10 mM ATP, pH 7.0 stored as single-use frozen stock, 0.5 mL of 200 mM DTT made fresh, 50 µL of 1 mM Taxol stored as frozen stock in dimethylsulfoxide, 1 mL of 100 mg/mL Na-ascorbic acid made fresh (optional antioxidant), 0.5 mL of "1×" PMEE plus 20 µM Taxol, 2 mL of Assay Buffer (1× PMEE plus 20 µM Taxol, 4 mM DTT, 2 mg/mL Na-ascorbic acid acid), and 0.5 mL of Blocking Buffer (Assay Buffer plus 5 mg/mL casein, Sigma Cat #C7078, undissolved casein removed by filtration).
2. Prepare 12 motility chambers coated with DEAE-dextran.
3. Dilute fluorescent microtubules, prepared as above, 1:50 (to 14 µg/mL, concentration will vary) in 1× PMEE plus Taxol, and mix gently (optional: use cut tips to prevent shearing of microtubules).

4. Add 5 μL of microtubules to the chambers, incubate for 3 min at room temperature, and wash 3× with Blocking Buffer and 2× with Assay Buffer. The Blocking Buffer (casein) is used to decrease the binding of vesicles directly to glass.
5. Thaw fluorescent vesicle aliquots and immediately add 5 μL of the vesicles to each chamber and incubate for 10 min at room temperature in a humid environment. Longer incubation times will decrease motility; shorter incubation times may not allow sufficient binding to microtubules. A drop of assay buffer at the edges can prevent dehydration and insure a nonoxidizing environment.
6. Wash off unbound vesicles with Assay Buffer and store chambers on ice in humid environment.
7. At the microscope, heated to 37°C, remove a single chamber, wash in Assay Buffer, and focus on a field of microtubule-bound vesicles, avoiding excess excitation light.
8. Initiate image acquisition and add 20 μL of 50 μM ATP in Assay Buffer. Capture images at 1 frame per 3 s for 120 s.
9. Repeat trafficking assay for all chambers.

Motility is sensitive to oxidative damage from the microscope excitation light. Fluorescence excitation should be minimized, for instance by lowering the intensity, using transmitted light or long-wavelength light when focusing, and by capturing at low frame rates. Many laboratories use oxygen-scavenging systems to reduce photodamage (e.g., 0.1 mg/mL glucose-oxidase, 0.018 mg/mL catalase, 2.3 mg/mL glucose (24, 25)). In our hands these components can bundle microtubules and do not increase motility during the short acquisition times of these assays. We do include the antioxidants, DTT and ascorbic acid. An ATP-regenerating system (e.g., 0.16 mg/mL creatine phosphokinase, 8 mM phosphocreatine) may be employed but did not increase the motility of our purified vesicles (5, 26). The above assay uses a low concentration of ATP (50 μM), which may limit extraneous ATPase activity. Many factors can inhibit motility and initial negative results should not be a cause for despair. Motor proteins are sensitive to salt, oxidation, ATPases, free phosphate, ADP, and other factors, and buffer stocks can go bad. It has been very helpful to have a source of active motor proteins or vesicles as a positive control. For instance, microtubule gliding assays (5) can be performed in the presence of isolated vesicles to determine if motility inhibitors are present. The endocytic vesicle preparation typically contains enough motors to cause some bending or sliding of microtubules even when attached by DEAE-dextran. Microtubule-bound vesicles also frequently exhibit short, back-and-forth diffusive movement prior to ATP addition. If vesicles show no motility, no microtubule bending, and no diffusive movement, they may be cross-linked to the microtubules. Higher salt or removal of the cross-linking molecule(s) may be required.

2.9. Trafficking Assay Adapted for Actin and Microtubules

Assays for vesicle trafficking along actin filaments can be performed in the same manner as for trafficking along microtubules with a few *key differences*. Because these filaments can be distinguished by their diameter and brightness (Fig. 2), both actin and microtubules can be imaged in the same field.

1. Prepare buffers as described in Vesicle Trafficking Assay (Sect. 2.8). Instead of "1×" PMEE plus Taxol, prepare 0.5 mL of *AM Buffer* (5 mM Tris, 50 mM KCl, 2 mM $MgCl_2$, 0.07 μM phalloidin, 20 μM Taxol, pH 8.3).
2. Prepare 12 motility chambers coated with 50 μg/mL *polylysine*. The polylysine can bundle microtubules. To prevent this, both actin and microtubules are diluted into the high pH AM buffer prior to their introduction into the microscope chambers.
3. Dilute fluorescent F-actin 1:150 (to 1.8 μg/mL or 41 nM) and fluorescent microtubules 1:200 (to 3.5 μg/mL, 64 nM tubulin) into AM Buffer. Actin and microtubules are not stable in this buffer and should be used immediately.
4. Add 5 μL of the microtubules and actin to the polylysine chambers and *immediately* wash 2× with AM Buffer followed by 2× with Blocking Buffer and 2× with Assay Buffer.
5. Incubate vesicles with the actin and microtubules for 5 min, wash, and store chambers on ice.
6. Initiate motility experiments with 1 mM ATP at 30°C. Higher ATP was found to be more effective for actin motility, and 37°C was found to be inhibitory.

With this assay we have found that vesicles isolated from rat liver move short distances (<2 μm) or release from actin filaments upon addition of ATP whereas microtubule-based movement lasts for 40–80 s, covering 10 to 100s of microns (5, 27). Therefore it may be preferable to add both ATP and vesicles during microscopy acquisition, rather than pre-binding the vesicles. In Fig. 2 microtubule and actin-based movement is observed in the same field of view. Microtubules appear as long, thick, filaments while actin is dim. A microtubule-bound vesicle moves left during the 14-s sequence while a second vesicle binds an actin filament at 9 s, moves up, and releases after 14 s. This indicates that endogenously expressed actin and microtubule motors are present on the vesicles.

2.10. Labeling Vesicle Proteins with Fluorescent Antibody

In addition to imaging fluorescent-labeled ligand, we typically quantify the localization and motility of various proteins that are present. This can be achieved by immunofluorescence labeling of epitopes that are found on the exterior (cytosolic) face of the vesicles.

Fixatives are usually avoided so that motile activity is retained and because they are unnecessary. Since the experimental volume and diffusion barriers are low, high antibody concentrations can be used with short incubation times, and typically vesicles can be imaged in less than an hour.

After step 5 of the Vesicle Trafficking Assay (i.e., after 10-min incubation of vesicles with the microtubules, Section 2.8):

1. Wash vesicles 3× 15 μL, using Blocking Buffer for this and subsequent washes and dilutions. This serves to remove unbound vesicles and to block potential nonspecific antibody-binding sites. Alternate blocking reagents may be used.
2. Perfuse 3× 15 μL of primary antibody, incubate for 6 min, and wash 5×. Suggested starting antibody concentration is 5 μg/mL; monoclonal antibodies frequently require higher concentrations. Incubations are typically performed on ice, although some may work better at room temperature.
3. Perfuse 3× 15 μL of fluorescent-labeled secondary antibody (e.g., Anaspec, Inc Cat #61056-1-H555), incubate for 6 min, and wash 5×; suggested antibody concentration is 10 μg/mL.
4. Continue with step 7 of the Vesicle Trafficking Assay (i.e., bring to microscope and initiate motility experiments).

Antibodies must be carefully evaluated. Western blots showing significant nonspecific reactivity should be excluded. Although western blots and unfixed vesicles presumably evaluate denatured and native protein, respectively, in our experience most antibodies with good signal by western blot also exhibit good signal for vesicle immunofluorescence, in contrast to whole cell or histological immunofluorescence where epitopes may be blocked or destroyed by the fixative. The primary and secondary antibodies should be titrated to include zero concentration and a specific increase in fluorescence should be observed. If possible, additional antibodies to the same protein should be used. It is also helpful to compare the immunofluorescence to that of "total membrane" in the field to determine if the signal is concentrated to particular vesicles. The total membrane can be estimated using high-contrast bright field or phase-contrast microscopy (28). We have also found that 2 μM bodipyFL C_5 ceramide (Life Technologies Cat #B22650) will label all the vesicles in a rat liver preparation with uniform intensity.

2.11. Microscope Systems

Many microscope systems will function for vesicle trafficking assays. The main components to consider are the following:

1. Low excitation intensity light, high signal detection: Laser scanning confocal systems have strong excitation light and confocal imaging is not necessary since the contents are perched on a single focal plane. Similarly, spinning disk confocal systems, although less photodamaging, may not have significant

advantage over non-confocal, wide-field imaging. If using confocal we recommend opening the pinhole and reducing laser intensity. Also because longer wavelength light has lower intensity, motility may be more robust in the red and far red wavelengths. Total internal reflection imaging can provide high signal-to-noise ratio and can allow high-resolution automated tracking (29), although switching between multiple wavelengths may be difficult and photodamage is a concern. The objective lens should have high numerical aperture (≥1.4) and high magnification (e.g., 60×). EMCCD cameras give the highest quantum efficiency (signal detected per photon). However their many settings can complicate image quantitation. Additionally, EMCCDs have large pixel size and lower spatial sampling. sCMOS cameras appear to offer many advantages as they are less expensive than EMCCD and have high quantum efficiency, and high frame rates. We have had success with wide field fluorescence imaging and cooled CCD camera that has approximately 60% quantum efficiency (CoolSnap HQ, Photometrics, Inc.).

2. Moderately fast, multichannel fluorescence image acquisition: The assays described have a recommended acquisition speed of 1 set of images per 3 s or faster with up to four fluorescence channels. At 200 ms exposure, this leaves 550 ms to switch channels and to store images. In practice, delays creep into the system and a goal of 100 ms channel switching capability is recommended. This can be accomplished with a white light source and filter wheels (excitation, dichroic, emission), with lasers, or notably with new solid-state excitation sources such as the SOLA Light Engine (Lumencor Inc.). Our system uses a DG-4 xenon 175 W excitation lamp (Sutter Instrument Company) with four excitation filters, a Sedat Quad dichroic mirror (Chroma Technologies), and a Lambda 10–20 emission filter wheel (Sutter Instrument Company, Novato, CA) with proper emission filters for DAPI, FITC, Cy3, and Cy5 or equivalent wavelengths. All imaging is directly to the camera (the eye piece has been removed) and displayed on the computer screen. The devices are run with Metamorph software (Molecular Devices, Sunnyvale, CA). It should be noted that very rapid vesicle movement may be missed at these acquisition speeds and that higher frame rates may reveal novel features of vesicle trafficking.

3. Heating the stage: Most experiments are performed at 37°C and any thermal gradients will alter the focal plane. An enclosure placed around the microscope stage and a hot air heater (Airthermy, World Precision Instruments) can suffice to equilibrate the stage; the temperature at the lens should be carefully monitored.

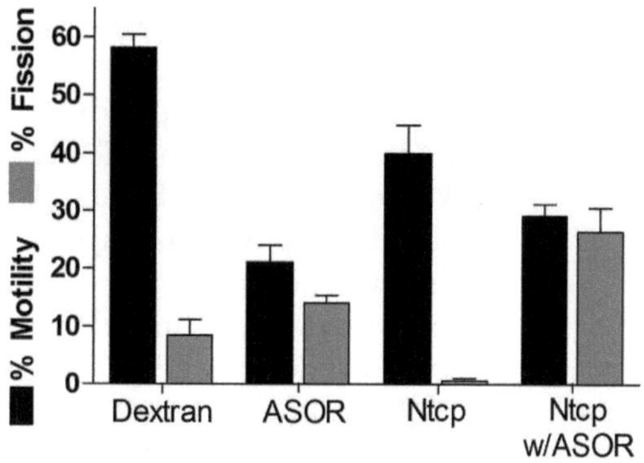

Fig. 3. Quantifying the motility and fission of different populations of vesicles. Vesicle trafficking assays were performed as described in Section 2.8. Different populations of vesicles (as indicated) exhibit different levels of motility and fission, and the level of fission is independently regulated from motility. Texas Red-labeled amino dextran and Texas Red ASOR endocytic vesicles were purified in separate isolations from rat liver as described (Sections. 2.6 and 2.7). Rat liver vesicles were labeled by immunofluorescence for the bile acid transporter, Ntcp, according to methods of Section 2.10, and vesicles were colocalized for the presence of both Ntcp and ASOR. Interestingly these vesicles had moderate motility (29%) but very high fission (26%), suggesting that vesicles containing proteins directed toward different cellular compartments, i.e., the plasma membrane (ntcp) and lysosome (ASOR), may have higher rates of fission. Error, standard error of the mean.

2.12. Data Analysis

Initially, an investigator may wish to know how much motility is observed for different populations of vesicles and under various treatments. This can be answered by identifying vesicles and manually scoring whether movement occurs, relative to the cytoskeleton, upon addition of ATP. It is helpful to establish rigorous distance and time intervals for scoring since back-and-forth diffusive movement is also observed. Gliding or bending or other factors may preclude scoring particular vesicles. The vesicles should be identified prior to scoring since their movement will increase their visibility and lead to overestimation of the motility. Some vesicles undergo fission, or splitting of their contents (30). With endocytic vesicles this can be manually scored as the separation of fluorescent puncta after addition of ATP. Fission has been shown to be a specific, independently regulated activity of endocytic vesicles and not a result of the movement overlapping vesicles (31). Figure 3 demonstrates the motility and fission of different populations of vesicles using this technique. Rat liver vesicles that contain the endocytic marker, fluorescent dextran, exhibit high motility (58%) with moderate fission (8%) while vesicles containing the ASOR ligand show moderate motility (22%) with relatively high fission (14%). Interestingly, vesicles that contain both lysosomally directed ASOR and a plasma membrane-directed bile acid transporter, ntcp, show highest fission (26%), suggesting that high rates of fission may correlate with vesicles that contain proteins that are destined for different cellular compartments.

Many detailed parameters of vesicle trafficking can also be quantified, and the public domain program, ImageJ, developed by Wayne Rasband at the US National Institutes of Health (http://rsbweb.nih.gov/nih-image/) is an excellent resource for quantification methods. For instance, the position of vesicles can be tracked using ImageJ Manual Tracking software "plugin" to reveal run length, instantaneous velocity, and movement periodicity. Additional technical advances using high-contrast, rapid image acquisition have allowed automated particle tracking with nanometer precision (4, 32) and combined with super-resolution techniques (33) may provide spatial–temporal "signatures" for specific vesicle trafficking events, although this has so far remained elusive. The direction of motility relative to microtubule polarity mark has been scored manually in trafficking assays and has revealed the bidirectional nature of early and late endocytic vesicles from rodent liver, HeLa, and other cell types (20, 34). Chemical inhibitors, when used appropriately, can help distinguish the activity of different motors. For instance, inclusion of 5 μM vanadate or 1 mM AMP-PNP along with ATP in the trafficking assays has been shown to inhibit dynein- and kinesin-based motility, respectively (1, 5, 35). We have also found that phosphoinositide 3-kinasese inhibitors will reduce vesicle motility, and that a specific protein kinase c zeta inhibitor will block motility of vesicles that contain the bile acid transporter, ntcp, but not late endosomes (35). Another means of determining motor specificity is the use of inhibitory antibodies. Specific antibodies to kinesin and dynein have been effective at inhibiting motility in vitro and when microinjected into cells (36). Typically these are incubated with the microtubule-bound vesicles for 5 min prior to washing and initiation of motility.

2.13. Protein Colocalization

A major advantage of vesicle trafficking assays is that they can be used to quantify protein colocalization. Scoring colocalization can be automated using macro programs within the ImageJ software. The strategy is to allow biochemically isolated organelles to attach either directly to the glass surface or else to cytoskeletal filaments that are bound to microscope chambers. Soluble material is washed away and immunofluorescence is performed. This can be repeated for multiple primary antibodies provided that they do not cross-react. Because the isolated vesicles are attached to a 2-dimensional substrate, image complexity and background intensity are minimized and it is straightforward to "segment" the image into fluorescent spots and background. To report colocalization, users can measure the total area of the spots that overlap, or they can count the percentage of spots that show fluorescence overlap (28).

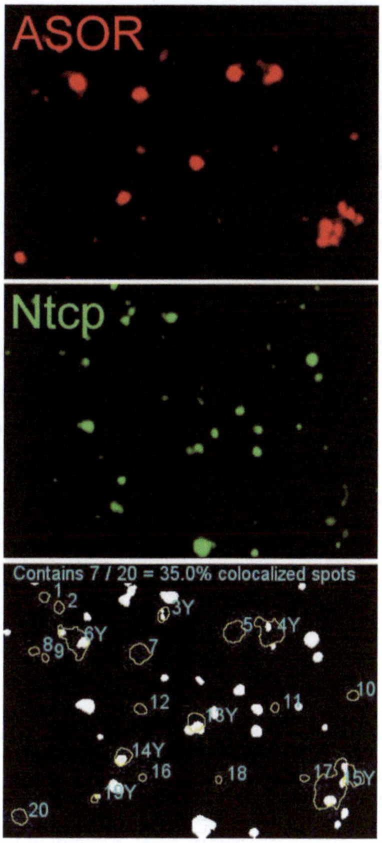

Fig. 4. Demonstration of the quantification of protein colocalization using endocytic vesicles and the "simple colocalization" macro presented in Sect. 3.1. Rat liver vesicles containing fluorescent-labeled ASOR (*top panel*) were incubated within an untreated microchamber. The vesicles bound to the chamber were co-stained for the bile acid transporter, Ntcp (*middle panel*), and imaged. An individual cropped, 9 × 13 μm field is presented for the two fluorescent channels as well as a final image (*bottom panel*) that results from running the macro within ImageJ, resulting in a score of 35% colocalization. Typically 1,000s of vesicles and numerous experiments are quantified, and image processing, thresholding, and error checking are performed to suit the image data sets.

Figure 4 shows an example of a field of ASOR vesicles that has been co-stained for the bile acid transporter, ntcp, and quantified for the percentage of colocalized (overlapping) vesicles using the "Simple Colocalization" macro presented below. This macro numbers the ASOR vesicles and displays a "Y" (yes) for any vesicle showing ntcp colocalization. The total is summarized directly on the image. The code for this Simple Colocalization macro is provided below to let the reader explore the ImageJ macro language in this context. The ImageJ Web site provides excellent help and many macro examples. Some familiarity with computer programming may be necessary.

3. Note

3.1. Simple Colocalization Macro Program for ImageJ to Demonstrate Colocalization of Fluorescent Spots

Download the ImageJ program at http://rsbweb.nih.gov/ij/download.html.

Start ImageJ and create a new text window (File→New→Text Window). Then copy and paste the text below into this window. Open two images of the same size. They should be 8-bit or 16-bit gray scale (Image→Type→8-bit). For instance one can copy and paste the images of Fig. 2 into ImageJ. Run the macro (in the text window: Macros→Run Macro). Note that "//" denotes a comment, is not read by the program, and instead provides a running description of the operations that are being performed.

```
macro 'Simple Colocalization [z]' // Press 'z' as shortcut to run the macro
{                                          // curly brackets enclose the macro operations
requires("1.44e");                // ImageJ version 1.44e required for 'Overlay' text functions
MinSpotSize = 30;                 // Spot must be greater than this number of pixels
MaxSpotSize = 2000;                        // Spot cannot be greater than this number of pixels
MinimumForColocalization = 2;              // Min % overlap to be scored as positive for colocalization
SpotsWithColocalization = 0;      // Initialize variable
setLineWidth(1);                  // Line width of drawings
setForegroundColor(255, 255, 0);  // Outline color, Yellow
setColor("Cyan");                 // Text color, Cyan
setFont("SanSerif", 20, "antialiased");
run("Set Measurements…", "centroid redirect=None decimal=2");
selectImage(2);                   // Select 2nd, "Green" image that was opened
GreenImage = getTitle();
setAutoThreshold("Triangle dark"); // Automatic thresholding using the 'Triangle' method
                  // Below: find the spots, make 'Mask' used for colocalization
run("Analyze Particles…", "minimum=" + MinSpotSize + " maximum=" + MaxSpotSize+ " bins=20
show=Masks display clear ");
MaskGreen = getTitle();
run("Invert");                             // Invert the image, black to white
run("RGB Color");                 // Convert to color image
selectImage(2);
close();                                   // Close "Green" image
selectImage(1);                   // Select 1st, "Red" image opened, base image
run("Set Scale…", "distance=0 known=0 pixel=1 unit=pixel"); // Remove any distance scaling
RedImage = getTitle();
setAutoThreshold("Triangle dark");
                  // Below: find the spots on base, Red image, make 'Mask' used for colocalization
run("Analyze Particles…", "minimum=" + MinSpotSize + " maximum=" + MaxSpotSize+ " bins=20
show=Masks display clear ");
MaskRed = getTitle();
run("Invert");
selectImage(1);
close();                                   // Close original "Red" image
NumberOfSpots = nResults;         // Stores the number of spots
for (i = 0; i < NumberOfSpots; i++) // For loop: Go thru each spot and check for colocalization
{selectWindow (MaskRed);
// Below: Use wand tool at positions from 'Analyze Particles' function
doWand(getResult("X", i), getResult("Y",i));
selectWindow (MaskGreen);
run("Restore Selection");
                  // Get number of black pixels, counts [0], white pixels, counts [255], in spots
getHistogram(values, counts, 256);
FluorescencePixelsInSpot =counts[255];
TotalPixelsInSpot = counts[255] + counts[0];
PercentColocalizationOfSpot = FluorescencePixelsInSpot / TotalPixelsInSpot * 100;
if ( PercentColocalizationOfSpot > MinimumForColocalization )
{IsSpotColocalized = "Y";         // Spot colocalizes? "Yes"
SpotsWithColocalization++;
} else
```

```
IsSpotColocalized = "";                  // Empty "" when not colocalized
run("Draw");                             // Draw selection
                                 //Below: Number the spots, "Y" if colocalized
Overlay.drawString( (i+1)+ IsSpotColocalized, getResult("X", i) + 10, getResult("Y",i) );
Overlay.show();
} // for loop
                  //Below: print % colocalization on the image
TotalColocalizedSpots = SpotsWithColocalization / NumberOfSpots * 100;
setColor("Black");                       // Draw black box beneath text for visibility
setLineWidth(10);
Overlay.drawRect(10, 5, 372, 10)
setColor("Cyan");
Overlay.drawString("Contains " + SpotsWithColocalization + " / " + NumberOfSpots +
" = " + d2s (TotalColocalizedSpots ,1) + "%" + " colocalized spots", 10 , 16 );
Overlay.show();
} // End Simple Colocalization macro
```

The macro may give error messages or problems with some images. Please contact the author with any questions. For actual analysis, alternate thresholding methods are employed along with image processing and error detection. Determining the threshold method is critical for quantifying data and this should be investigated by the user. Most commands within ImageJ can be automated and incorporated into macros using the "record" function (Plugins→Macros→Record). The ability to quantify protein colocalization in this manner has, for instance, allowed studies of protein sorting during endocytic fission in vitro and, when combined with proteomic analysis, has helped demonstrate an unexpected role for Rab1a in allowing early endosome trafficking progress toward protein degradation (31, 37).

4. Summary

Instructions for performing cytoskeletal-based in vitro vesicle trafficking assays have been provided along with an overview and discussion of how these can be quantified and interpreted. The goal has been to be as explicit as possible so that investigators with varying experience will find them helpful. The assays themselves are especially suitable for identifying the immediate and primary effects of protein or chemical treatments on vesicle trafficking, since reagents can be added directly to organelles and their effects studied instantaneously in an environment free of cytosol. The assays are also easy to quantify since they avoid the complicated 3-dimensional environment of the living cell. The primary difficulties in producing reliable data are the vesicle isolation procedure and the imaging system, which should be minimally damaging to proteins. These approaches should continue to be useful in answering important questions of cell biology such as how motor proteins are regulated and how populations of vesicles are differentiated.

References

1. Vale RD, Reese TS, Sheetz MP (1985) Identification of a novel force-generating protein, kinesin, involved in microtubule-based motility. Cell 42(1):39–50
2. Lye RJ, Porter ME, Scholey JM, McIntosh JR (1987) Identification of a microtubule-based cytoplasmic motor in the nematode *C. elegans*. Cell 51(2):309–318. doi:0092-8674(87)90157-7 [pii]
3. Waterman-Storer CM (2001) Microtubule/organelle motility assays. Curr Protoc Cell Biol 11 doi:10.1002/0471143030.cb1301s00 (Chapter 13:Unit 13)
4. Hendricks AG, Perlson E, Ross JL, Schroeder HW 3rd, Tokito M, Holzbaur EL (2010) Motor coordination via a tug-of-war mechanism drives bidirectional vesicle transport. Curr Biol 20(8):697–702. doi:10.1016/j.cub.2010.02.058, S0960-9822(10)00331-3 [pii]
5. Murray JW, Bananis E, Wolkoff AW (2000) Reconstitution of ATP-dependent movement of endocytic vesicles along microtubules in vitro: an oscillatory bidirectional process. Mol Biol Cell 11(2):419–433
6. Hyman A, Drechsel D, Kellogg D, Salser S, Sawin K, Steffen P, Wordeman L, Mitchison T (1991) Preparation of modified tubulins. Methods Enzymol 196:478–485
7. Pollock N, Koonce MP, de Hostos EL, Vale RD (1998) In vitro microtubule-based organelle transport in wild-type Dictyostelium and cells overexpressing a truncated dynein heavy chain. Cell Motil Cytoskeleton 40(3):304–314. doi:10.1002/(SICI)1097-0169(1998)40:3<304::AID-CM8>3.0.CO;2-C [pii]
8. Hyman AA, Salser S, Drechsel DN, Unwin N, Mitchison TJ (1992) Role of GTP hydrolysis in microtubule dynamics: information from a slowly hydrolyzable analogue, GMPCPP. Mol Biol Cell 3(10):1155–1167
9. Soppina V, Rai A, Mallik R (2009) Simple non-fluorescent polarity labeling of microtubules for molecular motor assays. Biotechniques 46(7):543–549. doi:000113124 [pii]10.2144/000113124
10. Kim T, Kao MT, Hasselbrink EF, Meyhofer E (2007) Active alignment of microtubules with electric fields. Nano Lett 7(1):211–217. doi:10.1021/nl061474k
11. Yokokawa R, Tarhan MC, Kon T, Fujita H (2008) Simultaneous and bidirectional transport of kinesin-coated microspheres and dynein-coated microspheres on polarity-oriented microtubules. Biotechnol Bioeng 101(1):1–8. doi:10.1002/bit.21874
12. Yanagida T, Nakase M, Nishiyama K, Oosawa F (1984) Direct observation of motion of single F-actin filaments in the presence of myosin. Nature 307(5946):58–60
13. Kron SJ, Spudich JA (1986) Fluorescent actin filaments move on myosin fixed to a glass surface. Proc Natl Acad Sci U S A 83(17):6272–6276
14. Geuze HJ, Slot JW, Strous GJ, Lodish HF, Schwartz AL (1983) Intracellular site of asialoglycoprotein receptor-ligand uncoupling: double-label immunoelectron microscopy during receptor-mediated endocytosis. Cell 32(1):277–287
15. Oka JA, Weigel PH (1983) Recycling of the asialoglycoprotein receptor in isolated rat hepatocytes. Dissociation of internalized ligand from receptor occurs in two kinetically and thermally distinguishable compartments. J Biol Chem 258(17):10253–10262
16. Steer CJ, Ashwell G (1980) Studies on a mammalian hepatic binding protein specific for asialoglycoproteins. Evidence for receptor recycling in isolated rat hepatocytes. J Biol Chem 255(7):3008–3013
17. Hubbard AL, Wilson G, Ashwell G, Stukenbrok H (1979) An electron microscope autoradiographic study of the carbohydrate recognition systems in rat liver. I. Distribution of 125I-ligands among the liver cell types. J Cell Biol 83(1):47–64
18. Weibel ER, Staubli W, Gnagi HR, Hess FA (1969) Correlated morphometric and biochemical studies on the liver cell. I. Morphometric model, stereologic methods, and normal morphometric data for rat liver. J Cell Biol 42(1):68–91
19. Hagiwara H, Yorifuji H, Sato-Yoshitake R, Hirokawa N (1994) Competition between motor molecules (kinesin and cytoplasmic dynein) and fibrous microtubule-associated proteins in binding to microtubules. J Biol Chem 269(5):3581–3589
20. Loubery S, Wilhelm C, Hurbain I, Neveu S, Louvard D, Coudrier E (2008) Different microtubule motors move early and late endocytic compartments. Traffic 9(4):492–509. doi:TRA704 [pii]10.1111/j.1600-0854.2008.00704.x
21. Soppina V, Rai AK, Ramaiya AJ, Barak P, Mallik R (2009) Tug-of-war between dissimilar teams of microtubule motors regulates transport and fission of endosomes. Proc Natl Acad Sci U S A 106(46):19381–19386. doi:0906524106 [pii]10.1073/pnas.0906524106

22. Ghosh RN, Gelman DL, Maxfield FR (1994) Quantification of low density lipoprotein and transferrin endocytic sorting HEp2 cells using confocal microscopy. J Cell Sci 107(Pt 8): 2177–2189
23. Schroer TA, Schnapp BJ, Reese TS, Sheetz MP (1988) The role of kinesin and other soluble factors in organelle movement along microtubules. J Cell Biol 107(5):1785–1792
24. Kishino A, Yanagida T (1988) Force measurements by micromanipulation of a single actin filament by glass needles. Nature 334(6177):74–76. doi:10.1038/334074a0
25. Kapoor TM, Mitchison TJ (1999) Allele-specific activators and inhibitors for kinesin. Proc Natl Acad Sci U S A 96(16):9106–9111
26. Bananis E, Murray JW, Stockert RJ, Satir P, Wolkoff AW (2000) Microtubule and motor-dependent endocytic vesicle sorting in vitro. J Cell Biol 151(1):179–186
27. Fort AG, Murray JW, Dandachi N, Davidson MW, Dermietzel R, Wolkoff AW, Spray DC (2011) In vitro motility of liver connexin vesicles along microtubules utilizes kinesin motors. J Biol Chem 286(26):22875–22885. doi:M111.219709 [pii]10.1074/jbc.M111.219709
28. Murray JW, Bananis E, Wolkoff AW (2002) Immunofluorescence microchamber technique for characterizing isolated organelles. Anal Biochem 305(1):55–67
29. Ross JL, Wallace K, Shuman H, Goldman YE, Holzbaur EL (2006) Processive bidirectional motion of dynein-dynactin complexes in vitro. Nat Cell Biol 8(6):562–570. doi:ncb1421 [pii]10.1038/ncb1421
30. Pucadyil TJ, Schmid SL (2008) Real-time visualization of dynamin-catalyzed membrane fission and vesicle release. Cell 135(7):1263–1275. doi:S0092-8674(08)01495-5 [pii]10.1016/j.cell.2008.11.020
31. Murray JW, Sarkar S, Wolkoff AW (2008) Single vesicle analysis of endocytic fission on microtubules in vitro. Traffic 9(5):833–847. doi:TRA725 [pii]10.1111/j.1600-0854.2008.00725.x
32. Hoffman MT, Sheung J, Selvin PR (2011) Fluorescence imaging with one nanometer accuracy: in vitro and in vivo studies of molecular motors. Methods Mol Biol 778:33–56. doi:10.1007/978-1-61779-261-8_4
33. Patterson G, Davidson M, Manley S, Lippincott-Schwartz J (2010) Superresolution imaging using single-molecule localization. Annu Rev Phys Chem 61:345–367. doi:10.1146/annurev.physchem.012809.103444
34. Nath S, Bananis E, Sarkar S, Stockert RJ, Sperry AO, Murray JW, Wolkoff AW (2007) Kif5B and Kifc1 interact and are required for motility and fission of early endocytic vesicles in mouse liver. Mol Biol Cell 18(5):1839–1849. doi:E06-06-0524 [pii]10.1091/mbc.E06-06-0524
35. Sarkar S, Bananis E, Nath S, Anwer MS, Wolkoff AW, Murray JW (2006) PKCzeta is required for microtubule-based motility of vesicles containing the ntcp transporter. Traffic 7(8):1078–1091. doi:TRA447 [pii]10.1111/j.1600-0854.2006.00447.x
36. Murray JW, Wolkoff AW (2003) Roles of the cytoskeleton and motor proteins in endocytic sorting. Adv Drug Deliv Rev 55(11): 1385–1403
37. Mukhopadhyay A, Nieves E, Che FY, Wang J, Jin L, Murray JW, Gordon K, Angeletti RH, Wolkoff AW (2011) Proteomic analysis of endocytic vesicles: Rab1a regulates motility of early endocytic vesicles. J Cell Sci 124(Pt 5):765–775. doi:jcs.079020 [pii]10.1242/jcs.079020

Part II

Interaction, Dynamics, and Proteomics

Chapter 3

Interaction of the Glycocalyx with the Actin Cytoskeleton

Mia M. Thi, Eno E. Ebong, David C. Spray, and Sylvia O. Suadicani

Abstract

In this chapter, we summarize and describe various methods that have been used to detect glycocalyx and to study the interactions between glycocalyx and actin cytoskeleton with the emphasis in the nervous tissue. We highlight two methodologies to analyze the direct interaction between glycocalyx and actin cytoskeleton: one method focusing on visualization using immunostaining and confocal microscopy techniques and the other focusing on direct protein–protein interactions using co-immunoprecipitation techniques.

Key words: Heparan sulfate, Chondroitin sulfate, Hyaluronic acid, Actin cytoskeleton, Astrocyte

1. Introduction

Cellular form is provided by both external and internal skeletons. The cellular exoskeleton consists of a complex macromolecular meshwork that comprises the extracellular matrix (ECM), also termed the cell surface coat or glycocalyx. The endoskeleton is formed by actin, intermediate filaments, and microtubules, which together with bundling elements (e.g., actinin, myosin, villin, etc.) form the cytoskeleton. The endo- and exoskeletons are linked both physically and functionally so that extracellular stimuli are transduced into intracellular responses. In this chapter we briefly review components of the two skeletons, focusing on the nature of the direct interaction between the actin cytoskeleton and glycocalyx in astroglia and other cell types and methods by which such interactions can be investigated.

1.1. The Glycocalyx

The glycocalyx provides a membrane-bound extracellular environment that creates, facilitates, and stabilizes molecular interactions regulating diverse cellular functions. The spatiotemporal expression of the glycocalyx components in the nervous system is highly

Rolf Dermietzel (ed.), *The Cytoskeleton: Imaging, Isolation, and Interaction*, Neuromethods, vol. 79,
DOI 10.1007/978-1-62703-266-7_3, © Springer Science+Business Media, LLC 2013

regulated and associated with their essential roles in proliferation and differentiation of neural progenitor cells, cell migration, axonal pathfinding, synaptogenesis, neural plasticity, and regeneration.

The molecular components of the glycocalyx include proteoglycans and hyaluronic acid (or hyaluronan). These proteoglycans consist of a core protein bound to one or more glycosaminoglycans (GAGs) that include heparin, heparan, dermatan, keratin, and chondroitin sulfate families. They are structurally diverse, owing to the large number of core proteins and the number, type, and stoichiometry of the GAG components. Certain proteoglycans also exhibit temporal diversity such that their association with the GAG chain can be altered by shedding or interaction with other ECM components. Some proteoglycans are secreted and contribute to the glycocalyx through binding to other structural components while others are cell surface bound or transmembrane. The transmembrane proteoglycans contain extracellular, transmembrane, and cytoplasmic domains. The extracellular domains interact with components of the exoskeleton, cell surface receptors, and signaling molecules, while the intracellular domains interact with the PDZ domains of several intracellular proteins and may also interact directly with kinases and cytoskeletal proteins (Fig. 1) (1–3).

Members of the lectican/hyalectan family (aggrecan, versican, neurocan and brevican) of secreted chondroitin sulfate proteoglycans (CSPG) bind prominently to hyaluronic acid in the nervous system ECM and are particularly abundant in this tissue (4, 5). Of the lectican members, neurocan, versican V2, and brevican are central nervous system specific (5). Phosphacan, a secreted CSPG-isoform of receptor-type protein-tyrosine phosphatase β (RPTPβ), is also nervous tissue specific and, like lecticans, has been shown to play major roles in the brain ECM (6–9).

The membrane-bound group of proteoglycans comprises a diverse set of membrane-spanning and membrane-anchored glycoproteins. Among the transmembrane proteoglycans are the melanoma-CSPG (NG2), neuropilin-1, betaglycan, CD44, members of the syndecan family (2), and neuroglycan-C, a brain-specific CSPG (10). The membrane-anchored proteoglycans include the members of the glypican family of heparan sulfate proteoglycans (HSPG), which bind to the outer cell membrane through their C-terminal glycosylphosphatidylinositol (GPI) anchors (11). The GPI-anchored HSPGs often function as coreceptors, interacting in particular with integrins and growth factor receptors. The transmembrane proteoglycans (TM-PG) have also the ability to independently signal through their cytoplasmic carboxyl terminal domains by way of intracellular interactions with cytoskeletal and other proteins (Figs. 1 and 3). Unlike typical cell surface receptors, the TM-PGs can bind to numerous and diverse ligands, such as growth factors, enzymes, cytokines, and chemokines, which considerably broadens their functional repertoire. In the case of the

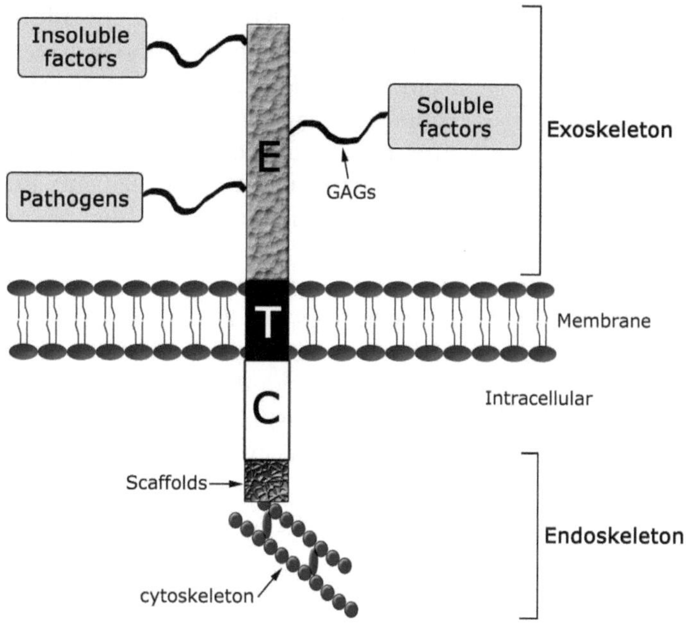

Fig. 1. Schematic diagram of transmembrane proteoglycan core protein, illustrating the extracellular (E), transmembrane (T), and cytoplasmic (C) domains. The extracellular domain is a major component of the glycocalyx, a complex macromolecular meshwork, where it binds to one or more glycosaminoglycans (GAGs). The GAGs attached to the extracellular domains of the proteoglycans interact with pathogens (bacteria, viruses, etc.) and insoluble (other GAGs, proteoglycans, collagen, integrins, etc.) and soluble factors (growth factors, enzymes, peptides, proteins, etc.), forming the exoskeleton. The transmembrane and cytoplasmic domains physically and functionally link the exoskeleton to the endoskeleton. Through this linkage extracellular stimuli are transduced into intracellular responses. The endoskeleton, formed by actin, intermediate filaments, and microtubules, is held together by bundling proteins such as actinin, myosin, and villin and binds to the cytoplasmic aspect of transmembrane proteoglycans through scaffolding elements (adapted from (1)).

syndecan family members (syndecan-1, syndecan-2, syndecan-3, and syndecan-4) this ability to respond to a wealth of ligands parallels the diversity of their extracellular domains and their GAG substitutions (1). In contrast, because of their highly conserved transmembrane and cytoplasmic domains, the multitude of extracellular signals that syndecans convey to the cells converges onto common intracellular transduction pathways. For example, the cytoplasmic domain of all members of the syndecan family of transmembrane proteoglycans can interact with proteins of the actin-associated cytoskeleton, supporting a key role for TM-PG in regulation of cytoskeleton-dependent cellular functions (2, 3, 12).

1.2. The Cytoskeleton

The cytoskeleton dictates cell shape and polarity, applies forces against substrates and adjacent cells to achieve locomotion, and anchors to and assists junction formation with substrate, extracellular matrix, and adjacent cells. In addition, it forms the spindles necessary for cell division and provides transport of proteins to and from the surface membrane participating in intracellular signaling, endocytosis, and secretion events. The cytoskeleton is made up of three distinct filament classes: microtubules, intermediate filaments, and actin filaments. These filaments are held together and linked to cellular organelles and to the plasma membrane by accessory proteins.

1.2.1. Microtubules

Microtubules are 25–30 nm rods formed by α- and β-tubulin heterodimers arranged as a tube with a cross section of 13 tubulin molecules (subunit Mr: 55 kDa). Microtubule-associated proteins (MAPs and tau) bind to microtubules and increase stability; stabilization efficiency is regulated by MAP/tau phosphorylation. Microtubule motor proteins, which carry vesicular cargoes along the microtubules, include kinesin and dynein.

1.2.2. Intermediate Filaments

Intermediate filaments (IF), so-called because they are intermediate in size (10 nm) between actin and tubulin polymers, are composed of a variety of approximately 35 kDa proteins; glial fibrillary acidic protein (GFAP), vimentin, and nestin are the primary IF proteins in astrocytes. Protein subunits are elongated in shape, with a central rod-shaped domain and globular head and tail domains. Functions of intermediate filaments include roles in migration and process formation.

1.2.3. Actin

Actin molecules are globular (G actin) 43 kDa proteins that polymerize end to end into filamentous (F) actin, forming cables 7 nm in diameter and many micrometers in length. Actin filaments form a prominent network below the membrane surface, mediating changes in cell shape, and generating the forces necessary for movement (13). Contractile bundles of actin filaments cross-linked by α-actinin (and thereby providing spaces in which myosins can bind, a process facilitated by lengthwise binding of tropomysin to actin filaments) form stress fibers that are attached at focal adhesions to integrins and include other proteins such as talin and vinculin. Actin is also attached to the membrane at adherens junctions through catenin attachment to cadherin. Actin networks form the cortical cytoskeleton, which is attached to membrane proteins including spectrin, ankyrin, and protein 4.1. Certain myosins (e.g., myosin V) function to carry cargoes along actin filaments, analogous to the vesicular transport along microtubules mediated by kinesins and dynein.

1.2.4. Interactions of the Exo- and Endoskeletons

The linkage that extracellular matrix components provide between the outside of the cell and the cytoskeleton is critical for both the functions that individual cells perform and also for those of the tissues that they comprise. Particularly noteworthy in this regard are the syndecans, the major TM-PG that interact with components of the actin cytoskeleton, and junctional proteins, which form multiprotein signaling complexes. Whereas other chapters in this volume (see Chapters 1, 4, 5, 6 this volume) discuss methods used for the study of the cytoskeleton, in the sections of this chapter we provide overviews and detailed protocols for identification of extracellular matrix components and their interactions with cytoskeletal elements.

1.3. Overview of Methods that Have Been Used for Glycocalyx Detection

1.3.1. Electron Microscopy Analysis

The existence of a glycocalyx in mammalian cells was first reported almost half a century ago by Gasic (14) in mouse ascites tumor cells fixed in 10% neutral phosphate-buffered formalin using Hale stain (ferric ferrocyanine precipitation) and electron microscopy. This study showed that pretreatment with neuraminidase (100 μg/ml) abolished the staining, thereby providing histochemical evidence that the extracellular matrix contains a sialic acid-rich layer. In subsequent studies Rambourg and coworkers (15) found glycoproteins and acidic carbohydrates within the extracellular matrix of adult rat tissues (fibrocytes, blood cells, epithelial and nervous system cells) using periodic acid-silver methenamine and colloidal thorium. For these studies, tissue pieces were fixed in 2.5% glutaraldehyde and postfixed in 2% osmium tetroxide to enhance the contrast between cellular components. Further electron microscopy analysis revealed that the glycocalyx of the nerve cells could be traced along the dendrites and axons, around glial processes, and even within synaptic clefts.

Concomitantly, the existence of the glycocalyx in endothelial cells was demonstrated using staining with ruthenium red, a dye that binds to negatively charged glycoproteins (16). Subsequent *in situ* and *in vitro* studies using electron-dense tracers such as cationized ferritin or Alcian Blue further characterized the structure and properties of the endothelial glycocalyx. Taking advantage of the large molecular size of cationized ferritin to outline glycocalyx in transmission electron micrographs of 2.5% glutaraldehyde fixed frog mesenteric microvessels, Adamson and Clough (17) demonstrated that integrity of endothelial glycocalyx depended on the presence of plasma proteins such as albumin. When Rostgaard and Qvortrup (18) used a new method of perfusion–fixation (2% glutaraldehyde dissolved in 13.3% oxygen-carrying fluorocarbon blood substitute and postfixation with 1% osmium tetroxide), they observed bush-like filament structures in the capillary fenestrae of the rat small intestine. However, the ultrastructural organization of the endothelial glycocalyx was not well known until the studies by Squire et al. (19) applying computational methods to deconvolve glycocalyx images obtained from conventional and freeze fracture electron microscopy. These studies revealed that the glycocalyx layer is a quasi-periodic bush-like structure. More recently, van den Berg and coworkers (20)

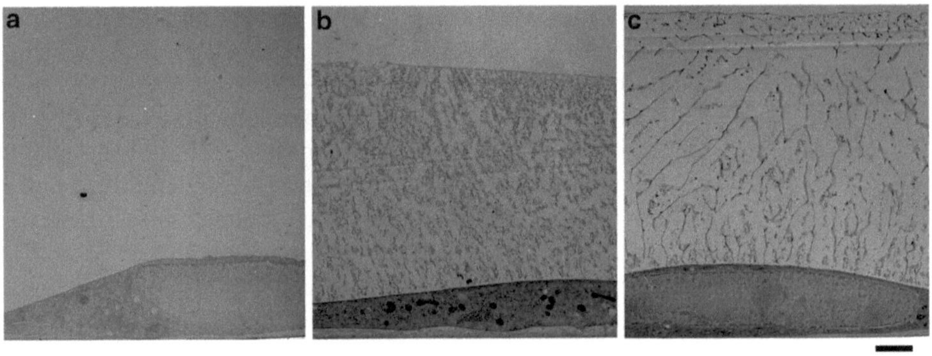

Fig. 2. Detection of endothelial glycocalyx after rapid freezing/freeze substitution. (**a**) Transmission electron microscopy (TEM) image of bovine aortic endothelial cell (BAEC) preserved using conventional paraformaldehyde/glutaraldehyde fixation, labeled with ruthenium red and osmium tetroxide, and alcohol dehydrated. The glycocalyx structure cannot be seen at this level of magnification. (**b**) TEM image of BAEC preserved using rapid freezing/freeze substitution (RF/FS) and stained with osmium tetroxide. A substantial glycocalyx layer can be seen, and its morphology is that of a fibrous mesh. (**c**) TEM image of RF/FS preserved BAEC treated with Haparinase III to degrade the heparan sulfate component of the glycocalyx. The Haparinase-treated BAEC glycocalyx is less dense and exhibits more elongated elements than its untreated counterpart. Bar = 2 μm.

stabilized the anionic carbohydrate structures in rat myocardial capillaries with Alcian blue 8GX, following perfusion with oxygenated calcium-free cardioplegic solution containing 0.1% BSA to remove blood prior to phosphate-buffered fixation with $MgCl_2$ followed by 0.05% Alcian blue staining, and postfixed with 1% osmium tetroxide and 1% lanthanum nitrate. Their results demonstrated that the luminal surfaces of myocardial capillaries were uniformly lined with dense hairlike structures, which were removed by pretreatment with the hyaluronic-specific degradation enzyme hyaluronidase (25 IU/ml).

In recent studies we used rapid freezing/freeze substitution (RF/FS) preservation and osmium tetroxide staining to evaluate the structure of the glycocalyx on the surface of bovine aortic endothelium (21). Whereas traditional transmission electron microscopy (TEM) techniques do not normally detect the glycocalyx in situ (Fig. 2a) due to dehydration artifacts, we found that RF/FS-TEM stably preserved the glycocalyx in its hydrated and protein-rich configuration and revealed a mesh structure (Fig. 2b) unlike the shrub-like appearance that has been previously reported (21). Pretreatment with the heparan sulfate-specific degradation enzyme heparinase (15 mU/ml, a concentration that does not completely remove the glycocalyx (21)) altered the ultrastructure of the glycocalyx (Fig. 2c), which became less dense and more elongated than in control untreated conditions (Fig. 2b). In short, RF/FS-TEM clearly discerns differences in glycocalyx structure due to altered local environment. Moving forward, we believe that this technique will be extremely useful to elucidate details in the linkage between the glycocalyx and the actin cytoskeleton.

1.3.2. Epifluorescence, Immunohistochemistry, and Confocal Microscopy Analysis

As indicated above, molecular components of the glycocalyx are now reasonably well characterized, and antibodies against many of the core glycoproteins are available. In one of the first studies to use such tools for detecting surface expression of HSPG, Rapraeger and coworkers (22) modified conventional immunostaining by incubating live epithelial cell cultures with primary and secondary antibodies at 4°C followed by sequential acetone fixation (50%, 1 min; 100%, 5 min; 50%, 1 min) on ice. They generated a rat monoclonal IgG2A that was specific for the core protein of the epithelial cell HSPG and found that HSPGs appeared as dots of epifluorescence uniformly dispersed over the cell surface. They concluded that this staining pattern was due to HSPG cross-linking, which was unaffected when the primary antibody incubation time was varied from 20 min to 1 h on ice. However, they found that when live cells were incubated with the antibody at 37°C, the distribution of HSPG clusters rapidly changed, being assimilated into larger sized aggregates on the apical surface and also lining the margins of the cells, an event that was observed as early as 10 min after incubation. At 1 h after incubation at 37°C they observed a marked loss of HSPG staining and appearance of fewer aggregates at the cell borders.

In subsequent studies Carey and Stahl (23) used a similar procedure to analyze the surface expression of HSPG in primary Schwann cells extracted from rat sciatic nerves. They incubated chilled live cells for 1 h with primary antibody prior to fixation. Their epifluorescence analysis indicated that HSPG was expressed in small spikelike, actin-rich structures that protruded from the surface of Schwann cells. We evaluated the distribution of cell surface HSPGs in cultured endothelial cells (13) using methods adapted from those described above (22, 23), noting that chilling live cells during primary antibody incubation was the crucial step in detecting cell surface expression of proteoglycans, which otherwise undergo internalization. For this we used Celltracker Orange (Invitrogen) to outline the cell cytosolic compartment, and cross-sectional confocal analysis to show the cell surface expression of HSPG in endothelial cells. Alternatively, slight modification in fixation also works well, as demonstrated in our recent studies using 2% paraformaldehyde/0.1% glutaraldehyde fixation prior to primary antibody incubation (21). Similar live immunostaining was also performed by Nishimura and coworkers (24) in hippocampal neuron cultures. Their technique of incubating live cells with primary antibody for a shorter time (15 min) at 37°C worked equally well. Using specific antibodies, they showed that CSPG expression was localized at focal contacts in axons and cell bodies while HSPG expression was ubiquitously distributed on the cell surfaces of dendrites and axons. Cell surface distribution of HSPG and CSPG is also observed in cortical astrocyte cultures following live cell immunostaining, as shown in Figs. 3 and 4. The use of this

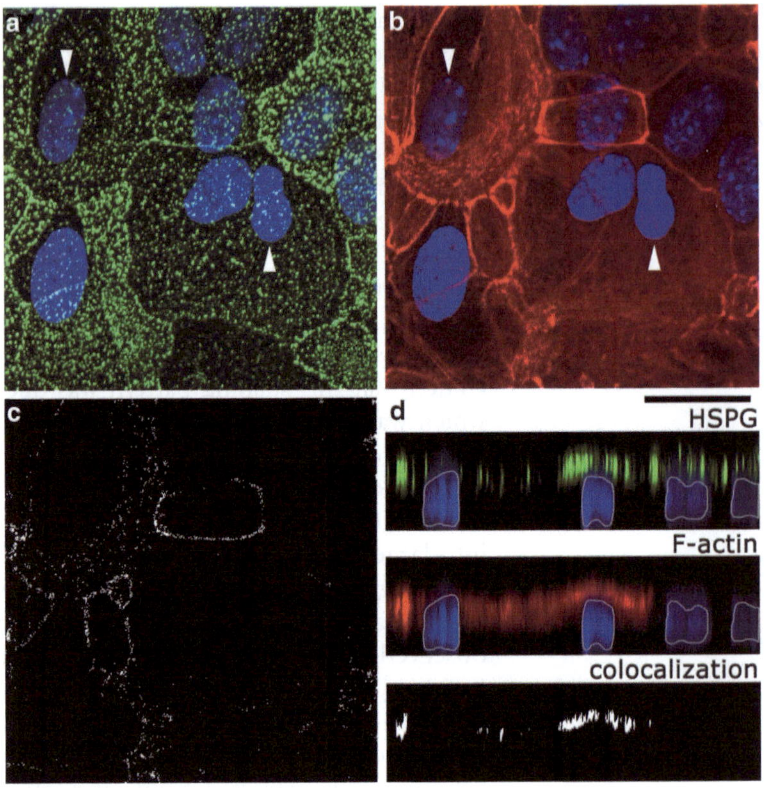

Fig. 3. Interaction of cell surface heparan sulfate proteoglycans (HSPGs) and actin cytoskeleton in mouse cortical astrocytes. An example of HSPG (**a**) and F-actin (**b**) distribution in the same stacked image acquired using a Zeiss confocal microscope. *Arrow heads* indicate astrocyte nuclei. Nuclei (*blue*), HSPG (*green*), F-actin (*red*). (**c**) Colocalization of cell surface HSPG and the underlying actin cytoskeleton was performed using Zeiss LSM Image Browser and ImageJ. (**d**) Cross-sectional views of the stacked confocal images from the region indicated by *white lines* in (**a**) and (**b**) showing the degree of colocalization between cell surface HSPG and actin cytoskeleton in astrocytes. Nuclei were outlined to distinguish cell surface expression of HSPG and underlying actin cytoskeleton. Bar = 20 μm.

technique combined with electron microscopy (EM) (25) provided early evidence that astrocytes could secrete an extracellular matrix containing HSPG and laminin, and that morphological differentiation and contact with neurites regulated the expression of the HSPG and laminin components of the astrocytic ECM. In this study the authors observed a fibrillar pattern of staining for HSPG and laminin in astrocytes subcultured for 5 or 9 days in serum-containing media, which was reduced with culture aging (30–42 days after subculturing). When astrocytes were subcultured for 5 days in serum-free defined media, the astrocytes displayed changes in morphology with increased formation of stellate processes and lacked an ECM visible by EM, showing no immunostaining for

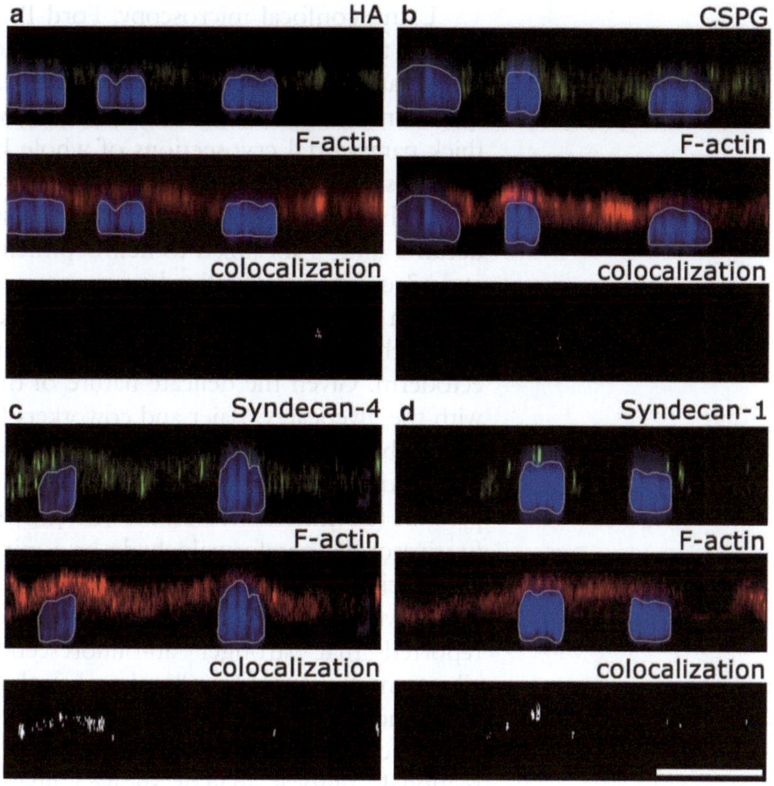

Fig. 4. Interaction between various cell surface proteoglycans and actin cytoskeleton. Cross-sectional views of stacked confocal images showing the various degrees of interaction between F-actin and (**a**) hyaluronic acid (HA), (**b**) chondroitin sulfate (CSPG) and two of the heparan sulfate proteoglycan (HSPG) family members, (**c**) syndecan-4 and (**d**) syndecan-1. Colocalization of cell surface proteoglycans and the underlying actin cytoskeleton was performed using Zeiss LSM Image Browser and ImageJ. Nuclei (*blue*), HA (**a**), CSPG (**b**), syndecan-4 (**c**), syndecan-1 (**d**) (*green*), F-actin (*red*). Bar = 20 μm.

either HSPG or laminin. Reduced staining for these ECM components and development of stellate processes were also observed in astrocytes that were in direct contact with the neurites from dorsal root ganglion explants. Studies that followed supported a key role for astrocyte proteoglycans in providing substrates for neuron adhesion, process extension, and other cell interactions (25–27).

On the tissue level, proteoglycan detection has been performed using either paraffin-embedded or cryosected samples. For example, Oohira et al. (28) demonstrated that CSPG (specifically neurocan) expression in the cerebrum was developmentally regulated, with detectable expression beginning as early as embryonic day 14 and disappearing around postnatal day 30. Expression levels of proteoglycans have also been shown to change in the CNS in response to injury, during regeneration and in pathological conditions (26, 29).

Using confocal microscopy, Ford-Perriss and colleagues (30) showed the expression of various components of HSPG during the proliferative phase of brain development. Immunolocalization of syndecan-1, -2, -3, and -4 and perlecan was performed on 10 μm thick parasagittal cryosections of whole brain from mouse E10.5 embryos using conventional staining at room temperature. Their comprehensive studies revealed that cell surface expression of syndecan-1 and -4 localized to neuroepithelial cells while syndecan-2 and -3 expression localized to neurons in the developing neural tube. In addition they showed that perlecan expression was localized to the basement membrane of the neural tube and the overlying ectoderm. Given the delicate nature of the components associated with the glycocalyx, Baier and coworkers (31) used various fixation and embedding procedures to better assess the distribution of HA expression in the developing mouse cerebellum. They used either paraffin (10 μm) or vibratome (50 μm) sections fixed in Bouin's fixative or 4% paraformaldehyde or a mixture of 95% ethanol and acetic acid (3:1) or 0.5% cetylpyridinium chloride (CPC), since the latter two reagents were shown to better preserve the HA. They reported that stronger autofluorescence was associated with vibratome sections. Overall, other than the stronger staining for HA in ethanol/acetic acid fixed samples, they did not observe obvious differences among the differentially fixed samples. Their cross-sectional confocal analysis showed fiber-like structures predominantly located in interneuron precursors and oligodendroglial cells.

In summary, these studies strongly indicate that fixation plays a critical role in preserving glycocalyx structure. The different microscopy methods summarized in this section reveal not only the structure but also the integrity of the glycocalyx in various cell types and tissues, and can be applied to assess the interaction between the glycocalyx and the actin cytoskeleton.

1.3.3. Direct Assessment of Interaction Between the Glycocalyx and Actin Cytoskeleton

The first indication that TM-PG interact with the actin cytoskeleton came from a series of studies in the mid 1980s in which Woods and coworkers (32, 33) analyzed the relationship of HSPGs to the cytoskeleton and extracellular matrix in fibroblasts. Double labeling studies for HSPG and either actin or fibronectin demonstrated a closer correlation between HSPG and actin than fibronectin during cell rounding or spreading. Their follow-up studies showed that after 0.2% Triton X-100 detergent extraction the cytoskeleton-matrix residues contained lipophilic HSPGs, which suggested that HSPG binds directly or indirectly to the actin cytoskeleton (33). Subsequent studies by Carey and Todd (34) using HSPG extracts from Schwann cells showed that the Triton-insoluble fraction contained three cytoskeletal proteins, namely, spectrin, vimentin, and actin. These findings strongly implied that HSPG is anchored to the Schwann cell actin cytoskeleton. Similar findings were reported by Rapraeger and coworkers (22) who showed detergent-soluble and -insoluble pools of cell surface HSPG in epithelial cells. In addition they demonstrated that as epithelial cells became polarized in vitro, mobile cell surface HSPGs become sequestered at the basolateral cell surface and bind to the actin cytoskeleton. Using the live staining procedure described above, they were the first to suggest colocalization of HSPG and F-actin fibers at the basolateral cell surface and at cell margins. To determine whether movement of HSPG clusters was cytoskeleton dependent they treated the cells with either 25 μM colchicine or 2 μM cytochalasin D to disrupt either microtubule or actin cytoskeleton. They observed that while colchicine had no effect on the HSPG movement, cytochalasin immobilized HSPG with occasional co-distribution of collapsed actin cytoskeleton aggregates and HSPG at basolateral regions of the cell. Based on these findings, the authors proposed that cell surface proteoglycans could function as matrix receptors to regulate the morphology and adhesion of epithelial sheets in response to extracellular cues.

Experimental evidence that TM-PG interact with the cytoskeleton and participate in regulation of cellular morphogenesis was obtained by Carey and coworkers (35) in studies with clonal Schwann cell lines derived after syndecan-1 transfection. They observed that stable ectopic expression of syndecan-1 in these cells resulted in changes in morphology, enhanced cellular spreading on fibronectin and laminin-coated substrata, and induced reorganization of actin filaments and formation of focal adhesions. Double immunolabeling of these cells with syndecan-1 antibody and phalloidin demonstrated that cell surface syndecan-1 expression colocalized with actin polymerization sites at the lateral borders of the cells. Colocalization was only observed at the time when the cells were spreading, which indicated that syndecan-1 was transiently interacting with the actin cytoskeleton and suggested that such interactions could be dynamically regulated. In subsequent experiments Carey and colleagues (36) began to disclose the nature of this interaction between syndecan-1 and microfilaments. Using point mutations and truncations of the syndecan-1 core protein they demonstrated that syndecan-1 colocalization with actin filaments required the presence of a conserved tyrosine residue localized on the central region of the syndecan cytoplasmic domain.

In a recent study we used live cell immunostaining for HSPG, CSPG, HA, and syndecan-1 and -4 to determine the extent of the interaction between astrocytic glycocalyx and actin cytoskeleton. We observed that both syndecan-1 and -4 colocalize with the cortical astrocytic actin cytoskeleton (Figs. 3 and 4c, d), while little or no interaction was detected with CSPG or HA (Fig. 4a, b). Further analysis using immunoprecipitation and Western blotting techniques (Fig. 5) confirmed the HSPG association with underlying actin cytoskeleton. Detailed methods used in this study are presented below.

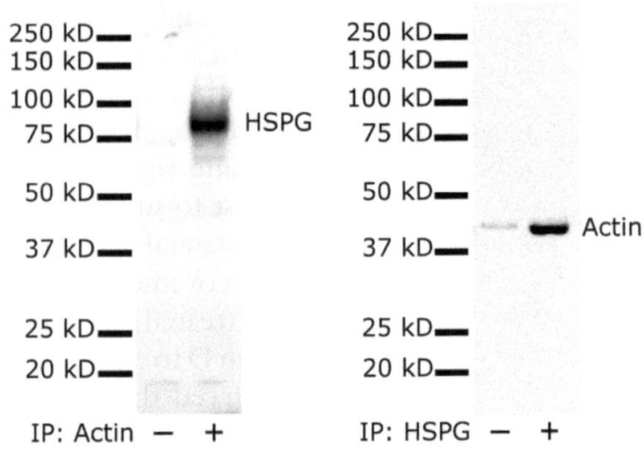

Fig. 5. Representative Western blots of the co-immunoprecipitation (IP) of neonatal mouse cortical astrocyte lysates with either heparan sulfate proteoglycan (HSPG) or actin antibody. This data demonstrates that there is a molecular complex consisting of HSPG (~100 kDa) and actin (~42 kDa) indicating the interaction between HSPG and actin cytoskeleton.

In endothelial cells the interaction between the glycocalyx and the cytoskeleton plays a crucial role in cellular remodeling in response to fluid shear stress (13, 37). In these experiments, cells were exposed to fluid shear stress of 10–15 dyn/cm^2 for 5–24 h and cell alignment in direction of flow as well as the intracellular distribution of molecular components of the endoskeleton were compared to control (no flow) conditions and to cells treated with Heparinase III to digest exoskeleton proteins. Strikingly, when glycocalyx was compromised, the shear stress-induced augmented disruption of the dense peripheral actin bands and stress fiber formation were completely abolished (Fig. 6) and alignment no longer occurred. These findings unequivocally confirmed the direct interaction between the endothelial glycocalyx and underlying cytoskeleton.

2. Materials

Prepare all solutions using double-distilled water. Store all media, reagents, and buffers according to manufacturers' instructions or store at room temperature. Aliquot and store all antibodies at -20°C to avoid repeated freezing and thawing, except for those from Santa Cruz Biotechnology (Santa Cruz, CA, USA) that are stored at 4°C. All chemicals are purchased from Sigma-Aldrich (St. Louis, MO, USA) unless otherwise stated.

2.1. Reagents, Buffers, and Materials for Immunofluorescence

1. Bovine serum albumin (BSA, A2153, Sigma-Aldrich).
2. 16% Formaldehyde solution (28908, Thermo Scientific, Rockford, IL, USA).

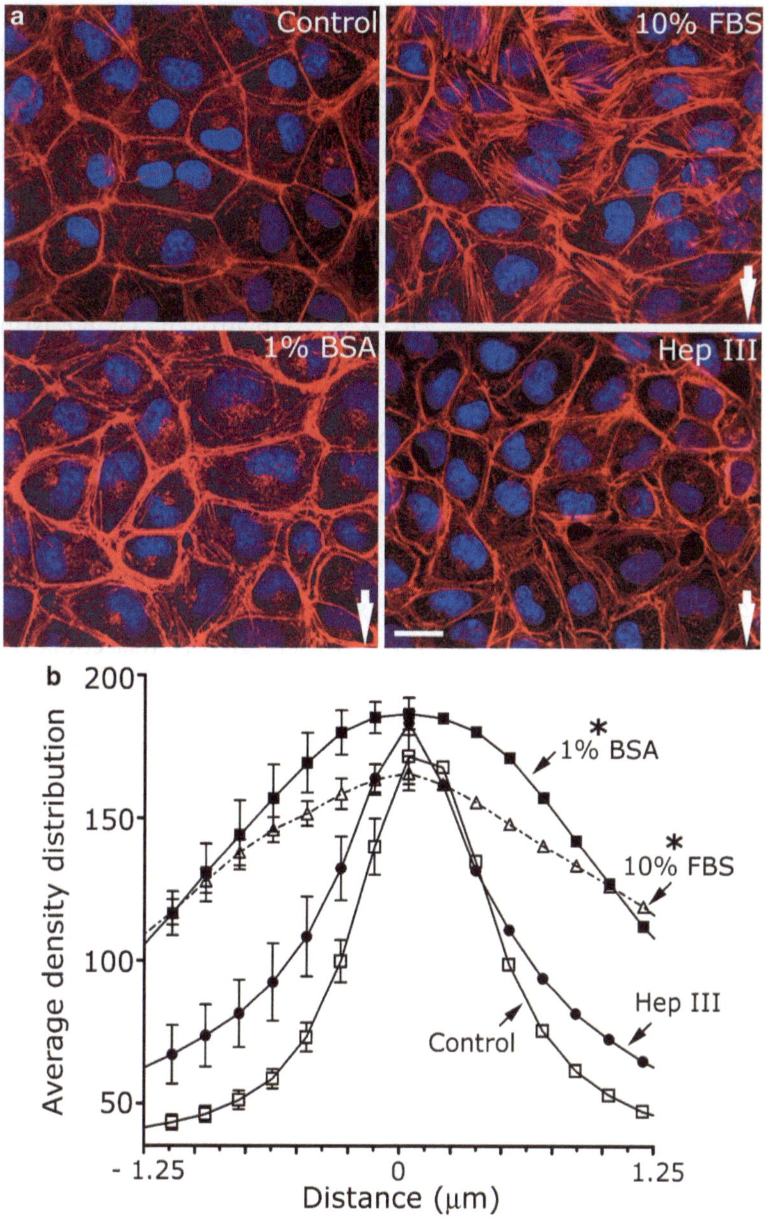

Fig. 6. (**a**) Reorganization of endothelial cytoskeleton in response to fluid shear stress of 10.5 dyne/cm^2 with various media compared to control (no flow) condition: intact glycocalyx media (DMEM + 10% FBS or DMEM + 1% BSA) and compromised glycocalyx media (Heparinase III + DMEM + 1% BSA). F-actin (*red*) and nuclei (*blue*). Bar = 20 μm. (**b**) Changes in the distribution of F-actin at cell–cell junction were plotted as average protein density profile from stacked images of various conditions and detected using Kurtosis analysis. All data sets are presented as the mean ± SE, $n = 160$. *$P < 0.05$ (reproduced with authorization from (13)).

3. Prepare blocking solution 1% BSA and fixative 4% formaldehyde in 1× Dulbecco's Phosphate-Buffered Saline (DPBS, pH 7.4), aliquot, and store at −20°C to avoid repeated freezing and thawing (see Note 1).

2.2. Antibodies for Immunofluorescence Microscopy

1. *Primary antibodies*: HSPG (10E4 epitope, H1890, US Biological, Swampscott, MA, USA); CSPG (clone CS-56, C8035, Sigma-Aldrich); biotinylated hyaluronic acid binding protein (B-HABP, 400763-1A, NorthStar BioProducts, East Falmouth, MA, USA); syndecan-4 (H140, sc-15350) and syndecan-1 (H174, sc-5632) (Santa Cruz Biotechnology).
2. *Secondary antibodies for immunofluorescence*: Alexa fluor 488 donkey anti-mouse IgG (A-21202), Alexa fluor 488 donkey anti-rabbit IgG (A-21206), Alexa fluor 564 phalloidin (A-22283) (Invitrogen, Grand Island, NY, USA); DyLight 488-conjugated IgG mouse anti-biotin (200-482-211, Jackson ImmunoResearch Laboratories, West Grove, PA, USA).
3. Prepare working dilutions for all antibodies in 1% BSA solution.

2.3. Reagents, Buffers, and Materials for Co-immunoprecipitation and Western Blots

1. Protein A-agarose immunoprecipitation reagent (sc-2001, Santa Cruz). Store at 4°C.
2. Immunoprecipitation buffer (IP buffer): 20 mM Tris–HCl (pH 8.0), 140 mM NaCl, 1% Triton X-100, 10% glycerol, 1 mM ethylene glycol tetraacetic acid (EGTA), 1.5 mM $MgCl_2$, 4 mM dithiothreitol (DTT), 1 mM phenylmethylsulfonyl fluoride (PMSF), 1 mM sodium orthovanadate, 1× protease inhibitor (0589297001, Roche Applied Science, Indianapolis, IN, USA). Store at 4°C (see Note 2).
3. IP wash buffer: 25 mM Tris–HCl (pH 8.0), 150 mM NaCl, 0.5% NP-40.
4. Laemmli's SDS-sample buffer (4×, nonreducing, BP-110NR, Boston BioProducts, Ashland, MA, USA).
5. Tris-buffered saline (1× TBS): 150 mM NaCl, 25 mM Tris–HCl (pH 7.4).
6. Wash buffer (TBST): 1× TBS with 0.05% Tween-20.
7. Blocking solution: 5% Skim milk in TBST. Make fresh for every experiment.
8. Immobilon Western HRP substrate peroxide solution (WBKLS0100, Millipore, Billerica, MA, USA).
9. Nitrocellulose transfer membrane (10401396, Whatman GmbH, Dassel, Germany).
10. Mini Trans Blot filter paper (1703932, Bio-Rad, Hercules, CA, USA).

2.4. Antibodies for Western Blots

1. Primary antibodies: Actin (A2066, Sigma-Aldrich), heparan sulfate (10E4, see Section 3.1, step 4).
2. Secondary antibodies: Goat anti-mouse IgG-HRP (sc-2005) and goat anti-rabbit IgG-HRP (sc-2004) (Santa Cruz Biotechnology).
3. Prepare working dilutions for all antibodies in blocking solution.

3. Methods

Use protocol described in (38) to isolate primary cortical astrocyte cultures from brains of neonatal (P0) wild-type C57BL/6 mice and plate astrocytes on 12 mm round coverslips (12-545-80, Fisher Scientific, Pittsburgh, PA, USA) or 100 mm culture dishes (see Note 4).

3.1. Immunofluorescence and Confocal Microscopy

To detect cell surface expression of proteoglycans and actin cytoskeleton on live cortical astrocytes in culture we use the protocol described in (13) which are methods adapted from Rapraeger and Carey's works (22, 23).

1. Chill confluent cortical astrocyte cultures on ice for 10 min in growth media (see Note 5).
2. Aspirate culture media and block cells with 1% BSA for 10 min on ice.
3. Wash cells briefly and gently for three times with ice-cold 1× DPBS. No agitation.
4. Incubate with primary antibodies for 1½ h on ice in an incubation chamber. Primary antibody concentrations: heparan sulfate, 1:200; B-HABP, 1:100; syndecan-4, 1:100; and syndecan-1, 1:100. For negative control, incubate a few coverslips with 1% BSA without the primary antibody (see Note 6).
5. Wash cells at 5-min intervals for three times with ice-cold 1× DPBS. No agitation.
6. Immediately fix cells with 4% formaldehyde at room temperature for 20 min.
7. Wash cells three times with 1× DPBS at 10-min intervals on a shaker.
8. Incubate with appropriate secondary antibodies against respective primary antibodies together with Alexa 546 conjugated to phalloidin (1:200) for 1½ h in an incubation chamber at room temperature. Secondary antibody concentrations: Alexa 488 against mouse or rabbit IgGs, 1:200.
9. Wash cells three times at 10-min intervals with 1× DPBS on a shaker.
10. Mount coverslips with cells on microscope slides with medium containing glycerol, 13.3% polyvinyl alcohol, 5% n-propyl gallate, and 20 μM of the nucleophilic dye DAPI (see Note 7).
11. Once mounting medium is dried, store the slides at 4°C (see Note 8).
12. Use confocal microscopy to acquire stacked images and analyze expressions of cell surface proteoglycans and underlying actin

cytoskeleton. We used the Zeiss LSM 510 Duo laser scanning confocal microscope (Carl Zeiss Microimaging GmbH, NY, USA) and a 63× objective at a zoom of 2 (see Note 9).

13. Take serial images from top to bottom of each image field. Typical image stacks consist of 18–20 optical sections per stack with each section collected at 0.29 μm *z*-axis step apart as shown in Fig. 3 for HSPG (A) and F-actin (B) (see Note 10).

3.2. Colocalization Analysis

1. Use Zeiss LSM Image Browser (Carl Zeiss Microimaging GmbH, NY, USA) to analyze multiplane and cross-sectional views of stacked images as shown in Figs. 3 and 4.
2. Save stacked images and cross-sectional images from plane of interest for each fluorophore (i.e., Alexa 488 or Alexa 564) as "tiff" files. Convert each tiff file to 8-bit grayscale file using Adobe Photoshop (Adobe Systems Incorporated, San Jose, CA, USA).
3. For colocalization analysis, open stacked images of two fluorophore images in ImageJ (NIH, USA). Under "Image" drop-down menu, select "Adjust" and "Threshold" images. Use "Default" and "B&W" setting. Take note of threshold values for each fluorophore image.
4. From the "Plugins" drop-down menu, choose "Colocalization." Select image corresponding to each channel (i.e., for either red or green). Keep ratio at 50 and enter threshold values acquired from previous step for respective channels. Select "colocalization point 8-bit" to obtain colocalization image such as the one shown in Figs. 3c, d and 4.

3.3. Co-immunoprecipitation and Western Blots

1. Reconstitute or collect samples in IP buffer.
2. Sonicate and centrifuge samples for 10 min, 20,000 × *g* at 4°C.
3. Determine protein concentration. Optimal concentration: 1–2 mg/ml.
4. Prepare duplicates for each sample.
5. Preclear samples (500 μl volume) with 30 μl of Protein-A agarose beads for 1 h at 4°C on a rotator.
6. Centrifuge samples for 10 min, 14,000 × *g* at 4°C.
7. Transfer supernatant to a new tube, add 0.5–2 μg of either actin or HSPG antibody, and mix overnight at 4°C on a rotator (see Note 11).
8. Add 30 μl of Protein-A beads to the samples and mix for 1 h at 4°C on a rotator.
9. Centrifuge samples for 10 min, 3,000 × *g* at 4°C.
10. Aspirate supernatant and gently wash five times with 500 μl of wash buffer. Centrifuge in between each wash (5 min, 3,000 × *g*).

11. Reconstitute beads with 30 μl of 4× Laemmli buffer with 100 mM DTT, centrifuge (5 min, 3,000×*g*), transfer the supernatant to a new tube, and centrifuge again (see Note 12).
12. Use general Western blotting protocol to electrophoretically separate proteins and transfer to nitrocellulose membrane.
13. Block membranes with 5% skim milk for 30 min at room temperature.
14. Incubate the actin immunoprecipitated blot with HSPG, 1:250, for 30 h at 4°C. Incubate the HSPG immunoprecipitated blot with actin, 1:5,000, overnight at 4°C.
15. Wash five times with TBST for 10 min each.
16. Use Immobilon HRP substrate solution to expose blots to X-ray films and visualize protein bands.

4. Notes

1. Shelf life for formaldehyde is very short. It becomes acidic and polymerizes over time at room temperature. To prevent this, once 4% formaldehyde is prepared, it is best to aliquot and freeze the aliquots, which can then be used for up to 1 year. If one encounters high background staining, instead of 1% BSA, either 10% goat serum or 2% donkey serum can also be used as blocking solution given that goat- or donkey-derived secondary antibodies were used.
2. IP buffer is prepared without DTT, PMSF, sodium orthovanadate, and protease inhibitor and stored at 4°C for not more than 3 months. DTT, PMSF, sodium orthovanadate, and protease inhibitor are added just before sample collection.
3. Adjust Immobilon Western HRP substrate peroxide solution if the resulting protein bands are too strong. Dilute the solution up to fourfold as needed.
4. Make sure that all animal procedures and experimental protocols are approved by your Institution's Association for Assessment and Accreditation of Laboratory Animal Care (AALAC) International and institutional guidelines.
5. There is an alternative way of visualizing cell surface proteoglycans as previously described in (13) using CellTracker orange (Molecular Probes) to identify proteoglycan distributions on the cell surface. We exclude this step since we are also visualizing the underlying actin cytoskeleton with Alexa 546. If coverslips are used for staining, be as gentle as possible when removing coverslips from the dishes so as not to damage the confluent cell monolayer.

6. Extra care should be taken while handling the coverslips at this point of the protocol because cells are still alive in this step. The temperature (4°C) is crucial during this step given that the glycocalyx is a very delicate layer and the proteoglycans associated with it tend to become internalized at 37°C, when cells are alive. It is therefore critical that the live cell incubation with primary antibody be performed at 4°C. As an alternative, a recent study (21) has shown that 2% paraformaldehyde/0.1 glutaraldehyde fixation prior to staining also works well with detecting cell surface proteoglycan expression.
7. As an alternative, VECTASHIELD mounting medium with DAPI (H-1200, Vector Laboratories, Inc., Burlington, CA, USA) can be used.
8. Microscope slides can be stored at 4°C for a long term, but imaging should be done within a few weeks to obtain optimal results.
9. Use objective 40× and above to attain optimal imaging. Use optical zoom function as needed.
10. The goal of this study is to analyze the colocalization of two fluorophores and therefore it is necessary to use the optical z-step size recommended by the microscope software. Use cell surface expression of proteoglycans to determine the topmost apical section and nuclei (DAPI) to determine the basal section.
11. Always have negative control for each sample (i.e., one tube of sample without receiving antibody of interest).
12. Some Protein-A beads can be still at the bottom of the tubes. Avoid pipetting all the way to the bottom of the tube when loading the gel. Bead contamination while resolving protein will damage the gel and subsequently ruin the protein separation.

5. Summary

In this chapter, we have attempted to summarize tools and reagents that are available for studies of glycocalyx–cytoskeleton interactions. Despite advances in these techniques, there remains no clear resolution of the interactions, their dynamic nature, or their functional relevance.

Acknowledgement

This work was supported by NIH grants DK091466, HL094889, and DK081435.

References

1. Choi Y, Chung H, Jung H, Couchman JR, Oh ES (2011) Syndecans as cell surface receptors: unique structure equates with functional diversity. Matrix Biol 30:93–99
2. Couchman JR (2010) Transmembrane signaling proteoglycans. Annu Rev Cell Dev Biol 26:89–114
3. Yoneda A, Couchman JR (2003) Regulation of cytoskeletal organization by syndecan transmembrane proteoglycans. Matrix Biol 22:25–33
4. Bandtlow CE, Zimmermann DR (2000) Proteoglycans in the developing brain: new conceptual insights for old proteins. Physiol Rev 80:1267–1290
5. Zimmermann D, Ma D-Z (2008) Extracellular matrix of the central nervous system: from neglect to challenge. Histochem Cell Biol 130:635–653
6. Barnea G, Grumet M, Milev P, Silvennoinen O, Levy JB, Sap J, Schlessinger J (1994) Receptor tyrosine phosphatase beta is expressed in the form of proteoglycan and binds to the extracellular matrix protein tenascin. J Biol Chem 269(20):14349–14352
7. Margolis RU, ReK M (1997) Chondroitin sulfate proteoglycans as mediators of axon growth and pathfinding. Cell Tissue Res 290:343–348
8. Maurel P, Rauch U, Flad M, Margolis RK, Margolis RU (1994) Phosphacan, a chondroitin sulfate proteoglycan of brain that interacts with neurons and neural cell-adhesion molecules, is an extracellular variant of a receptor-type protein tyrosine phosphatase. Proc Natl Acad Sci U S A 91:2512–2516
9. Shitara K, Yamada H, Watanabe K, Shimonaka M, Yamaguchi Y (1994) Brain-specific receptor-type protein-tyrosine phosphatase RPTP beta is a chondroitin sulfate proteoglycan in vivo. J Biol Chem 269:20189–20193
10. Watanabe E, Maeda N, Matsui F, Kushima Y, Noda M, Oohira A (1995) Neuroglycan C, a novel membrane-spanning chondroitin sulfate proteoglycan that is restricted to the brain. J Biol Chem 270:26876–26882
11. Filmus J, Capurro M, Rast J (2008) Glypicans. Genome Biol 9:224
12. Multhaupt HA, Yoneda A, Whiteford JR, Oh ES, Lee W, Couchman JR (2009) Syndecan signaling: when, where and why? J Physiol Pharmacol 60(Suppl 4):31–38
13. Thi MM, Tarbell JM, Weinbaum S, Spray DC (2004) The role of the glycocalyx in reorganization of the actin cytoskeleton under fluid shear stress: a "bumper-car" model. Proc Natl Acad Sci U S A 101:16483–16488
14. Gasic G, Berwick L (1963) Hale stain for sialic acid-containing mucins. J Cell Biol 19:223–228
15. Rambourg A, Leblond CP (1967) Electron microscope observation on the carbohydrate-rich cell coat present at the surface of cells in the rat. J Cell Biol 32:27–53
16. Luft JH (1966) Fine structures of capillary and endocapillary layer as revealed by ruthenium red. Fed Proc 25:1773–1783
17. Adamson RH, Clough G (1992) Plasma proteins modify the endothelial cell glycocalyx of frog mesenteric microvessels. J Physiol 445:473–486
18. Rostgaard J, Qvortrup K (1997) Electron microscopic demonstrations of filamentous molecular sieve plugs in capillary fenestrae. Microvasc Res 53:1–13
19. Squire JM, Chew M, Nneji G, Neal C, Barry J, Michel C (2001) Quasi-periodic substructure in the microvessel endothelial glycocalyx: a possible explanation for molecular filtering? J Struct Biol 136:239–255
20. van den Berg BM, Vink H, Spaan JAE (2003) The endothelial glycocalyx protects against myocardial edema. Circ Res 92:592–594
21. Ebong EE, Macaluso FP, Spray DC, Tarbell JM (2011) Imaging the endothelial glycocalyx in vitro by rapid freezing/freeze substitution transmission electron microscopy. Arterioscler Thromb Vasc Biol 31:1908–1915
22. Rapraeger A, Jalkanen M, Bernfield M (1986) Cell surface proteoglycan associates with the cytoskeleton at the basolateral cell surface of mouse mammary epithelial cells. J Cell Biol 103:2683–2696
23. Carey DJ, Stahl RC (1990) Identification of a lipid-anchored heparan sulfate proteoglycan in Schwann cells. J Cell Biol 111:2053–2062
24. Nishimura K, Ishii M, Kuraoka M, Kamimura K, Maeda N (2010) Opposing functions of chondroitin sulfate and heparan sulfate during early neuronal polarization. Neuroscience 169:1535–1547
25. Ard MD, Bunge RP (1988) Heparan sulfate proteoglycan and laminin immunoreactivity on cultured astrocytes: relationship to differentiation and neurite growth. J Neurosci 8:2844–2858
26. Dow KE, Wang W (1998) Cell biology of astrocyte proteoglycans. Cell Mol Life Sci 54:567–581

27. Johnson-Green PC, Dow KE, Riopelle RJ (1991) Characterization of glycosaminoglycans produced by primary astrocytes in vitro. Glia 4:314–321
28. Oohira A, Matsui F, Watanabe E, Kushima Y, Maeda N (1994) Developmentally regulated expression of a brain specific species of chondroitin sulfate proteoglycan, neurocan, identified with a monoclonal antibody IG2 in the rat cerebrum. Neuroscience 60:145–157
29. Matsui F, Oohira A (2004) Proteoglycans and injury of the central nervous system. Congenit Anom 44:181–188
30. Ford-Perriss M, Turner K, Guimond S, Apedaile A, Haubeck HD, Turnbull J, Murphy M (2003) Localisation of specific heparan sulfate proteoglycans during the proliferative phase of brain development. Dev Dyn 227:170–184
31. Baier C, Baader SL, Jankowski J, Gieselmann V, Schilling K, Rauch U, Kappler J (2007) Hyaluronan is organized into fiber-like structures along migratory pathways in the developing mouse cerebellum. Matrix Biol 26:348–358
32. Woods A, Höök M, Kjellén L, Smith CG, Rees DA (1984) Relationship of heparan sulfate proteoglycans to the cytoskeleton and extracellular matrix of cultured fibroblasts. J Cell Biol 99:1743–1753
33. Woods A, Couchman JR, Höök M (1985) Heparan sulfate proteoglycans of rat embryo fibroblasts. A hydrophobic form may link cytoskeleton and matrix components. J Biol Chem 260:10872–10879
34. Carey DJ, Todd MS (1986) A cytoskeleton-associated plasma membrane heparan sulfate proteoglycan in Schwann cells. J Biol Chem 261:7518–7525
35. Carey DJ, Stahl RC, Cizmeci-Smith G, Asundi VK (1994) Syndecan-1 expressed in Schwann cells causes morphological transformation and cytoskeletal reorganization and associates with actin during cell spreading. J Cell Biol 124:161–170
36. Carey DJ, Bendt KM, Stahl RC (1996) The cytoplasmic domain of Syndecan-1 is required for cytoskeleton association but not detergent insolubility. J Biol Chem 271: 15253–15260
37. Yao Y, Rabodzey A, Dewey CF Jr (2007) Glycocalyx modulates the motility and proliferative response of vascular endothelium to fluid shear stress. Am J Physiol: Heart Circ Physiol 293:H1023–H1030
38. Scemes E, Dermietzel R, Spray DC (1998) Calcium waves between astrocytes from Cx43 knockout mice. Glia 24:65–73

Chapter 4

Cytoskeleton Dynamics and Binding Factors

Elena E. Grintsevich and Emil Reisler

Abstract

The dynamic nature of cellular cytoskeleton is vital to many cell functions. It was shown that all three filamentous components of the cytoskeleton (F-actin, microtubules, and intermediate filaments) share dynamic behavior. In this chapter we address briefly some basic features of cytoskeletal dynamics and then focus on the actin filament system. In this context we review the major classes of actin-interacting partners and their effects on actin cytoskeleton.

Key words: Cytoskeletal dynamics, Actin-binding proteins, Capping proteins, G-actin sequestering proteins, Actin-stabilizing proteins, Cross-linking proteins, Motor proteins

1. Introduction

Actins represent a conserved family of proteins essential to all eukaryotes. In cells, actin filaments can be organized into higher order short- or long-lived structures such as lamellipodial dendritic arrays, filopodial protrusions, stress fibers, etc. (1). Microtubules (MTs) are also key components of cytoskeleton, playing an important role in spatial distribution of organelles and chromosomes, and assisting in cell division, intracellular transport, and motility of cilia and flagella (2). Actin and tubulin share some common features despite the morphological differences between their polymers. Both proteins bind tightly nucleotides-ATP and GTP for actin and tubulin, respectively-and their *in vivo* polymer assembly involves energy consumption through ATP or GTP hydrolysis. *In vitro* experiments showed that ATP hydrolysis is not an absolute requirement for F-actin assembly, as filaments can be formed from the monomers containing a non-hydrolyzable ATP analog-ANP-PNP or just ADP. In contrast to that, MT cannot be assembled from GDP-tubulin (3). Polarity is another common feature of MTs and

Rolf Dermietzel (ed.), *The Cytoskeleton: Imaging, Isolation, and Interaction*, Neuromethods, vol. 79,
DOI 10.1007/978-1-62703-266-7_4,

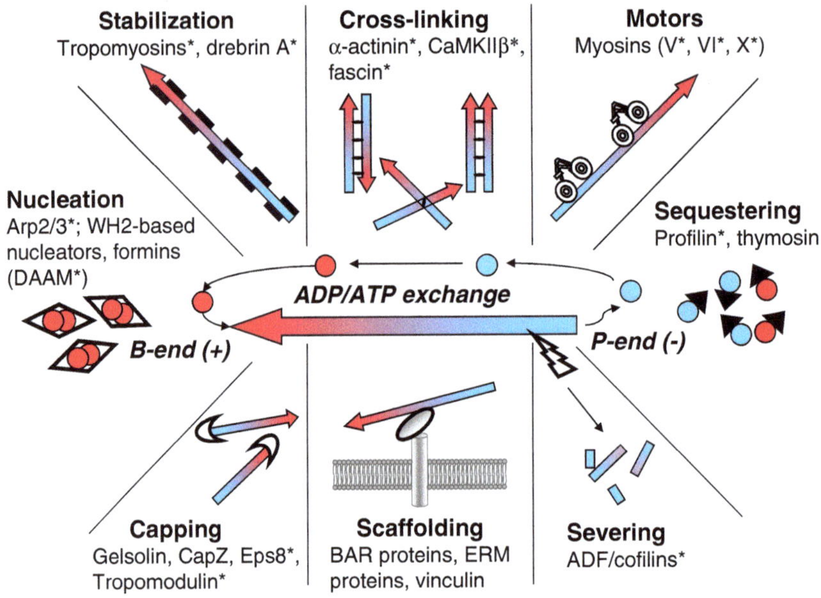

Fig. 1. Schematic representation of filament treadmilling (in the *center*) and examples of actin-binding proteins. Actin filaments are represented as *arrows* with *arrowheads* indicating the fast-growing ends (B-ends). ATP- and ADP-bound free monomers (*circles*) and in stretches of filaments are colored in *red* and *blue*, respectively. Mixed-state regions undergoing Pi release are colored in light purple. Major classes of actin-binding proteins are given with some examples. *Asterisks* identify examples of actin-binding proteins important to the neuronal function (10, 22, 85).

F-actin: in both polymers one of their ends (plus end) grows faster than the other end (minus end). The polymerization cycle of actins and tubulin involves the addition of an ATP/GTP-containing monomer to the plus end of the polymer, followed by nucleotide hydrolysis, Pi release, and the dissociation of ADP/GDP-containing monomer. This monomer undergoes then nucleotide exchange (ADT-ATP or GDP-GTP) and can re-associate with the plus end of the polymer. This process, called treadmilling, is characteristic of pure F-actin filaments at their steady-state condition (Fig. 1, center). In the absence of interacting factors F-actin treadmills very slowly, as the rate of this process is limited by ADP-protomer dissociation from the minus end (~0.2 s^{-1}) (4). The dynamic behavior of MT is quite different from that of F-actin. Dynamic instability is a unique and fascinating feature of MT. It was originally proposed by Mitchison and Kirchner (5) and then confirmed by direct microscopic observations. This phenomenon involves MT growth, followed by random shrinkage called "catastrophe," and then by the "rescue" which initiates a new phase of growth. It was shown that MTs spend significant time in the pause stages, which were shown to represent unproductive growth due to repetitive switching between growth and shortening events within up to five tubulin layers (6). It was shown also that tubulin dimers incorporate into MT at an average rate comparable to that observed for F-actin (association constant for the plus end = 4.6–8.9 $\mu M^{-1}/s^{-1}$). However, compared to actin, tubulin dissociation is significantly faster ($k^{-}_{(+end)} = 14.1–44\ s^{-1}$) and accelerates dramatically during the random shrinkage ("catastrophe") event $k^{-}_{s\ (+end)} = 733–853\ s^{-1}$, resulting in complete disassembly of MT (7–9). How the dynamic instability of MTs is regulated *in vivo* remains an open and important question. The stability of MT depends to a great extent on a cell cycle or a cell compartment; MT can be very stable in cilia and very dynamic in the mitotic spindle. Clearly, *in vivo* changes in MT dynamics are mediated by a variety of microtubule-associated proteins (plus-end-tracking, MT stabilizing, destabilizing, severing, and motor proteins). For a more detailed discussion of such factors and effects we refer the reader to a recent review by Dent et al. on MT regulators that contribute to axon outgrowth and guidance (10).

Direct microscopic *in vitro* observations revealed significant differences in the homogeneity of actin and tubulin polymers. In TIRF assays, elongation of individual actin filaments proceeds at a similar rate, leading to the assumption of relative homogeneity of actin filaments in terms of their assembly kinetics (see (11) for an example). A different picture arises from the data accumulated on MT growth. Significant variations in the dynamic parameters were shown for individual MT within the same preparation. These findings decrease the value of bulk solution experiments, making microscopy techniques more suitable for following MT dynamics (3, 6).

The dynamics of actin and MT has been well established by solution studies and direct microscopy observations. In contrast to that it was believed for a long time that intermediate filaments (IFs) were a static cytoskeletal component, preventing cell deformation. However, it was shown that IFs are dynamic as well. Fluorescence recovery after photobleaching (FRAP) in living neurons revealed that IF subunit exchange takes place, but at much slower rates than in F-actin and MT, yielding a recovery half-time ~40 min (12).

2. General Aspects of Actin Biology

2.1. Intrinsic F-Actin Dynamics and Effects of Cofactors

Actin monomer binds two interrelated cofactors—a divalent cation to a high-affinity site, and a nucleotide. Cation binding to low-affinity sites on actin has been also demonstrated. The high-affinity cation-binding site can coordinate Ca, Mn, Cd, Mg, Zn, or Ni (in decreasing affinity) (13). The fact that Pi release is blocked in F-actin assembled from Cr-ATP-G-actin was used successfully in *in vitro* experiments to obtain filaments mimicking the ADP-Pi state of a native F-actin (3). Considering that the physiological concentration of Mg^{2+} is much higher than that of all the above cations, it is accepted that *in vivo* the cation-binding site of actin will be occupied by magnesium. Actin filaments exist typically with mixed nucleotide states: they contain ATP-cap at the fast growing end, followed by an ADP-Pi-enriched region, and ADP actin protomers in the rest of the filament. It is widely accepted that the properties of the actins filaments depend on the nature of the bound nucleotide and that ATP hydrolysis upon polymerization affects the strength of protomer–protomer contacts. Early solution work indicated that interprotomer contacts are weakened by ATP hydrolysis, resulting in lower stability of ADP-F-actin compared to its ATP- and ADP-Pi forms (3). The fact that tight binding of Pi analogs, AlF_4^- and BeF_3^-, to actin filaments stabilizes them supports the view on ATP and ADP.Pi role in strengthening of intraprotomer contacts. Experiments with Cr-bound actin, which does not release phosphate after ATP hydrolysis, yielded similar results (3). However, it must be noted that the structure of F-actin with the bound AlF_4^-, BeF_3^-, and Cr is undefined, leaving open the possibility of additional effects.

The accepted view of nucleotide-dependent actin dynamics was challenged by the electron microscopy (EM) reconstruction of newly formed filaments (~2 min), presumably containing mostly ATP and ADP-Pi protomers. The EM imaging revealed the presence of less stable tilted state in "young" filaments which would result in weaker interprotomer contacts, contradicting the nucleotide-based view of F-actin stability (14). Moreover, quite unexpected results were reported recently by T. Mitchison's group: using single filament imaging and bulk solution experiments they showed increased stability of aged actin filaments compared to the newly formed ones (15). This adds new complexity to the current understanding of actin behavior, namely, not only cation- and nucleotide-bound states but also the age of filaments may affect actin dynamics. If confirmed, the discovery of time-dependent changes in actin structure might be physiologically significant. As mentioned above, the life span of actin-based structures varies significantly. In cellular environments, requiring very fast filament turnover (such as lamellipodia), these filaments may have limited time to "mature." Therefore, it is possible that actin structures observed after short polymerization time are physiologically relevant. This possibility calls for further investigation even though uncomplexed, free actin filaments may be seldom present in a cellular environment in which multiple actin-binding factors modulate actin structure and dynamics (see below).

The polymorphism of actin filaments formed *in vitro* from G-actins in the same cation- and nucleotide-bound state shows that the filaments are not rigid structures; they undergo nanoscale motions and fluctuations that can be compared to breathing. Such structural plasticity of actin was defined by Kueh and Mitchison as "change in the structural state of a polymer without change in the chemical state of its bound nucleotide" (2). Multiple lines of evidence support the view of structural plasticity of actin filaments and MT. For actin, structural polymorphism was detected in solution experiments (16, 17) as well as electron microscopy image analysis (18). More specifically, chemical cross-linking of multiple yeast actin mutants, combined with negative staining (NS EM), electron paramagnetic resonance (EPR), and light scattering data, revealed a remarkable plasticity of the interprotomer interface in F-actin. Moreover, cross-linking results showed that internal actin dynamics allowed the formation of actin states that destabilized the filament structure (16). A detailed cryo-EM microscopy analysis of *in vitro* formed actin filaments has revealed that blocks of structurally different states were present within the same filament. At least six different states have been recently identified: with about ~24% of filament segments adopting a "tilted conformation" (T-state) and five other substates assigned based on the structural states of a highly mobile subdomain 2. Notably, the fact that these multiple states could be assigned based on image analysis of the short F-actin segments suggested that the transitions between such states should be highly cooperative (19). Although plasticity and polymorphism of actin filament structure are still debated in the field (18, 20) the combined results of high-resolution EM and solution studies strongly support the structural plasticity view.

In the living cell the variability of F-actin helical twist is suggested to be a part of the regulatory mechanism modulating its interactions with multiple proteins. Below we discuss the effects of binding factors on actin dynamics in more details.

3. Actin-Binding Proteins

The number of identified actin-binding proteins (ABPs) (>150) (21) is still increasing and it is beyond the scope of this review to describe/discuss all of them individually. Instead, these proteins can be subdivided into groups based, for example, on the following features: (1) sequence homology (ADF/cofilins, WH2 domain, calponin-homology domain proteins, formins, etc.); (2) function (severing, capping, bundling proteins, etc.); (3) ability to bind monomeric or filamentous actin (G- and F-binding proteins); (4) structural fold, etc. None of these classifications is ideal because some ABPs can be remarkably multifunctional, and in many cases the physiological relevance of their documented activities is yet to be clarified. Here we subdivide ABPs according to their main *in vivo* function (Fig. 1).

3.1. Actin Nucleators and Elongation Factors

It is well established that actin nucleation is the rate-limiting and unfavorable step in actin polymerization (4). Thus, actin dynamics in cells cannot rely on a spontaneous formation of unstable actin dimers and trimers. Moreover, in cells, most of the G-actin pool is complexed with profilin and other sequestering proteins that block the spontaneous nucleation and control monomer incorporation into filaments (4). Despite their structural differences, all nucleators serve a similar purpose—the formation and/or stabilization of a viable nucleus for actin polymerization. The nucleators are a heterogeneous group of proteins that can be subdivided into three major classes: Arp2/3 and its activators, WH2 domain-containing nucleators, and formins. It should be noted that despite a large number of studies aiming to elucidate the mechanism of the nucleation activity, we still lack a detailed understanding of how formins and WH2-based nucleators exert their function *in vivo*, i.e., when most of the actin monomers are sequestered by profilin or thymosins. Actin nucleators themselves need to be tightly controlled to prevent random polymerization, which would be energetically costly and most likely damaging to the cell. Accordingly, most (but not all) nucleators exist in an auto-inhibited conformation and need to be activated by either nucleation-promoting factors (NPF) (in the case of Arp2/3) or other factors/mechanisms such as small GTP-ase (for formins) and phosphorylation (23, 24).

3.1.1. Arp2/3

The morphology, strength, and lifetime of dendritic actin arrays promoting lamellipodia extensions depend on the Arp2/3-mediated branching of actin filaments. Arp2/3 complex consists of seven subunits and two of them (Arp2 and 3) bind ATP and are related structurally to actin. On its own, Arp2/3 is a weak nucleator that undergoes conformational changes upon its interaction with NPFs. These include WASP/SCAR/WAVE family proteins containing the WCA motif (WH2 domain followed by a central/connecting and acidic regions) (25). According to a current model, upon activation by NPF, Arp2 and 3 adopt the conformation mimicking that of protomers in F-actin. This facilitates the attachment of actin monomer (bound to the WH2 domain within WCA motif of NPF) and nucleates the formation of a daughter filament branched at an angle ~70° to the mother filament (26). In this way, Arp2/3 serves also as a P-end capper of a daughter filament. It was shown that ATP hydrolysis—followed by Pi release—occurs with a delay on the Arp2 subunit and promotes filament debranching. Indeed, in *in vitro* assays, Arp2/3-induced branches dissociate spontaneously after a few minutes ($t_{1/2}$ ~ 800 s), which is a long time on the scale of filament's life span in cells (27).

It is important to consider the Arp2/3 branching in the context of rapid actin dynamics in the cell. The lifetime of branched actin arrays is regulated by many factors, besides ATP hydrolysis, and more are yet to be identified. An ABP cortactin and a phallotoxin phalloidin were shown to stabilize filament branches without affecting Pi release from Arp2 (27). In contrast to this, proteins such as coronin 1B and glia maturation factor (GMF) promote filament debranching (28, 29). Tropomyosin was shown to reduce the branch density on F-actin by competing with Arp2/3 (30). Thus, functional differences among Tm isoforms should be considered in the context of this effect. On the other hand, actin-interacting factor cortactin stabilizes the Arp2/3 branches by interacting with both F-actin and Arp2/3 (31).

Recent report from Small's group (32) altered the view on actin's ultrastructure at the cell's leading edge. Presented data suggested some deviations from the canonical model and call for a deeper investigation of the role of filament debranching factors, cross-linking proteins, and other actin effectors in modulating the *in vivo* morphology of actin dendritic arrays. According to the dendritic nucleation model of filament formation, supported by electron microscopy data, the branch density in lamellipodia is high. Essentially every new filament originates as a branch of a preexisting one, which results in dense, highly branched, deformation-resistant F-actin networks (33, 34). Recent electron tomography observations of frozen cells appeared to be inconsistent with this view (32). In that study, lamellipodia was reported to consist of unbranched actin filaments, questioning the hypothesis of dendritic nucleation and contradicting the well-characterized activity of Arp2/3. It was suggested that branches detected in earlier studies were artifacts of the critical-point drying method. However, independent reevaluation of the electron tomography data by T. Svitkina's group resolved this contradiction by finding the F-actin branch points in the original raw images from J. Small's group (35). Nevertheless, Urban et al. still observed longer and less branched filaments in lamellipodia than expected from earlier data, which calls for a refinement of the existing model ((32) also see (36)).

3.1.2. Tandem WH2 Domain-Based Nucleators

This group includes rather specialized nucleators such as Spire (required for egg and embryo development), Cordon-bleu (neuron specific), Leiomodins (muscle specific), and others. Nucleators that belong to this group contain helical WH2 actin-binding domains (ABDs) (as was found also in NPFs activating Arp2/3). These domains are arranged in tandems and are separated by linkers of different length. It was reported that the length and amino acid sequence of these linkers often define the mechanism and strength of their nucleation activity. Most of the tandem WH2-based factors do not seem to interact with Arp2/3 and represent independent class nucleators.

Several activities are attributed to WH2-based nucleators. For example, Spire and Cordon-bleu (Cobl) nucleate actin at low mole ratios and sequester monomers at high Spire/actin ratios (37–39). Additionally, weak P-end capping activity has been suggested for Spire (37). Another report indicated that human Spire severs F-actin (38) and interacts with their B-ends, inhibiting their elongation from profilin–actin. Despite the different results on interactions with filament ends, the *in vitro* nucleation effect of Spire appears established now. However, the available data is insufficient to understand how Spire functions *in vivo*. The reported inability of Spire to utilize profilin–actin complex is especially interesting, considering that most of the monomeric actin in cells is bound to profilin (38). Additional studies are needed to define which of the biochemical activities of Spire is dominant under physiological conditions. The above activities (nucleation, sequestering, and severing) were recently reported for the neuron-specific actin nucleator Cordon-bleu (39). However, similar to Spire, this protein is also unable to utilize the physiologically relevant profilin–actin complex for filament nucleation. This led Husson et al. to the conclusion that *in vivo* Cordon-bleu may be a severing agent rather than a nucleator. However, in order to assess the physiological significance of a severing activity of WH2-based nucleators, additional experiments are needed to compare them with efficient severing factors such as cofilin.

3.1.3. Formins

In contrast to Arp2/3, formins promote the formation of unbranched actin structures, such as found in filopodia protrusions. So far, among all actin nucleators only formins were reported to regulate both F-actin and MT dynamics. In most cases formins were found to stabilize MT (24). The presence of conserved C-terminal formin homology domains (FH1 and FH2) is a common feature of this class of nucleators. The unstructured proline-rich FH1 domain interacts with G-actin–profilin complexes and the FH2 domain forms homodimer that nucleates filaments probably by stabilizing transient actin dimers or trimers. Formins are also processive elongation factors: they remain bound to the filaments, accelerating their elongation through addition of G-actin–profilin complexes captured by FH1 at the B-end. An important difference between the processive movement of formins and motor proteins is that formins most likely do not hydrolyze nucleotides. Proteins belonging to the formins group can be also seen as leaky F-actin cappers, as they allow B-end elongation even in the presence of actin-capping proteins. Additional activities of some formins include bundling, severing/depolymerization of actin filaments *in vitro*, but the physiological relevance of such activities is yet to be clarified (reviewed in (24)).

Clearly, filaments need to be restricted to a certain length that fits a specific actin-based structure. An important question related to formin regulation is how it is displaced from the growing end of the filament to avoid its uncontrolled elongation. Although few formin-inhibiting factors have been documented, including Dia-interacting protein (DIP) (40), yeast Bud14 (41), and Spire (in oocytes) (42), no clear mechanism for formin displacement has emerged yet. Possible mechanisms of formin regulation by tropomyosin are discussed below (see Section 3.5 this chapter).

3.1.4. Actin Elongation Factors: Ena/VASP

Proteins of the *Ena*bled/*Va*sodilator-*s*timulated *p*hosphoprotein family (Ena/VASP)—including factors such as Mena, VASP, and EVL—can modulate the morphology of filopodia and lamellipodia, thereby affecting cell motility (see (43) for review). It was reported that Ena/VASP family proteins antagonize the capping proteins but cannot uncap the filaments (43). Thus, proteins of this family can be called B-end protectors. Direct TIRF microscopy observations revealed that the lifetime of VASP at the B-end is ~3 orders of magnitude shorter than that of a capping protein (CP). Nevertheless, VASP is able to decrease the association rate of CP due to its very rapid binding to the filament ends (44). This explains the inability of VASP family proteins to uncap the B-ends. Ena/VASP contain a polyproline region that often correlates with the ability of proteins to bind the profilin–actin complex. Similar to formins, Ena/WASP proteins stay bound to B-end, accelerating actin polymerization in the presence of profilin, but do not appear to nucleate at physiological salt concentrations (43). It was suggested that Ena/VASP activity is important for the elongation of actin bundles rather than individual filaments. The fact that Ena/WASP proteins exist as tetramers, multimerizing through their coil-coiled region and interacting with B-ends through their C-terminal EVH2 domains, allows for a simultaneous elongation of four actin filaments.

It should be noted also that different nucleators are specifically recruited to different cell sites, such as dendritic arrays in lamellipodium (Arp2/3), filopodium, endocytic sites, or yeast cables and contractile rings (formins), to promote the formation of distinct actin structures. Recently, the question about possible mechanisms that recruit particular nucleators, along with sets of other ABPs, to specific cell sites has attracted attention. An interesting hypothesis links the nucleator-induced changes in the filament structure to the recruitment of ABPs that interact preferentially with these filament conformations (45). However, more structural work on F-actin remodeling by different interacting factors is needed to test this hypothesis.

3.2. Capping Proteins

Capping proteins interact with either B- or P-end of actin filaments, usually with high affinity. Because of the kinetic differences between actin polymerization and depolymerization at the two ends of the filament, it is clear that the effects of barbed-end and pointed-end cappers on actin dynamics will be different. A common feature of the capping proteins is that they interact with the two last subunits at either end of the filament, preventing assembly/disassembly reactions at that end. Intuitively, it appears possible that the ability to stabilize the terminal protomers in actin filaments would suggest some nucleation activity of the cappers. This appears to be the case for at least some of these proteins, such as gelsolin (B-end) and Leiomodin (P-end) (46). Such activity can be explained by the cappers' weak binding to transient actin dimers or trimes, resulting in their stabilization. This appears to be a likely scenario in the case of Lmod, but the mechanism of gelsolin action may be more complex. It was shown that gelsolin seeds (a complex of gelsolin with two actins) contain monomers arranged in an antiparallel orientation. Because antiparallel dimers do not support filament nucleation, the exact mechanism of gelsolin's nucleation activity remains unclear (47).

3.2.1. Barbed-End Filament Capping

Barbed-end filament capping was shown to be important in the formation of dendritic actin arrays in lamellipodia. This was confirmed in a reconstituted actin-based motility system in which a heterodimeric actin capping protein (CP) was found to be one of the five proteins required to sustain *Listeria*-like movement (48). In agreement with this, the ability of cultured mammalian cells to form lamellipodia protrusions was found to depend on CP (49). It was shown that CP binds to actin filaments near membranes, and then gets released due to either severing-induced depolymerization or displacement (49). According to the current view, B-end filament capping serves two major purposes: (1) maintaining the pool of monomeric actin by preventing subunits from incorporation at the B-end. This speeds up the growth at the free, uncapped ends (rapid polymerization of the free ends will push the membrane forward); (2) regulation of actin filament length. In addition to B-end cappers, such as gelsolin and CP, a small ligand—cytochalasin—was shown to have capping activity along with monomer binding (13). This compound is widely used in *in vivo* experiments to induce F-actin depolymerization.

B-end capping proteins are abundant in cells and their inhibition is needed for the formation of actin structures containing long filaments (1). Some actin interacting factors can interfere with the activity of B-end capping proteins due to their competitive binding (formins and Ena/VASP) (24, 44) or direct displacement of CP (CARMIL) (50).

3.2.2. P-End Capping: Tropomodulin

Most of what is known about tropomodulin (Tmod) effects on actin dynamics was gained from the muscle cell system. Actin filaments in the sarcomere are capped at their P-end with Tmod, and with CapZ (B-end capper) at the other end, for a tight control of the thin filament length (51). One of the major differences between Tmods and other B-end cappers is the relatively low affinity of the former for the P-end of the filament (low μM range). However, Tmod affinity for actin is significantly increased by its interaction with tropomyosins (Tms) ($K_d \sim 50$ pM). It was shown that Tmod interacts with the N-terminus of Tm in an isoform-specific manner, and binds to two Tm molecules attached to the sides of the same filament (52).

It was presumed that P-end cappers do not affect significantly actin dynamics in lamellipodia because Tmod expression was mostly linked to highly differentiated, nonmotile cells (51). However, one of the Tmod isoforms (Tmod3) has been identified in endothelial cells. Increased expression of Tmod3 results in fewer free P-ends as well as a decreased number of B-ends. Based on the assumption that depolymerization of the aged (ADP) actin is a rate-limiting step in the treadmilling cycle at the leading edge, this unexpected result was interpreted in terms of lowering the local supply of actin monomers. This could also be due to Tmod interference with cofilin-induced depolymerization or—jointly with tropomyosin—the inhibition of both Arp2/3 and cofilin activity (53). It is tempting to speculate that in some cell types the Tm–Tmod complex may participate in a rearrangement of dendritic actin networks. By stabilizing short filament fragments that would move towards lamella and bind formins, the Tm–Tmod complex would promote the assembly of long, unbranched filaments.

3.3. Severing by ADF/Cofilins

Cofilin is a classical example of F-actin-severing protein. Yeast has only one cofilin isoform which is essential for cell vitality. Three cofilin isoforms (ADF, cofilin 1 and 2) are present in mammalian cells and their local distribution depends on the cell type and compartment. siRNA silencing of either one of the three isoforms leads to reduced cell motility and cytokinesis (54). Cofilins are critical to the assembly of various cellular actin-based structures, especially those that depend on high rate of filament turnover, such as lamellipodium and clathrin–actin endocytic sites. Similarly, in dendritic spines, where actin turnover rate is very fast (90% of actin fluorescence recovery within 2 min, as estimated by FRAP experiments), cofilin 1 is required for proper spine morphology and function (55).

Most ADF/cofilins can interact with both, G- and F-actin. Several lines of evidence support the assumption that ADF/cofilin effects *in vivo* depend on its local concentration. *In vitro* assays have shown that at sub-saturating concentrations cofilin severs actin filaments (56, 57). The ability of cofilin to accelerate pointed-end depolymerization of actin filaments has been documented (56, 58). Also, differential scanning calorimetry (DSC) and microscopy data suggest that high cofilin concentrations can lead to filament stabilization (59). Some reports suggest also that at high concentrations cofilin favors the nucleation of new filaments, probably by stabilizing a longitudinal actin dimer, i.e., the first intermediate in a spontaneous assembly of actin (56). Current models of lamellipodia (leading edge) dynamics view the net effect of cofilin as that of acceleration of filament turnover. Since cofilin has much higher affinity for ADP actin than ATP/ADP-Pi actin, it was proposed to disassemble aged, ADP actin filaments, mostly at their pointed ends. It should be noted that actin filaments turnover at the leading edge is faster than Pi release from pure F-actin. As suggested by *in vitro* experiments, physiological concentrations of cofilin speed up the Pi release, helping to "age" the filaments (4). Then, due to its severing activity, cofilin increases the number of filament ends accelerating the depolymerization process. The released actin monomers undergo nucleotide exchange (ADP to ATP) and this recycled actin is then incorporated at the growing end of the filament.

As evident from the above, cofilins are remarkably multitasking proteins. Thus, considerable efforts were made to dissect cofilins' activities and test for their significance *in vivo*. It was reported that cofilin-mediated severing and depolymerization can be uncoupled by mutations (60). Moriyama et al. showed that the expression of either pig cofilin mutant S120A (with diminished severing activity in the *in vitro* assays) in fission yeast or the corresponding yeast cofilin mutant 104A led to severe growth defects. In contrast to this, the pig cofilin mutant Y82F, or its homologous substitution in yeast cofilin that impaired only the depolymerization activity, had no apparent growth defects. It is also important to assess the role of cofilin-induced severing in relation to specific actin-based structures in cells. Recently, Chen and Pollard addressed the role of cofilin severing activity in cytokinesis by expressing mutants with normal monomer binding activity but impaired ability to sever filaments. They showed that cells dependent on mutant cofilins have slower growth rates and unusual forms of contractile ring assembly (61).

While cofilin-induced severing is recognized now as important for a number of cell functions, its mechanism is still poorly understood. At a structural level, cofilin served for a long time as the main example of an ABP that changes dramatically the F-actin morphology. Electron microscopy image reconstruction, supported by solution studies, revealed the unique ability of ADF/cofilins to increase the helical twist of actin filaments by ~5° per protomer (62). Considering the high cooperativity reported for cofilin-F-actin interactions, it was suggested that upon its binding cofilin induces (or captures/stabilizes) this twisted filament conformation. This conformation was shown to weaken protomer–protomer contacts and was suggested therefore to cause the fragmentation of actin filaments near the boundaries between the decorated and undecorated segments (63). Somewhat more complex picture arises from our recent study using several actin and cofilin isoforms. We have shown that non-muscle human cofilin 1 binds tightly to yeast actin filaments but does not sever them when added at substoichiometric ratios (64).

Examination of yeast actin filaments decorated with human cofilin revealed that despite its inability to sever these filaments it changes their twist similar to other cofilins (our unpublished observations). These results call for explanation of cofilin-induced severing unconnected to changes in the helical pitch of actin. We also found that the ability of cofilin to sever actin filaments correlates with the cofilin-induced flexibility changes (64). Considering previously reported differences in the dynamic properties of muscle and non-muscle actin isoforms (65), it is tempting to speculate that isoform-specific cofilin–actin interactions may determine or regulate the local severing activity *in vivo*. Taken together, the present data show that more information is needed on the structure–function relationship in F-actin–cofilin complexes to gain deeper understanding of the cofilin-modulated actin dynamics.

Multiple factors have been identified that synergize with or antagonize cofilin in modulating actin dynamics. The increasing number of reports on complex mixtures of actin-modulating factors stems from continuous improvements in methods and tools available for probing protein interactions and dynamics. Several binding factors and/or their combinations were found to regulate cofilin-induced F-actin disassembly. For example, *in vitro* experiments with newly formed actin filaments have shown that cofilin-induced disassembly is accelerated in the presence of coronin and actin interacting protein 1 (Aip1). Moreover, previously unknown type of filament disassembly was observed in this system: abrupt, infrequent bursts resulting in the loss of filament mass, mostly from their ends (66). It was suggested that the synergism of cofilin, coronin, and Aip1 leads to the weakening of lateral contacts in F-actin, resulting in strand separation and rapid depolymerization. Detailed mechanism of such "burst disassembly" and its physiological relevance is yet to be clarified. Recently, cofilin synergism with coronin was investigated in greater detail (67). It was shown that yeast coronin modulates cofilin-induced actin disassembly in a nucleotide-sensitive manner: it blocks cofilin severing of ADP-Pi/ATP actin and accelerates the severing of ADP-filaments. It should be noted that in the above studies by Kueh et al. and Gandhi et al. coronins from different sources were employed (mammalian and yeast, respectively) and therefore the reported effects may be isoform specific (66, 67). Detailed structure–function studies of the coronin–actin system are needed to explain the observed effects.

Important questions about cofilin recycling were addressed in the *in vitro* system containing multiple ABPs. A complex study on F-actin regulation by cofilin, profilin, Srv2/CAP, and Abp1 suggested that the Srv2/CAP complex interacts with ADP-actin–cofilin and catalytically detaches ADP-actin monomers from cofilin (68). This activity is physiologically important for two reasons: (1) the need for ADP–ATP exchange in monomers which is greatly inhibited in their complex with cofilin and (2) the need to recycle cofilin for a next round of filament severing/depolymerization, to promote overall filament turnover. Mammalian CAP was shown to catalyze nucleotide exchange in G-actin on its own, but yeast Srv2 had to synergize with profilin to assist in the exchange. The observed activity was consistent with *in vivo* observations that a loss of Srv2 reduces the rates of F-actin disassembly in yeast.

3.4. G-Actin-Sequestering Proteins

The recently discovered sequestering activity of some multidomain actin nucleators (Spire and Cobl), together with the fact that proteins such as ADF/cofilins can bind actin monomers as well, makes it somewhat difficult to discuss their effects on actin dynamics. For simplicity, in this review we define a sequestering protein as one that contains a single actin-interacting domain, binding exclusively to monomeric actin and blocking filament nucleation upon complex formation. Using this definition, we end up with heterogeneous—in terms of sequence homology—and unrelated group of proteins. However, the common structural aspect of G-actin interaction with sequestering proteins is that they block/interfere sterically with the attachment of another actin monomer to form the actin nucleus/polymer. For example, profilin, vitamin D-binding protein (DBP), twinfilin, and Ciboulot bind between actin SD1 and SD3 blocking monomers' B-end. Similarly, DNAse I interacts with SD2, blocking the P-end (however, physiological relevance of this interaction is not well understood). Monomer B-end blockers listed above have no sequence homology but share a common structural element—short α-helixes characterized by the presence of exposed and conserved hydrophobic side chains that bind in the hydrophobic cleft of actin (between SD1 and SD3) (reviewed in (69)). In cells, most of the monomeric actin exists in a complex with profilin. Profilin is a unique example of actin-sequestering protein because despite its inhibition of actin's nucleation it supports filament elongation at rates almost identical to that of free G-actin (70). At the same time, the actin–profilin complex does not support P-end elongation, making only B-end elongation physiologically relevant. Multiple actin-binding factors contain prolyne-rich regions to recruit the G-actin–profilin complex to the polymerization site.

3.5. Actin Stabilizing Factors

Tropomyosin (Tm) is a classic example of actin-stabilizing proteins. It is difficult to discuss specific effects of Tm on actin dynamics due to a large number of tropomyosin isoforms. For instance, mammals utilize four Tm genes to generate more than 40 functionally distinct isoforms (reviewed in (71)). Thus, the effects of Tm on actin dynamics and its cross talk with other binding factors cannot be generalized.

The functional form of Tm molecules is a hetero- or homodimer composed of two Tm polypeptide chains of the same size. Tm interacts with a number of actin protomers equal to the number of its helical quasi-repeats, which varies depending on the isoform. The affinity of a single Tm molecule for F-actin is low; however, its binding is highly cooperative and strongly increased due to the tail-to-head interactions of Tm molecules on the filament. Based on biochemical studies on F-actin–Tm complexes, the net effect of Tm on F-actin can be defined as that of overall stabilization of filament structure and the slowing down of its dynamics. It was shown with multiple Tm isoforms that actin filaments decoration with Tm inhibits their cofilin- and gelsolin-induced severing/depolymerization. Tm prevents also Arp2/3-mediated filament branching, which is important for a number of actin-based structures, and among them actin cables in yeast (72). It was shown that the loss of Tm leads to a rapid disappearance of such cables (73). The ability of some Tm to promote annealing of capped actin filaments suggests that these Tms may assist in the displacement of capping proteins from the filament ends (74). It was also documented that yeast SpTm isoform synergizes with Cdc12p to promote the elongation of actin filaments (11).

Drebrin is another protein that belongs to the category of actin-stabilizing factors. Drebrin A is highly enriched in dendritic spines and is known to compete with Tm *in vitro* (75). Drebrin and Tm do not share any structural homology. Drebrin contains the N-terminal ADF-homology domain, followed by a unique charged region which shares some homology with a mammalian actin-binding protein 1 (mAbp1). Drebrin binds at 1:5 mole ratio to F-actin and EM image analysis suggested several modes of its attachment to actin (76). Similar to Tm, drebrin competes *in vitro* with actin's cross-linking proteins such as α-actinin and fascin (75). Using AFM imaging we showed recently that drebrin increases actin filament stiffness, which is consistent with their overall stabilization by this protein (77). The AFM imaging revealed also that drebrin is one of a few actin-interacting factors modulating actin filament morphology, which was never reported for any of the Tm isoforms. In contrast to cofilin and fascin (62, 78) which shorten the helical pitch of actin filaments ("overtwisting" them), drebrin induces filament "untwisting," which translates into an increased (~40 nm) helical pitch of F-actin (77). It appears that the drebrin-decorated dendritic spine's cytoskeleton may represent a unique actin-based structure with still unknown properties. We are currently exploring in-depth drebrin effects on actin dynamics and clarifying its cross talk with other ABPs to gain a new insight into actin dynamics in dendritic spines. However, why neuronal cells need high levels of drebrin along with multiple Tm isoforms is an open and intriguing question that should be addressed in future work.

3.6. Cross-Linking Proteins

The best known examples of actin cross-linkers are the spectrin family proteins (such as α-actinin, spectrin, dystrophin, etc.) and, structurally unrelated to them, fascin. Proteins of this group cross-link F-actin into bundles or networks. Actin cross-linkers are often localized at the leading edge and focal adhesions, and act frequently as scaffolds connecting actin cytoskeleton to extracellular matrix proteins. In skeletal myoblasts, neurons, and fibroblasts the cross-linking protein fascin assists in cell migration/growth cone extension and the formation of actin-based protrusive structures (79).

The structure of cross-linking proteins defines the molecular architecture of actin bundles. For example, the spectrin family cross-linkers contain N-terminal ABD composed of two calponin homology domains (CH) and a rod domain containing a different number of spectrin repeats (4 (α-actinin) to ~24 (dystrophin)). These proteins dimerize through their rod domains to form antiparallel dimers. The length of rod domains and the mode of attachment to actin determine the spacing between filaments cross-linked by N-terminal ABDs. It was suggested by solution studies that cross-linkers such as α-actinin, L-plastin/fimbrin (80), fascin, and filamin (81) slow down F-actin depolymerization. Notably, the density and morphology of bundles/networks can regulate the access of regulatory factors that can, in turn, rearrange the bundles and/or move cargo.

The architecture of bundles would suggest an overall decrease in actin dynamics in these structures due to a more limited access for destabilizing factors (such as cofilin and monomer sequestering proteins). However, recent reports on cofilin–fascin cross talk suggest a more complex mechanism of bundles' rearrangement than originally thought. It was revealed by direct TIRF microscopy observations that fascin cross-linking aids in cofilin severing (82). Earlier study by Schmoller et al. concluded the opposite: fascin-cross-linked filaments were more cofilin resistant (81). Structural studies on actin bundles showed that some cross-linking proteins can alter filaments' morphology by increasing the variations in their helical pitch (scruin) (83) or filaments' "overtwisting" (fascin) (78). It is possible that cofilin may take advantage of such conformational changes (see Sect. 3.3), yielding accelerated severing. To resolve this contradiction, a more detailed investigation of cofilin's cross talk with actin cross-linking proteins is needed now. Factors such as isoform specificity and experimental approach limitations should be taken into consideration. The fact that cross-linkers can compete with each other suggests that assembly/disassembly of actin bundles may be fine-tuned by their exchange.

Intriguingly, isoform-specific effects of myosin on actin bundles have been documented. It was shown that myosin X but not myosin V recognizes selectively fascin-bundled actin filaments and uses them for cargo transport to filopodial tips (84). In contrast to this, myosin II causes filamin bundles' disassembly (81). In line with the hypothesis advanced for actin nucleators (45), these cross-linker-induced changes in F-actin may help in recruiting a set of appropriate ABPs in order to build/remodel actin structures.

3.7. Motor Proteins

This class of protein converts the energy of ATP hydrolysis into mechanical force that powers their movements. These proteins employ actin filaments or MT as biopolymer-based "highways" for cargo transport. Molecular motors can be divided into three superfamilies depending on the "roads" they take: myosin (F-actin), kinesin (MT), and dynein (MT). We refer the reader to a recent review by Hirokawa et al. for a detailed discussion of molecular motors in neuronal cells (85).

Myosin motors are critical for a wide variety of cell movements, from muscle contraction to cargo transport. These proteins undergo cyclical interactions with actin in order to produce force and movement. Although the actomyosin ATPase cycle includes multiple nucleotide-linked steps (see (86) for a recent review), myosin state intermediates can be defined based on their affinity to actin (weak or strong). Weak-affinity intermediates contain myosin attached to actin in the ATP- or ADP-Pi states and do not bear force. Force generation (power-stroke) coincides with Pi release and the transition from a weakly to strongly bound actomyosin intermediates—in the ADP- and nucleotide-free (rigor) states. More than 35 classes of myosins can be subdivided based on their structural similarities (single- or double-headed, etc.) or functional features (duty ratio, speed and direction of movement, etc.). All myosins have the same basic ATP hydrolysis cycle, but the lifetimes and rates of transitions between the states vary, yielding different duty ratios (defined as the fraction of ATPase cycle spent in the strongly bound state). It should be noted that duty ratios are load dependent and tend to increase under intermediate loads but decrease under high loads (reviewed in (86)). Low-duty-ratio myosins (such as myosin I) are usually single-headed and work at high concentrations to sustain the movement. In contrast to this, high-duty-ratio myosins (such as myosin V) function as processive dimers and have at least one of their heads attached strongly to the filament for efficient cargo transport. Most of the myosins establish directional movement towards the fast-growing end (B-end) of the filament, with the exception of myosin VI which is reported to move towards the pointed end.

4. Brief Summary and Perspectives

Multiple lines of evidence suggest that the dynamic plasticity of actin filaments is an important factor in their interactions with many binding factors. The classification of these factors became challenging with the increasing number of reports on their remarkable multitasking. This leads to several questions: #1: What is the physiological relevance of the multiple activities of ABPs? For many ABPs, answering this question will require genetic uncoupling of their multiple activities and assessing their role *in vivo*. In parallel, a question about the significance of these activities for specific, localized actin-based structures should be addressed as well. #2: How actin receives or finds its binding partners? In other words, how different structures are formed in the same cytoplasm and what are the mechanisms of actin-binding factors' sorting and segregation? A growing number of studies suggest that along with the cell signaling modulation of F-actin morphology and dynamics by nucleators, bundling and stabilizing proteins may help in recruiting a particular set of ABPs to the actin-based structures. Additional structural studies are needed to test this hypothesis. #3: Why some of the ABPs have many isoforms? What is the division of tasks among these isoforms? We are becoming now more careful in generalizing the effects of ABPs on actin dynamics. A point in case could be the very different severing activities of cofilin/ADF isoforms, which can be tailored for local needs of specific cell environments. The complex mechanisms and pathways of actin regulation that arise from isoform-oriented studies will probably be the next fast-developing and challenging direction in actin research.

Acknowledgments

We recognize that due to space limitations many important and relevant studies were not cited directly in this review and we apologize for that to the authors of those publications. This work was supported by USPHS grant R01 GM 077190.

References

1. Chhabra ES, Higgs HN (2007) The many faces of actin: matching assembly factors with cellular structures. Nat Cell Biol 9:1110–1121
2. Kueh HY, Mitchison TJ (2009) Structural plasticity in actin and tubulin polymer dynamics. Science 325:960–963
3. Carlier MF, Gutfreund H, Bayley PM (1992) Nucleotide hydrolysis regulates the dynamics of actin filaments and microtubules [and discussion]. Philos Trans R Soc Lond B Biol Sci 336:93–97
4. Pollard TD, Borisy GG (2003) Cellular motility driven by assembly and disassembly of actin filaments. Cell 112:453–465
5. Mitchison T, Kirschner M (1984) Dynamic instability of microtubule growth. Nature 312:237–242
6. Schek HT III, Gardner MK, Cheng J et al (2007) Microtubule assembly dynamics at the nanoscale. Curr Biol 17:1445–1455
7. Pedigo S, Williams J (2002) Concentration dependence of variability in growth rates of microtubules. Biophys J 83:1809–1819
8. Chrétien D, Fuller SD, Karsenti E (1995) Structure of growing microtubule ends: two-dimensional sheets close into tubes at variable rates. J Cell Biol 129:1311–1328
9. Walker RA, O'Brien ET, Pryer NK et al (1988) Dynamic instability of individual microtubules analyzed by video light microscopy: rate constants and transition frequencies. J Cell Biol 107:1437–1448
10. Dent EW, Gupton SL, Gertler FB (2011) The growth cone cytoskeleton in axon outgrowth and guidance. Cold Spring Harb Perspect Biol 3(3): a001800
11. Skau CT, Neidt EM, Kovar DR (2009) Role of tropomyosin in formin-mediated contractile ring assembly in fission yeast. Mol Biol Cell 20:2160–2173
12. Okabe S, Miyasaka H, Hirokawa N (1993) Dynamics of the neuronal intermediate filaments. J Cell Biol 121:375–386
13. Sheterline P, Clayton J, Sparrow JC (2002) Actin, 4th edn. Oxford University Press, Oxford
14. Orlova A, Shvetsov A, Galkin VE et al (2004) Actin-destabilizing factors disrupt filaments by means of a time reversal of polymerization. Proc Natl Acad Sci U S A 101:17664–17668

15. Kueh HY, Brieher WM, Mitchison TJ (2008) Dynamic stabilization of actin filaments. Proc Natl Acad Sci 105:16531–16536
16. Oztug Durer ZA, Diraviyam K, Sept D et al (2010) F-actin structure destabilization and DNase I binding loop fluctuations: mutational cross-linking and electron microscopy analysis of loop states and effects on F-actin. J Mol Biol 395:544–557
17. Scoville D, Stamm JD, Toledo-Warshaviak D et al (2006) Hydrophobic loop dynamics and actin filament stability. Biochemistry 45:13576–13584
18. Galkin VE, Orlova A, Schroder GF et al (2010) Structural polymorphism in F-actin. Nat Struct Mol Biol 17:1318–1323
19. Orlova A, Prochniewicz E, Egelman EH (1995) Structural dynamics of F-actin: II. Cooperativity in structural transitions. J Mol Biol 245:598–607
20. Fujii T, Iwane AH, Yanagida T et al (2010) Direct visualization of secondary structures of F-actin by electron cryomicroscopy. Nature 467:724–728
21. Dos Remedios CG, Chhabra D, Kekic M et al (2003) Actin binding proteins: regulation of cytoskeletal microfilaments. Physiol Rev 83:433–473
22. Fortin DA, Srivastava T, Soderling TR (2012) Structural modulation of dendritic spines during synaptic plasticity. Neuroscientist 18(4):326–341
23. Chesarone MA, Goode BL (2009) Actin nucleation and elongation factors: mechanisms and interplay. Curr Opin Cell Biol 21:28–37
24. Chesarone MA, DuPage AG, Goode BL (2010) Unleashing formins to remodel the actin and microtubule cytoskeletons. Nat Rev Mol Cell Biol 11:62–74
25. Xu XP, Rouiller I, Slaughter BD et al (2011) Three-dimensional reconstructions of Arp2/3 complex with bound nucleation promoting factors. EMBO J 31(1):236–247
26. Rouiller I, Xu XP, Amann KJ et al (2008) The structural basis of actin filament branching by the Arp2/3 complex. J Cell Biol 180:887–895
27. Le Clainche C, Pantaloni D, Carlier MF (2003) ATP hydrolysis on actin-related protein 2/3 complex causes debranching of dendritic actin arrays. Proc Natl Acad Sci 100:6337–6342
28. Cai L, Makhov AM, Schafer DA et al (2008) Coronin 1B antagonizes cortactin and remodels Arp2/3-containing actin branches in lamellipodia. Cell 134:828–842
29. Gandhi M, Smith BA, Bovellan M et al (2010) GMF is a cofilin homolog that binds Arp2/3 complex to stimulate filament debranching and inhibit actin nucleation. Curr Biol 20:861–867
30. Bugyi B, Didry D, Carlier MF (2010) How tropomyosin regulates lamellipodial actin-based motility: a combined biochemical and reconstituted motility approach. EMBO J 29:14–26
31. Weaver AM, Karginov AV, Kinley AW et al (2001) Cortactin promotes and stabilizes Arp2/3-induced actin filament network formation. Curr Biol 11:370–374
32. Urban E, Jacob S, Nemethova M et al (2010) Electron tomography reveals unbranched networks of actin filaments in lamellipodia. Nat Cell Biol 12:429–435
33. Mullins RD, Heuser JA, Pollard TD (1998) The interaction of Arp2/3 complex with actin: nucleation, high affinity pointed end capping, and formation of branching networks of filaments. Proc Natl Acad Sci 95:6181–6186
34. Svitkina TM, Borisy GG (1999) Arp2/3 complex and actin depolymerizing factor/cofilin in dendritic organization and treadmilling of actin filament array in lamellipodia. J Cell Biol 145:1009–1026
35. Yang C, Svitkina T (2011) Visualizing branched actin filaments in lamellipodia by electron tomography. Nat Cell Biol 13:1012–1013
36. Ydenberg CA, Smith BA, Breitsprecher D et al (2011) Cease-fire at the leading edge: new perspectives on actin filament branching, debranching, and cross-linking. Cytoskeleton 68:596–602
37. Quinlan ME, Heuser JE, Kerkhoff E et al (2005) Drosophila Spire is an actin nucleation factor. Nature 433:382–388
38. Bosch M, Le KHD, Bugyi B et al (2007) Analysis of the function of Spire in actin assembly and its synergy with formin and profilin. Mol Cell 28:555–568
39. Husson C, Renault L, Didry D et al (2011) Cordon-Bleu uses WH2 domains as multifunctional dynamizers of actin filament assembly. Mol Cell 43:464–477
40. Eisenmann KM, Harris ES, Kitchen SM et al (2007) Dia-interacting protein modulates formin-mediated actin assembly at the cell cortex. Curr Biol 17:579–591
41. Chesarone M, Gould CJ, Moseley JB et al (2009) Displacement of formins from growing barbed ends by bud14 is critical for actin cable architecture and function. Dev Cell 16:292–302
42. Quinlan ME, Hilgert S, Bedrossian A et al (2007) Regulatory interactions between two

actin nucleators, Spire and cappuccino. J Cell Biol 179:117–128
43. Bear JE, Gertler FB (2009) Ena/VASP: towards resolving a pointed controversy at the barbed end. J Cell Sci 122:1947–1953
44. Hansen SD, Mullins RD (2010) VASP is a processive actin polymerase that requires monomeric actin for barbed end association. J Cell Biol 191:571–584
45. Michelot A, Drubin DG (2011) Building distinct actin filament networks in a common cytoplasm. Curr Biol 21:R560–R569
46. Chereau D, Boczkowska M, Skwarek-Maruszewska A et al (2008) Leiomodin is an actin filament nucleator in muscle cells. Science 320:239–243
47. Hesterkamp T, Weeds AG, Mannherz HG (1993) The actin monomers in the ternary gelsolin: 2 actin complex are in an antiparallel orientation. Eur J Biochem 218:507–513
48. Loisel TP, Boujemaa R, Pantaloni D et al (1999) Reconstitution of actin-based motility of Listeria and Shigella using pure proteins. Nature 401:613–616
49. Iwasa JH, Mullins RD (2007) Spatial and temporal relationships between actin-filament nucleation, capping, and disassembly. Curr Biol 17:395–406
50. Uruno T, Remmert K, Hammer JA (2006) CARMIL is a potent capping protein antagonist. J Biol Chem 281:10635–10650
51. Fischer RS, Fowler VM (2003) Tropomodulins: life at the slow end. Trends Cell Biol 13: 593–601
52. Kostyukova AS (2008) Tropomodulin/tropomyosin interactions regulate actin pointed end dynamics. Adv Exp Med Biol 644:283–292, Landes Bioscience and Springer Science + Business Media
53. Fischer RS, Fritz-Six KL, Fowler VM (2003) Pointed-end capping by tropomodulin3 negatively regulates endothelial cell motility. J Cell Biol 161:371–380
54. Hotulainen P, Paunola E, Vartiainen MK et al (2005) Actin-depolymerizing factor and cofilin-1 play overlapping roles in promoting rapid F-actin depolymerization in mammalian nonmuscle cells. Mol Biol Cell 16:649–664
55. Hotulainen P, Llano O, Smirnov S et al (2009) Defining mechanisms of actin polymerization and depolymerization during dendritic spine morphogenesis. J Cell Biol 185:323–339
56. Andrianantoandro E, Pollard TD (2006) Mechanism of actin filament turnover by severing and nucleation at different concentrations of ADF/cofilin. Mol Cell 24:13–23
57. Pavlov D, Muhlrad A, Cooper J et al (2007) Actin filament severing by cofilin. J Mol Biol 365:1350–1358
58. Carlier MF, Laurent V, Santolini J et al (1997) Actin depolymerizing factor (ADF/cofilin) enhances the rate of filament turnover: implication in actin-based motility. J Cell Biol 136:1307–1322
59. Bobkov AA, Muhlrad A, Pavlov DA et al (2006) Cooperative effects of cofilin (ADF) on actin structure suggest allosteric mechanism of cofilin function. J Mol Biol 356:325–334
60. Moriyama K, Yahara I (1999) Two activities of cofilin, severing and accelerating directional depolymerization of actin filaments, are affected differentially by mutations around the actin-binding helix. EMBO J 18:6752–6761
61. Chen Q, Pollard TD (2011) Actin filament severing by cofilin is more important for assembly than constriction of the cytokinetic contractile ring. J Cell Biol 195:485–498
62. McGough A, Pope B, Chiu W et al (1997) Cofilin changes the twist of F-actin: implications for actin filament dynamics and cellular function. J Cell Biol 138:771–781
63. Suarez C, Roland J, Boujemaa-Paterski R et al (2011) Cofilin tunes the nucleotide state of actin filaments and severs at bare and decorated segment boundaries. Curr Biol 21:862–868
64. McCullough B, Grintsevich E, Chen C et al (2011) Cofilin-linked changes in actin filament flexibility promote severing. Biophys J 101: 151–159
65. Kim E, Miller CJ, Reisler E (1996) Polymerization and in vitro motility properties of yeast actin: a comparison with rabbit skeletal α-actin. Biochemistry 35:16566–16572
66. Kueh HY, Charras GT, Mitchison TJ et al (2008) Actin disassembly by cofilin, coronin, and Aip1 occurs in bursts and is inhibited by barbed-end cappers. J Cell Biol 182:341–353
67. Gandhi M, Achard V, Blanchoin L et al (2009) Coronin switches roles in actin disassembly depending on the nucleotide state of actin. Mol Cell 34:364–374
68. Balcer HI, Goodman AL, Rodal AA et al (2003) Coordinated regulation of actin filament turnover by a high-molecular-weight Srv2/CAP complex, cofilin, profilin, and Aip1. Curr Biol 13:2159–2169
69. Dominguez R (2004) Actin-binding proteins—a unifying hypothesis. Trends Biochem Sci 29:572–578
70. Gutsche-Perelroizen I, Lepault J, Ott A et al (1999) Filament assembly from profilin-actin. J Biol Chem 274:6234–6243

71. Gunning P, O'Neill G, Hardeman E (2008) Tropomyosin-based regulation of the actin cytoskeleton in time and space. Physiol Rev 88:1–35
72. Lindberg U, Schutt CE, Goldman RD et al (2008) Tropomyosin regulate the impact of actin binding proteins on actin filaments. Adv Exp Med Biol 644:223–231
73. Pruyne DW, Schott DH, Bretscher A (1998) Tropomyosin-containing actin cables direct the Myo2p-dependent polarized delivery of secretory vesicles in budding yeast. J Cell Biol 143:1931–1945
74. Nyåkern-Meazza M, Narayan K, Schutt CE et al (2002) Tropomyosin and gelsolin cooperate in controlling the microfilament system. J Biol Chem 277:28774–28779
75. Ishikawa R, Hayashi K, Shirao T et al (1994) Drebrin, a development-associated brain protein from rat embryo, causes the dissociation of tropomyosin from actin filaments. J Biol Chem 269:29928–29933
76. Grintsevich EE, Galkin VE, Orlova A et al (2010) Mapping of drebrin binding site on F-actin. J Mol Biol 398:542–554
77. Sharma S, Grintsevich EE, Phillips ML et al (2010) Atomic force microscopy reveals drebrin induced remodeling of F-actin with subnanometer resolution. Nano Lett 11:825–827
78. Claessens MM, Semmrich C, Ramos L et al (2008) Helical twist controls the thickness of F-actin bundles. Proc Natl Acad Sci 105: 8819–8822
79. Hashimoto Y, Kim DJ, Adams JC (2011) The roles of fascins in health and disease. J Pathol 224:289–300
80. Lebart MC, Hubert F, Boiteau C et al (2004) Biochemical characterization of the L-plastin–actin interaction shows a resemblance with that of alpha-actinin and allows a distinction to be made between the two actin-binding domains of the molecule. Biochemistry 43:2428–2437
81. Schmoller KM, Semmrich C, Bausch AR (2011) Slow down of actin depolymerization by cross-linking molecules. J Struct Biol 173:350–357
82. Breitsprecher D, Koestler SA, Chizhov I et al (2011) Cofilin cooperates with fascin to disassemble filopodial actin filaments. J Cell Sci 124:3305–3318
83. Schmid MF, Sherman MB, Matsudaira P et al (2004) Structure of the acrosomal bundle. Nature 431:104–107
84. Nagy S, Rock RS (2010) Structured post-IQ domain governs selectivity of myosin X for fascin-actin bundles. J Biol Chem 285: 26608–26617
85. Hirokawa N, Niwa S, Tanaka Y (2010) Molecular motors in neurons: transport mechanisms and roles in brain function, development, and disease. Neuron 68:610–638
86. Sweeney HL, Houdusse A (2010) Structural and functional insights into the myosin motor mechanism. Annu Rev Biophys 39:539–557

Chapter 5

Principles of Proteomic Approaches to the Cytoskeleton

Christina Loosse, Katrin Marcus, and Bodo Schoenebeck

Abstract

In combination with bioinformatics and molecular biological techniques proteomic approaches become more and more indispensable in order to deepen our understanding of cellular functions. Since the cytoskeleton is presented by a highly dynamic network, engaged in many basic cellular functions like cell growth, migration, or intracellular transport mechanisms, many open questions remain to be clarified. Moreover, this concerns triggers of cytoskeletal remodeling or dynamics of membranous interaction partners. A proteomic description should exceed the pure listing of its constituents but rather should include functional proteomics as well as the description of protein interaction networks. Due to its mediating nature between cytosolic and membranous compartments of the cell different techniques are necessary to complete the investigation of the neuronal cytoskeleton. Within this article, we present a set of state-of-the-art approaches for further proteomic research of the cellular cytoskeleton and beyond.

Key words: Proteome, Membrane proteins, 1D-PAGE, 2D-PAGE, 2D-DIGE, Mass spectrometry, Relative quantification, Absolute quantification, ICPL, SILAC, iTRAQ, AQUA

1. Introduction

In analogy to the genome the proteome was defined as the quantitative expression of genes within, e.g., a cell, a tissue, or a body fluid at specific time points under defined circumstances (1). This definition implies a highly dynamic nature of the proteome according mainly to variations in (patho)physiological states. Alterations in the proteome's composition are expressed by shifts in protein levels and a multitude of posttranslational modifications (PTMs) including phosphorylation, glycosylation, or ubiquitination (2–8). Furthermore, it becomes more and more evident that alterations like thiol modifications play additional functional roles in cells (9, 10).

Katrin Marcus and Bodo Schoenebeck contributed equally.

Rolf Dermietzel (ed.), *The Cytoskeleton: Imaging, Isolation, and Interaction*, Neuromethods, vol. 79,
DOI 10.1007/978-1-62703-266-7_5, © Springer Science+Business Media, LLC 2013

In addition, mRNA levels fail in many cases to predict corresponding protein abundances (11, 12). Therefore, the proteome resembles merely a snapshot of inter- and intracellular dynamics (1). Since nervous tissue is highly complex with regard to its vast number of expressed genes and proteins (13, 14) proteomic studies are indispensable to gain a deeper insight of neuronal function.

In general, proteomic approaches imply protein profiling, functional proteomics, as well as description of protein interaction networks. Protein profiling defines protein expression in, e.g., diverse tissues or body fluids. Functional proteomics involves characterization of proteins and their function including, e.g., elucidation of PTMs (15). In a last consequential step, interactions between proteins have to be monitored to build up a description of interaction networks. This is of particular interest for clinical research as differences found between pathological and normal physiological states may help to discover biomarkers applicable for diagnosis and/or an eventual early clinical intervention (16–20).

To conquer difficulties arising from brain complexity, the generation of sub-proteomes has been proven to be beneficial. This can be achieved either by pre-analytical macroscopic anatomical preparations (21, 22) or by the separation of discrete cellular organelles. Thus, the proteomes of different cellular organelles have extensively been studied (23–27).

The cytoskeleton spans the entire cell and anchors at both cellular organelles and the outer plasma membrane. Therefore, a reasonable proteomic description of the neuronal cytoskeleton should include cytosolic as well as membrane proteins. Due to manifest differences in solubility between membrane and cytosolic proteins, diverse methods for protein detection have been established. For a detailed description of membrane proteins, see Section 3 this chapter<<<. The field of proteomics holds an immense inventory of different methods. In this article, we review the main proteomic approaches useful for investigations of the neuronal cytoskeleton. This rather should serve as decision guidance for the design of approaches than as a detailed description of each method listed below.

2. Materials and Methods

2.1. Sample Preparation

The first prerequisite in proteomic approaches is the adequate preparation of samples in order to reproducibly make proteins amenable for further analysis by, e.g., mass spectrometry (MS). This includes the disintegration of cellular membranes, either by mechanical cell disruption with, e.g., cell grinders or sonification or by the use of protein-denaturizing chemicals often in combination with detergents. Therefore, the choice of protein extraction is of critical importance for the subsequent analysis. In some cases, a pre-fractionation or enrichment of proteins or cellular organelles is advisable. This permits a stable measurement of even scarce proteins in samples like cerebrospinal fluid (CSF) (28) or serves as a means for confining the protein inventory of distinct cellular organelles. In a second step, proteins might either be analyzed directly by, e.g., MS, separation via gel electrophoresis, and subsequent detection via Western blotting or be digested into peptides. As illustrated later in this article, the latter procedure substantially improves protein identification via MS. Basically, two different strategies can be distinguished for protein digestion: in-gel digestion and in-solution digestion approaches. In both cases digestion is carried out with endoproteases offering specific cleavage into definite peptides. For most applications, trypsin is the protease of choice since it generates most frequent suitable peptides for MS analysis (29, 30). For specific questions the digestion with alternative proteases like GluC, LysC, AspN, or chymotrypsin may enhance the identification of proteins. By the combination of different proteases characterized by specific cleavage sites differing from each other, overlapping peptides are generated resulting in enhanced sequence coverage (31). For in-gel digestion, proteins are first solubilized with a detergent like SDS or CHAPS, then separated along a polyacrylamide gel, and subsequently digested within the gel bands. This method was first described by Shevchenko and coworkers in 1996 (32). The advantage of this approach is its rather robustness against impurities which would interfere with the digestion process. Thus, spurious amounts of detergents or salts can be washed out prior to final protein digestion. On the other hand, peptide recovery may be hampered especially in highly concentrated gels. For in-solution digestion, proteins are preferentially extracted in the presence of chaotropic reagents like urea or thiourea, respectively, and digestion takes place in solution after protein precipitation (33). This method can easily be automated but holds the drawback that substances impeding the digestion process are not precluded from the solution. Often, substances needed for protein extraction and digestion are not compatible with MS and one has to get rid of these substances before MS analysis. SDS as well as CHAPS result in strong dominance of MS spectra leading to underrepresentation of medium and low-abundance proteins. Therefore, exhaustive cleaning steps are mandatory for good MS results. These steps are effective for soluble proteins but can rather be challenging for membrane proteins as these proteins are mainly hydrophobic in nature. Therefore, their solubilization and digestion often need a special treatment (34) which is addressed later in Section 3 this chapter. Due to sample preparation being the first step in proteomic analyses its reproducibility in any approach is decisive for all next analytical steps. In general, the number of preparative steps should be restricted to an absolute minimum since otherwise variations due to methodological processing become greatly enhanced resulting in falsified results.

2.2. Gel-Based Proteomic Approaches

Proteomic investigations are challenging tasks because of their everywhere complexity. Many of the known genes in the human genome are expected to be expressed as different protein species due to multiple splice variants and an overwhelming amount of possible PTMs. In addition, the concentration range of proteins spans up to ten orders of magnitude, especially in easily accessible samples like blood (35). Therefore, gel-based approaches are often used as a pre-fractionation step. By this, protein detection in consecutive analyses via, e.g., MS becomes markedly increased in complex samples.

The simplest approach is the one-dimensional polyacrylamide gel electrophoresis (1D-PAGE) according to Laemmli (36). Gel-based separation of proteins is normally performed by sodium dodecyl sulfate-polyacrylamide gel electrophoresis (SDS-PAGE). Prior to separation, SDS is added as a negatively charged detergent to the samples in order to both retain a denaturated state and impart a uniformly negative charge to all proteins in proportion to their masses (36, 37). Thus, proteins migrate according to their different molecular masses along an applied electric field. In general, protein species of low molecular weight move faster through the porous gel-matrix than those bearing a higher molecular weight. Finally, proteins can be visualized in the gel with several staining methods, e.g., colloidal Coomassie staining or more sensitive silver staining (38–42). The choice of staining methods depends on the protein concentration expected or on appending methods selected for further characterization of the samples. For example, common silver staining protocols show benefits regarding an enhanced sensitivity but often fail to be applicable to further MS analysis. After transfer of the proteins from the gel-matrix to a nitrocellulose or a PVDF-membrane, a restricted subset of proteins is often analyzed by Western blotting (43, 44). Here, the protein of interest is selectively marked with an antibody which in turn is visualized either directly by an attached fluorophore or by an enzymatic driven reaction. In addition, staining of 1D-SDS-PAGE is used as a pre-fractionating method prior to further MS analysis. For this, colored protein bands are selected and cut and undergo an enzymatic digestion directly in the gel with subsequent extraction of the resulting peptides. In a final step, extracted peptides are identified via MS (45) (see Fig. 2 and Section 2.4 this chapter).

Whereas 1D-PAGE separates proteins simply due to their electrophoretic mobility, thus, mainly by molecular weight, the resolution in protein separation can considerably be enhanced by means of two-dimensional polyacrylamide gel electrophoresis (2D-PAGE) (46, 47). Here, the analytes pass two independent gel runs to utilize the protein characteristics of molecular weight with differences in their netto charge for improved separation. In detail, all proteins are subjected in the first dimension to isoelectric focusing (IEF), hence conducting a separation according to their respective isoelectric points (pI). In the second dimension, proteins run perpendicular to the first dimension according to their molecular weight via conventional SDS-PAGE. Thus, up to several thousand protein spots can be displayed or their respective PTMs (48, 49). IEF can be performed either using carrier-ampholyte (CA) (46, 47) or the so-called immobilized pH gradient (IPG) (50–52) gels. For the visualization of protein spots different staining methods are available which vary in sensitivity and dynamic range (see above). 2D-PAGE allows the differential proteome analysis by comparing spot patterns of different samples. Protein spots which show differences in intensity can be relatively quantified by analysis software tools (53, 54). Subsequent MS analysis gives rise to the protein identity of the spots of interest. Because of their hydrophobicity it is rather difficult to analyze membrane proteins or hydrophobic proteins with conventional 2D-PAGE approaches. Different strategies for the investigation of membrane proteins can be taken into account that omit IEF in the first dimension but include other gel-based approaches, e.g., native gel electrophoresis like Blue Native PAGE (55) or protein separation under denaturing conditions like urea-16BAC (56–58), CTAB (59, 60), or SDS/SDS (61).

A further development in 2D-PAGE is two-dimensional difference gel electrophoresis (2D-DIGE) (62) which is based on fluorescent labeling of the samples. This technique allows direct comparison of different samples within one gel. In this approach, each sample (e.g., disease vs. healthy control) is labeled with a distinct cyanine dye (Cy2, Cy3, or Cy5), each of it showing distinguishable fluorophores and thus preventing an overlap of corresponding detection wavelengths. Proteins can be labeled via two different labeling strategies called minimal labeling and saturation labeling, the second for faint protein abundance. During minimal labeling the dyes react with the amino group of a small portion of lysine residues (in average one lysine per protein) whereas in saturation labeling all cysteine residues are labeled allowing the detection of only 3 μg total protein amount per gel (59, 60). The labeled samples are mixed together and subjected to 2D-PAGE so that both samples are resolved on the same gel. In addition, a mixture of all samples labeled with a third fluorophore serves as an internal standard (63). Thus, accurate determination of protein expression changes is possible leading to relative quantification between the groups. After separation, the proteins are detected by scanning the gel at the respective excitation wavelength of each dye. Resulting images are further processed with automated background subtraction and normalization among all gels conducted by analysis software. The images within one gel are merged and compared between all gel runs. By this, differentially expressed proteins are detected and quantified (63). Whereas minimal labeling stains only about 3.5% of all proteins, saturation labeling of proteins has the drawback that only two dyes, Cy3 and Cy5, exist.

For this reason, only one condition (disease or healthy control) can be analyzed simultaneously with the internal control. Figure 1 gives an overall impression of the above-mentioned gel-based techniques.

2.3. Liquid Chromatography

Besides above-mentioned gel-based approaches liquid chromatography (LC) systems are state-of-the-art to conquer sample complexity in proteomic studies by pre-fractionation (64–66). Alternatively, gel-based techniques can easily be combined with LC by sequential loading of in-gel digestions of consecutively cut 1D-PAGE bands or 2D-PAGE spots (67, 68). For, e.g., samples of reduced complexity or the subsequent use of LC-MS setups characterized by a high-resolution power, gel-based techniques are often omitted. Here, in-solution digested protein samples can be loaded directly on an LC system. Figure 2 gives a comprehensive overview of typical workflows.

The LC system generally used in proteomics is high-pressure/performance liquid chromatography (HPLC) separation (64). The basic principle of this technique relies on an elution over time of peptides or proteins in solution (mobile phase) displaying different binding properties to a stationary phase in columns. Due to various chemical and physical characteristics the analyte molecules interact differently with the two phases. In detail, the netto charge of peptides and proteins defines their propensity to bind to the stationary phase. In proteomic research, reversed-phase (RP) HPLC serves as a standard method in LC (69). Here, the analytes bind to a nonpolar stationary phase, mostly C_2–C_{18} chains, coupled to silica-particles. The aqueous mobile phase is mixed with a polar solvent containing small amounts of acids and, e.g., acetonitrile or methanol (69). According to rising hydrophobicity, peptides elute from the column over time by gradually increasing the amount of the organic compound. The time point when the molecules elute from the column is called retention time. The eluting peptides pass a UV-detector which determines the total amount of peptides eluting at a given time point.

Usually, the HPLC setup is fully automated (70) meeting with benefits in reproducibility and time saving. The sample is automatically loaded on the column and eluted step by step by the programmed gradient. For higher resolutions, the total time for performing the gradient can be enhanced and longer columns can be used, respectively (71). In addition, the HPLC can be coupled directly to the electrospray ionization (ESI)-source of a mass spectrometer so that the eluting peptides are automatically measured referred to as online LC-MS (72). For subsequent detection via matrix-assisted laser desorption/ionization time of flight (MALDI-TOF) the separation is performed off-line and peptide-fractions collected are manually or automatically (64) spotted on an appropriate MALDI-target plate.

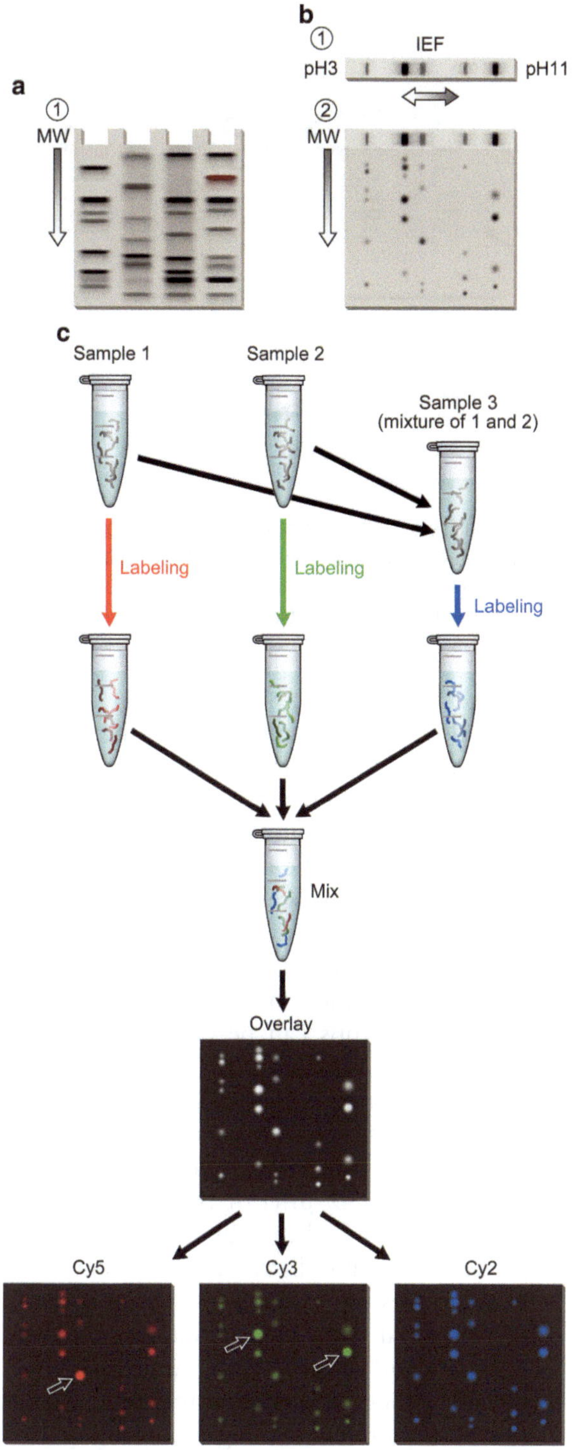

Fig. 1. Examples of gel-based proteomic techniques: In 1D-PAGE, lyzed and denaturated proteins are applied to an electrical field as the drifting force for the separation of proteins due to their electrophoretic mobility, i.e., mainly by molecular weight (MW) (**a**). In 2D-PAGE, proteins are separated in a first dimension by isoelectric focusing (IEF) according to the respective isoelectric point (pI) of proteins on strips of immobilized pH-gradients. In a second dimension, pre-separated proteins of IEF run perpendicular to their pI-values according to their electrophoretic mobility. This results in a two-dimensional spot pattern (**b**). In 2D-DIGE-technology, different samples are labeled with isobaric fluorophores and run as 2D-PAGE protected from sunlight. After gel run is completed, differential regulated proteins (*arrows*) can be visualized by comparing the respective fluorophore channels of each sample. Standardization is possible by interpolation with a third fluorophore corresponding to mean signals of a mixture of all samples included in the study (internal standard) (**c**).

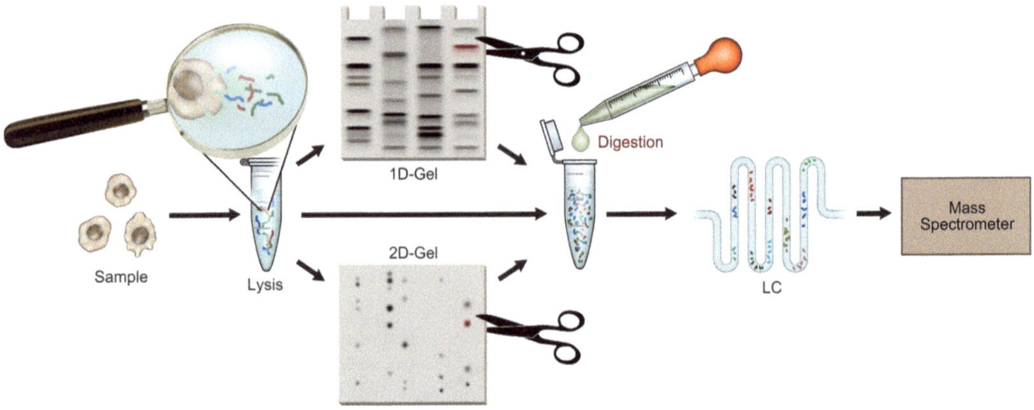

Fig. 2. Typical workflow for proteomic analyses: First, the sample is lyzed by appropriate buffers and detergents. Secondly, proteins are either directly subjected to enzymatic digestion (in-solution) or separated by gel-based techniques. Protein bands or even spots can be cut out from the gels, subsequently digested (in-gel digestion), extracted from the gel pieces, and optionally passed through a liquid chromatographic system (LC) before measured by mass spectrometry.

Another opportunity given by the automation is the application of an alternative separation technique, like multidimensional protein identification technology (MudPIT) (73). Here, the sample is separated in two dimensions on columns according to different characteristics. In the first dimension the samples can be separated via, e.g., ion exchange, hydrophilic interaction liquid chromatography (HILIC), or size exclusion followed mostly by RP-HPLC in the second dimension (74). This setup coupled to MS analysis allows identification of over 2,000 proteins per run (75).

HPLC setups can be downscaled to nano-size applications. Therefore, the material for the stationary phase is packed into capillary columns with a smaller inner diameter (76). This leads to increased sensitivity and less sample consumption. In order to gain even better separation efficiencies one possibility is to decrease the particle size (≤2 μm) of the packed material which improves the peak shape and thereby the resolution of the column. Smaller particle sizes also increase the backpressure and need a higher working pressure within the system that cannot be accomplished by conventional nano-HPLC-systems. Therefore, ultra-high performance liquid chromatography (UHPLC) (77–79) systems have been developed allowing for increased separation and shorter analysis times paired with a higher throughput of samples.

2.4. Mass Spectrometry

All techniques described above are useful for the pre-fractionation of protein mixtures in order to decrease sample complexity. However, these approaches do not allow for the identification of the separated proteins. Except for studies of confined interest in a sparse amount of proteins detectable by antibody-driven approaches, the vast amount of proteins inherent in cell lysates needs an accurate

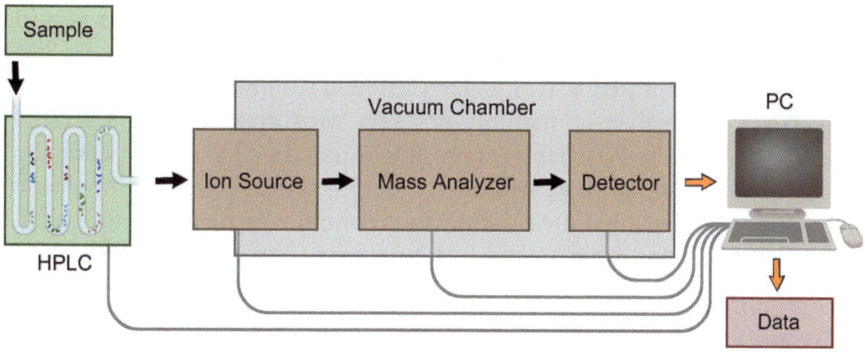

Fig. 3. Basic setup of a mass spectrometer. Protein or peptide samples enter the mass spectrometer at the ion source for initial ionization of the analytes. Charged proteins or peptides are passed within a vacuum device through a mass analyzer before subsequently reaching the detector. Signals induced by the detector are further analyzed by a computer system. Optionally, an HPLC-system can be coupled online to the mass spectrometer.

identification of as many proteins as possible within the same experiment. Since MS is capable to precisely detect proteins and peptides and eventually can be combined with upstream LC systems, MS is the method of choice for protein identification.

2.4.1. General Principle

Basically, three main components build up a typical mass spectrometer: an ion source, a mass analyzer, and a detector (80). The ion source creates charged peptide ions as a prerequisite for subsequent separation and detection. In detail, the mass-to-charge (m/z) ratios of peptide ions determine their further separation or even fragmentation via downstream mass analyzers. Finally, the ions are measured by a detector which is connected to a computer with appropriate software for data analysis. In order to reduce sample complexity, a separation of peptides prior to MS via HPLC is often used (81, 82). Figure 3 shows a simplified description of a general construction of a mass spectrometer.

2.4.2. Ionization Methods

For analysis via MS, it is inevitable to produce ions from peptides or proteins, since only ions correspond to the basic principles of MS detection, i.e., alternating electrical or magnetical stimulation. Basically, two methods have been established in proteomics for initial ionization of proteins and peptides, namely, matrix-assisted laser desorption/ionization (MALDI) (83) and ESI (84, 85). In MALDI, laser energy is transferred to the sample molecules which are immobilized by co-crystallization in the presence of various organic compounds (86). By this, incoming energy conducted by a laser beam is transferred on the embedded sample. This triggers desorption and subsequent ionization of analyte molecules in a hot plume derived from the crystallized surface (87). This results mainly in single charged analyte molecules. In proteomics, MALDI is preferably used in combination with TOF mass analyzers (83, 88, 89).

In contrast to MALDI, ESI generates charged peptide ions in solution (85). Liquid samples are dispersed by a capillary needle towards a strong electrical field resulting in ultra-fine dispersion of analyte-containing droplets with concomitant ionization events (90–93). Samples pre-fractionated by HPLC can easily be coupled online to an MS device. This is of fundamental importance for the sensitive measurement of complex protein mixtures with concurrent time saving. Alternatively, higher amounts of solutions of reduced complexity can be administered directly with a syringe to the system over a longer time period in order to ensure more robust measurements. Both ionization techniques described above are combined with different types of mass analyzers. Depending on the application desired, each combination holds its advantages and drawbacks concerning eminent differences in mass accuracy, sensitivity, dynamic range, or resolving power. Therefore, a battery of diverse mass spectrometers has been established for best performances on the respective demands (94).

After ionization of the incoming analyte ions the molecular masses are measured in mass analyzers. Basically, different types of analyzers can be described. TOF analyzers essentially are long vacuum tubes in which ions are given a defined kinetic energy at the ion source following a drift through a field-free region. The basic principle is the measurement of the time the ions need to reach the detector. Since TOF analyzers are often combined with MALDI, each protein or peptide is usually ionized by a single proton. As the velocity of ions depends on their m/z ratio heavy particles reach lower speeds at a given charge (95). Depending on, e.g., the length of the field-free drifting distance, TOF analyzers can reach moderate to high resolving powers. In addition, a mass accuracy of up to 2 ppm can be attained (94–96).

In contrast to TOF analyzers, quadrupole (Q) analyzers consist of four cylindrical arranged metal rods serving as stab electrodes (97, 98). This assembly is responsible for filtering the sample ions according to their m/z values by alterations of the applied electrical field. Thus, the m/z values can be calculated by varying the applied voltage. Only ions of defined m/z values oscillate through the rods before finally reaching the detector. All other ions run on unstable trajectories, thus failing to reach the detector. Furthermore, ions can be trapped in the electric fields for fragmentation. Q mass analyzers are often parts of the so-called hybrid mass spectrometers consisting of series-connection of diverse types of mass analyzers. This can be advantageous in order to combine a selection of precursor ions with subsequent fragmentation for the sake of enhanced mass accuracy and high-resolution power. This is the operation principle of Q-Q-TOF instruments where two Q units are combined with a TOF analyzer (94). The special combination of three Qs or the arrangement of two Qs followed by a linear ion trap (LIT) permits new scanning methods like product ion scanning, parent ion scanning (99, 100), neutral loss scanning (101, 102), or multiple reaction monitoring (103). These scanning methods use tandem MS analyses (MS/MS or MS^n, see also Sect. 2.4.3) for more detailed characterization of peptides. Due to the combination of mass analyzers as mass filters and subsequent collision cells, restricted groups of preselected peptide ions can be characterized in more detail.

In "ion trap" (IT) analyzers ions are trapped and with this are accumulated over a given time in a vacuum system. Nonlinear, three-dimensional ITs were first described by Paul et al. (104). For efficient trapping of ions dynamic electric fields are established between two adversely arranged hyperbolic electrodes and a corresponding ring electrode. In comparison to three-dimensional ITs LITs possess higher injection efficiencies and greater storage capacities. LITs are composed as a quadruple device confining ions by using an oscillating potential. With the progressive change of co-current and alternating current flow ratios during the scan period, the mass range of stable oscillation becomes shifted resulting in mass discrimination (105). By joined formation of static constant currents and oscillating alternating currents a three-dimensional electric field is capable to trap ions of a defined m/z range. Once focused in the center of the trap, the ions selectively can be ejected towards the detector by gradual changes of the electric field. ITs are typically filled with helium gas to a pressure of about 1 mtorr in order to enhance storage capacity by reducing the kinetic energy of all incoming ions. In general, ITs can perform MS/MS scans. Again, these additional further fragmentations lead to more detailed characterization of the peptides. This is typically done by collision-induced dissociation (CID) or electron transfer dissociation (ETD). This leads to unmatched sensitivity and fast data acquisition. In CID a neutral gas like helium, nitrogen, or argon is injected into the collision chamber. Alternatively, ETD defined on the basis of an electron transfer from a radical anion (fluoranthene) to protonated peptides is performed (106). One benefit of ETD is the increased preservation of PTMs in most cases, thus excellently complementing the characteristics of CID. Fragmentation patterns of CID and ETD differ substantially and a combination of both methods provides more comprehensive information about the analyzed peptides. For these reasons, both techniques are often combined in alternating scans yielding more confidential information about PTMs like phosphorylation or glycosylation (107–110). A detailed description concerning CID and ETD methodology can be found in (111–115).

Fourier transform ion cyclotron resonance (FT-ICR) mass spectrometers are special forms of ITs producing a strong magnetic field perpendicular to an electric field derived from trapping plates (116, 117). FT-ICR was invented in 1974 by Comisarow and Marshall (117). In this kind of mass analyzer, ions run in a circular path while passing nearby detection plates. Masses are resolved by their rotational frequency (cyclotron). m/z values are calculated by Fourier transformation of the ion cyclotron giving rise to remarkable mass accuracy of up to 1.0 ppm (118). Nevertheless, FT-ICR spectrometers are less common than other types because of their big dimensions and high operation costs.

A relatively new kind of mass analyzer is the so-called Orbitrap. Orbital trapping was first implemented by Kingdon in 1923 and was refined by Makarov in 2000 (119–121). The Orbitrap is featured by a coaxial spindle-shaped electrode surrounded by an outer cylinder. Due to sufficient tangential velocity of the incoming ions centrifugal forces are in balance with electrostatic attraction. Therefore, ions are orbiting around the central electrode. In addition, ions oscillate on the axis of the central electrode from which in turn the m/z ratios can be derived by fast Fourier transformation algorithms. Orbitrap analyzers present a high dynamic detection range and show excellent resolving power and mass accuracies comparable to those of FT-ICR instruments (119). Furthermore, the coupling of an LIT as a pre-selective device to an Orbitrap even augmented the above-mentioned pros. For this reason, Orbitrap instruments become increasingly popular in proteome analysis (122).

2.4.3. Identification of Proteins by Mass Spectrometry

In general, the sensitive measurement of m/z values generated by mass spectrometers is pivotal for the identification of proteins, peptides, or even peptide fragments. In particular for the discovery and accurate localization of PTMs or chemical groups covalently bound to the proper protein backbone the analysis of peptide fragments is required. Therefore, different scanning methods have been described suitable for both peptide mass and peptide fragmentation analysis, each of them suitable for a special focus of investigation. These scanning methods include peptide mass fingerprinting (PMF) (123), post-source decay (PSD) (124), MS/MS (also called MS^2), product ion scanning (30, 94, 125), neutral loss scanning (NL) (101, 102), precursor ion scanning (100, 126, 127), and selected/multiple reaction monitoring (SRM/MRM) (94, 103, 128).

PMF or peptide mass mapping is well adapted for samples of low complexity. Thus, this scanning method is best applicable in the further analysis of 2D-gel spots, e.g., by MALDI-TOF MS (30). Due to prior digestion by sequence-specific endoproteases a mixture of predictable peptide ions is generated. This is reflected by a characteristic peptide pattern depicted by the PMF. The peptide signals generated are matched with databases for final protein identification (129).

For protein mixtures of higher complexity, a more detailed analysis is necessary and fragmentation of peptides is the method of choice.

In MALDI-TOF instruments a proportion of desorbed proteins and peptides decay spontaneously along the field-free drifting distance, referred to as PSD (124). Since fragmentation occurs preferentially at the peptide bonds additive information of MS/MS mass spectra allow for the identification of the respective peptide or protein. In order to enhance both selectivity and sensitivity of MS measurements, mass analyzers can be connected in series to each other (130). By this, mass spectra can be confined by focusing of the first mass analyzer to a subset of ions to be precisely analyzed.

Usually, the information attained by MS/MS spectra is sufficient for confident peptide identification. For some reasons of in-depth analyses of fragment ions especially detection of PTMs further fragmentations are possible resulting in MS^n spectrometry (131, 132). Due to extended fragmentation steps in MS^n a clear identification of even those peptides becomes feasible whose parent ions share almost identical m/z-values.

In other applications it might be of interest to specifically detect (pre)defined peptide species from a (complex) mixture. Especially Q-Q-Q mass spectrometers are used for such specialized MS/MS applications called product ion scanning (30, 94, 125), precursor ion scanning (127), NL (30, 133), and SRM/MRM (128, 134, 135). All these applications have in common to use the sequential arrangement of Qs to combine selection, fragmentation, and scanning features within one experiment. In principle, each Q within the chain operates in selection, fragmentation, or scanning mode resulting in a task sharing. In general, Q2 serves as a collision chamber in order to produce fragment ions. A selection of ions to be analyzed can be achieved by the fixation of the electrical field of either Q1 (product ion scanning, SRM/MRM) or Q3 (precursor ion scanning). Alternatively, the concurrent scan of Q1 and Q3 allows for the confinement of ions reaching the detector (NL, SRM/MRM). Figure 4 gives an overview of typical approaches to monitor a subset of protein/peptide ions in a triple Q arrangement.

Product ion scanning and precursor ion scanning are appropriate methods when measuring PTMs which may be characterized by their product ions in a rather complex protein mixture (136, 137). NL is a scanning method for the indirect detection of uncharged molecules. This means the verification of a mass difference between parent ions in Q1 and product ions in Q3 resulting from loss of a neutral particle after collision in Q2 (30, 133). In other words, only fragment ions that differ from their parent ion by a characteristic mass difference will be detected. NL is applicable particularly for the identification of phosphorylated and glycosylated proteins since these chemical derivatives appear mostly in an uncharged form.

MRM is especially suitable for hypothesis-driven approaches in complex protein or peptide mixtures. This implies the preselection

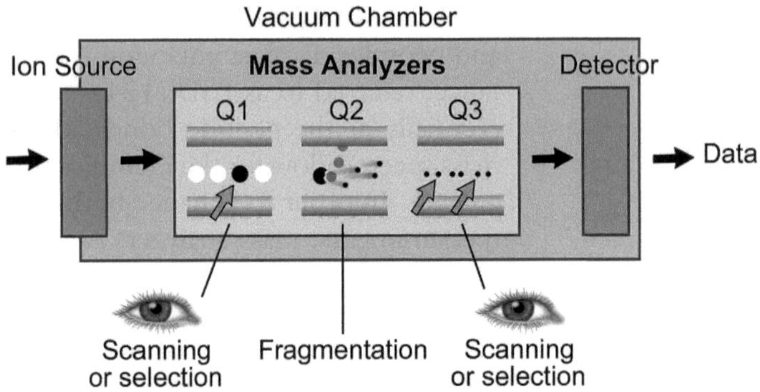

Fig. 4. Setup of a triple-quadrupole (Q1–Q3) for in-depth analyses of fragment ions. After initial ionization, proteins or peptides pass through a sequential arrangement of Qs. Either Q1 can be fixed to a defined electrical field resulting in the selection of precursor ions (designated by the *orange arrows*) or Q1 serves as a scanning device which monitors all incoming ions (sum of *white* and *black circles*). In Q2, fragmentation of the (preselected) analyte molecules occurs by collision with an inserted gas. Finally, Q3 serves again either for the scanning of all or the selection of product ions generated (the latter depicted by *orange arrows*). By diverse combinations of scanning, fragmentation, and selection features of the respective Qs different methods including product ion scanning, precursor ion scanning, NL, or SRM/MRM are generated. All of them end up with a focusing of a before complex protein or peptide mixture to a defined subset which greatly enhances the sensitivity of measurements. As illustrated as one possible method above, the predefined selection of a precursor ion with subsequent fragmentation in Q2 leads to a series of daughter ions reaching Q3. In Q3, again, a second selection of peptide fragments for final detection is included. By simultaneous scan of both Q1 and Q3 very specific SRM/MRM transitions can be detected.

of peptide ions with predefined m/z values for both precursor ions and product ions (MRM transitions). By monitoring these MRM transitions, only those ions can reach the detector which share both the predefined precursor ion m/z and their specific product(s) m/z ratio(s). By focusing on these MRM transitions, the specificity of even low-abundance proteins or peptides is markedly enhanced. With MRM relative or absolute quantification is feasible (135).

Together, all these applications end up in the focused analysis of a subset of proteins and peptides of (a rather) complex sample and consequently in improved detection of even those proteins and peptides characterized by a low abundance.

The attained raw data in MS imply information about m/z values, their respective intensities, as well as information resulting from MS^n scans. The actual identification of proteins or peptides finally results from the comparison of peak lists with databases. Typical search engines like MASCOT (138) or Sequest (139) use different algorithms to calculate a score or probability factor indicating the chance of correct protein assignment. Considering, e.g., prior digestion steps or further fragmentation of known selected precursors, the peak list generated is compared with a hypothetical digestion or fragmentation (140). In detail, characteristics of sample preparation and separation as well as individual settings of the mass spectrometer have to be taken into account. All these information give rise to a solid peptide assignment (138, 139, 141). Ideally, peptides should be unique for proper assigned proteins in order to circumvent false positive results. Especially when information of accessible databases is insufficient, manual description of MS/MS spectra for sequence generation might be performed (142). Figure 5 shows typical diagrams for an LC chromatogram with consecutive MS and MS/MS spectra and final database search.

2.4.4. Quantification of Proteins by Mass Spectrometry

In order to compare the protein composition of different samples it is not sufficient to simply identify the proteome of certain cell types or tissues as a pure list of proteins. Beyond that, it is crucial to monitor changes in protein abundances in the respective groups (e.g., healthy vs. disease, treated vs. non-treated). Most notably, protein quantification via 2D-PAGE holds limitations in displaying hydrophilic membrane proteins. Furthermore, only proteins in a mass range of ~5–150 kDa within a restricted pH range of 3.5–10 are properly presented. Due to these limitations, gel-free approaches based on MS are developed allowing for relative and absolute quantification in order to overcome the above-mentioned disadvantages of gel-based quantitative proteomics. Different approaches encompass quantification via various labeling methods as well as label-free approaches which can be used for relative and, respectively, in part for absolute quantification (143, 144). Each method has its own limitations and benefits (145). Isotope-coded labels can be introduced by chemical means, metabolically, or enzymatically into a sample. Alternatively, they are spiked into the sample as an internal standard (146). Metabolic labeling often shows the most accurate results. That is because the label is introduced at the earliest time point in sample preparation, thus minimizing the variance in sample processing. For chemical labeling, one has to keep in mind that the labeling reaction might be incomplete which would result in falsified results. Label-free approaches are less accurate than approaches that are based on internal standards. Nevertheless, they offer the benefit to be time- and cost-saving because no labeling reaction has to be carried out.

2.4.5. Relative Quantification

This section briefly describes most frequently used methods for MS-based relative protein/peptide quantification.

2.4.6. Labeling Approaches

For relative quantification, the introduction of stable isotopes into the sample is a common technique to mark either proteins or peptides. For this, two basic principles are state of the art. Chemical labeling is performed by a chemical reaction leading to covalent bonds of isotope-encoded tags with amino acid residues of the

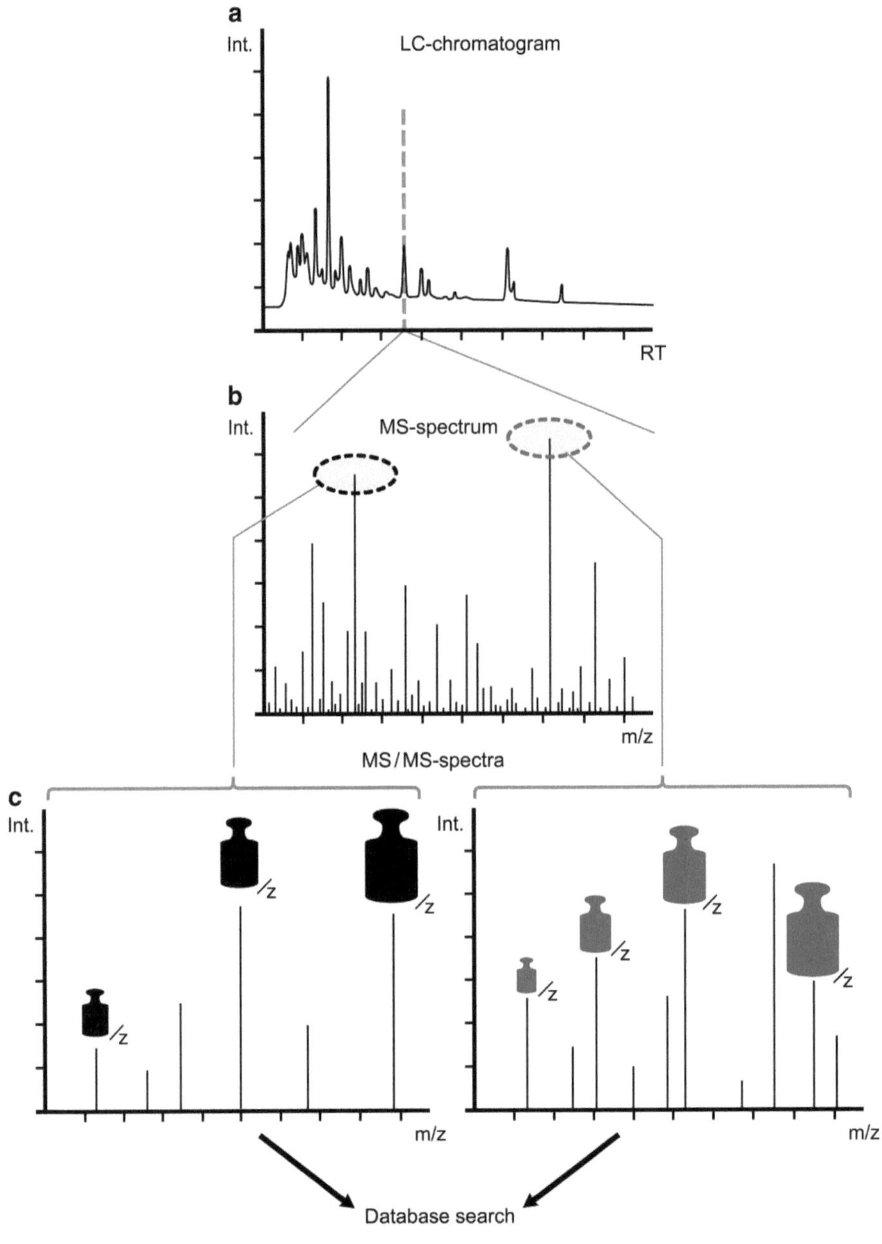

Fig. 5. Characteristic data hierarchy detected in MS-runs: (**a**) Peptides or proteins which elute from LC are measured via UV-detection and are depicted as a chromatogram over retention time (RT). (**b**) Eluted peptides of a defined RT are plotted as MS-spectrum, showing multiple intensity (Int.) peaks along a mass-to-charge (*m/z*) axis. (**c**) The most intense peaks are subsequently selected for MS/MS fragmentation (enclosed by scattered *ovals*) resulting in a characteristic set of *m/z* fragmentation products defined by both *m/z* values and intensity. Weight icons exemplify for definite fragmentation products of peptides separately for both MS/MS scans. (**d**) Analytical data measured in MS or MS/MS scans are compared online with databases comprising a list of possible peptide fragments. The proteolytic enzyme(s) as well as further chemical modification(s) can be predefined. By this, the measured *m/z* values are assigned to definite peptide fragments (exemplified by *black* and *grey* weight icons originating from the respective MS/MS scans). In a second run, determined peptide fragments give rise to peptides (scattered *ovals*) and furthermore proteins (p1; p2) by searches in a protein database. Here, proteins are selected which show preferable matches between their composition of peptides and those peptides found in the analytical data. Peptide fragments which fail to be detected in an MS/MS scan are depicted by colorless weights.

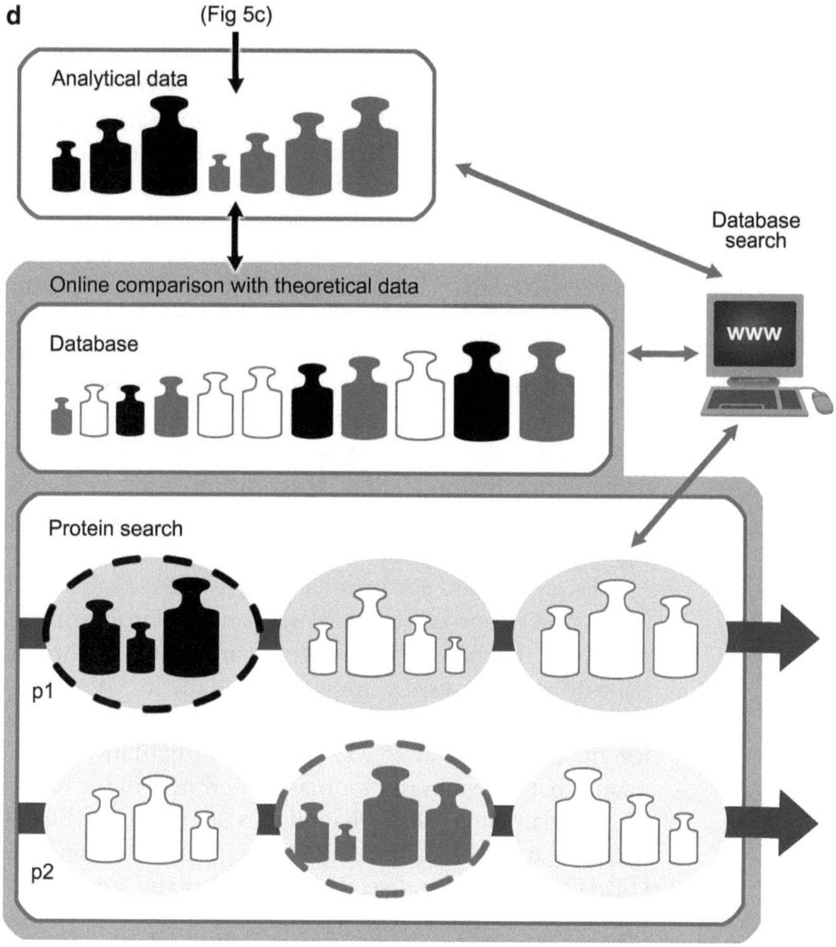

Fig. 5. (continued)

sample molecules. In contrast to this, in metabolic labeling the stable isotopes are in vivo incorporated in the proteins during growth of, e.g., cells in culture. The growth medium is supplemented with isotopically labeled amino acids or nutrients instead of unlabeled additives resulting in their integration during cell growth. After MS analysis labeled peptides can easily be distinguished from their original counterparts by a defined mass shift.

Among the list of quantification methods using chemical labeling, the isotope-coded affinity tags (ICAT) (147), isotope tags for relative and absolute quantification (iTRAQ) (148), isotope-coded protein labeling (ICPL) (149), and the chemical labeling with heavy water ($^{16}O/^{18}O$) (150, 151) have gained marked relevance. Depending on the label used peptides and/or proteins can be labeled.

ICAT is used for labeling the sulfhydryl group in cysteine residues. This reagent consists of three components: a biotin tag for affinity purification of labeled peptides, a thiol-specific reactive group interfering with cysteine residues, and an isotopically labeled linker region that allows for the discrimination between a light ICAT- (containing ^{12}C-atoms) and a heavy ICAT-form (containing ^{13}C-atoms). For relative quantification, same amounts of two protein samples are labeled with the light or heavy ICAT-form, respectively. Subsequently, the samples are pooled together and are subjected to enzymatic digestion. Resulting peptides are further enriched via their biotin tag, separated by HPLC, and subsequently analyzed by MS. The basic principle is that differentially labeled peptides co-elute in HPLC and can be measured in the same MS scans with a defined mass shift of 9 Da. By determining the ratio of the signal intensities of un/labeled peptide pairs the corresponding proteins are relatively quantified (152). Since only cysteine residues are labeled, the ICAT approach does not consider cysteine-free proteins for quantification. In addition, protein isoforms, PTMs, or degradation products, which show differences only in cysteine-free peptides, cannot be detected.

Tandem mass spectrometry tags (TMTs) (153) and iTRAQ (148, 154) approaches are based on the same principle chemical structure. These isobaric reagents are made of three elements: a peptide reactive group for covalent labeling of amine-groups in lysine residues and/or the N-terminus, a reporter group allowing for quantification after MS-driven fragmentation, and a balancer group that compensates for mass differences in the reporter groups. The fundamental idea is that all tags are isobaric which implies that each tag as a whole shows the same mass. To date, eight different iTRAQ mass tags and six TMTs are available which allow for multiplexing of samples. After labeling same amounts of different samples (mainly peptides) identical peptides labeled with different tags are isobaric. Labeled samples are mixed and respective peptides co-elute in HPLC. Since their isobaric nature, identical peptides labeled with different tags result in the same m/z ratios in the MS scan. Upon fragmentation of the peptides the low molecular reporter ions are released. Therefore, the different mass tags can be distinguished in the MS/MS-spectra according to different masses of the generated reporter ions. Again, relative amounts of corresponding peptides can be determined by calculating relative intensities of the reporter ions (155, 156). Besides the identification of the reporter ions via the MS/MS spectra they also allow for the identification of peptides (155, 156). Finally, iTRAQ and TMT can also be used for absolute quantification (157). Because the generation of MS/MS spectra is indispensable for quantification, iTRAQ and TMT approaches in complex samples are rather amenable for high-abundance proteins since less intense MS signals of low-abundance proteins will not automatically be chosen for MS/MS fragmentation. A further limitation of this technique lies in the generation of low molecular reporter ions which cannot be detected with all kinds of mass spectrometers.

Within ICPL all free amino groups in proteins are labeled (149). In this technique, up to four protein samples are reduced, alkylated, and subsequently labeled with one of the four different ICPL reagents which show mass shifts between 2 and 10 Da. The mass shifts arise from different degrees of ^{13}C incorporation. Samples are pooled and, if desired, subjected to chromatographic or electrophoretic separation in order to reduce complexity of the mixture. Alternatively, the mixture is directly enzymatically digested and subsequently analyzed by MS. Identical peptides from differentially labeled proteins differ in mass and are observable as quadruplets in the MS spectra. From the ratios of respective peak intensities the relative abundance of the corresponding protein can be concluded (158).

Via enzymatic labeling with heavy water ($^{16}O/^{18}O$ method) the C-termini of peptides are added either by ^{16}O- or ^{18}O-atoms (150, 151). This labeling strategy requires enzymatic digestion with trypsin, Glu-C, or Lys-C. With this method, two protein samples can be compared but no multiplexing of the method is possible. In detail, one protein sample is digested in $H_2{}^{16}O$-containing buffer whereas the other protein digestion buffer contains the heavy isotope $H_2{}^{18}O$. Thereby, up to two O-atoms are incorporated into the peptides leading to a mass shift of 2 or 4 Da for the isotopically labeled peptides. Prior to chromatographic separation and MS the samples are mixed together in equal amounts. Quantification is performed on signal intensities of peptide pairs in MS scans. Major drawbacks of this method are the potential back exchange of the O-atoms and inconsistent rates of isotopic incorporation depending, e.g., on size of the peptide or the amino acid sequence.

In contrast to the analysis of lyzed tissue or liquid samples like CSF or blood, in vivo samples can undergo metabolic labeling via stable isotope labeling by amino acids in cell culture (SILAC) (159). This is possible especially for cell culture (160–162) or simple organisms as *C. elegans* (163, 164), *D. melanogaster* (163, 165), and even mice (166, 167). This method is based on the incorporation of isotopically labeled amino acids into proteins. In cell culture approaches, one cell population is grown in medium with normal (light) amino acids while a second cell population is grown in medium supplemented by isotopically labeled (heavy) amino acids. After several cell cycles incorporation of light and heavy amino acids is completed and harvested cells can be mixed in equal ratios. Proteins are extracted, optionally fractionated and/or purified, proteolytically digested, and subjected to LC-ESI-MS(/MS). Again, light/heavy peptide pairs co-elute in HPLC and show fixed mass differences depending on amino acids and isotopes used (^{13}C, ^{15}N, ^{2}H, and ^{18}O). Relative amounts of detected peak pairs are reflected by the ratios of measured peak intensities.

Similar to the SILAC approach a labeling via ammonium salts in cell culture is possible. Here, the "light" medium contains ^{14}N whereas the "heavy" medium contains ^{15}N-enriched ammonium salts (168). Cells grown in these media fully incorporate the respective N-atoms in their proteins after some proliferation cycles. Differentially labeled cells are measured as mentioned above. Whereas in SILAC the peptide pairs show constant mass differences, ^{15}N labeling results in varying masses leading to a more complicated quantification procedure. Variable mass shifts depend on peptide length and amino acid composition and are caused by the incorporation of the respective N-atoms in each amino acid.

2.4.7. Label-Free Approaches

Unlike quantification via labeling approaches, where samples are pooled and analyzed in one LC-MS run, label-free approaches are restricted to separate measurements of the respective samples. Thus, the comparison of two and more samples is more prone to run-to-run variations. By omitting labeling of the samples, drawbacks like incomplete labeling or amino acid-dependant access to peptides are prevented. In general, relative quantification can be performed according to two different strategies, comparison of the so-called extracted ion chromatograms (XIC) on the MS level or comparison of the number of acquired spectra for each protein on the MS/MS level (81, 169, 170).

Spectral counting is based on the comparison of the number of identified MS/MS spectra assigned to each protein (45, 169, 171). This is feasible because the number of spectra assigned to one peptide correlates with its concentration (172). Thus, high-abundance proteins cause high amounts of peptides belonging to this protein equivalent with longer elution times from the HPLC column. As a consequence, spectra from these peptides are measured more often in the MS analysis. Indeed, several MS/MS spectra need to be generated in order to obtain reliable results. This is achieved by modern mass spectrometers featuring fast scan cycles that enable sufficient number of MS and MS/MS spectra. The major drawback of spectral counting is a rather poor quantification of low-abundance proteins, especially in complex protein samples. This can be explained by the fact that separation via HPLC is not able to completely circumvent the co-elution of different peptides. As an answer to this problem repeated analysis of the same sample is essential to identify as much peptides as possible (172). Moreover, only the most abundant ions in the MS survey scan are chosen for MS/MS spectra generation. Thus, less intense peptides are not taken into account for analysis. Latest mass spectrometers are able to generate MS/MS spectra for the 20 most intense ions in parallel which can minimize this problem to some extent. Regardless of these disadvantages spectral counting can be alternatively used as an inexpensive method for relative quantification, especially for high-abundance proteins.

Apart from relative quantification via MS/MS spectra the peak intensity of LC-MS analysis can be considered for quantification (173–176). Eluting peptides are subjected to MS analysis generating an MS spectrum in every scan cycle. Plotting the signal intensities of a specific *m/z* value (belonging to a specific peptide) against its innate retention time results in the so-called XIC. Because signal intensity correlates with ion intensity peak area values for each mass peak/peptide can be used for relative quantification. In order to obtain reliable results samples should be measured on highly reproducible HPLC systems as well as high-resolution mass spectrometers. Besides adequate hardware, appropriate analysis software is also required that detects and integrates peaks, removes isotope patterns, performs normalization, and after peak alignment matches corresponding peaks from several LC-MS runs. The advantage of this method is that all signals generated in the MS scans are considered for quantification and no information, e.g., on less intense signals is lost. However, compared to spectral counting this approach is less reproducible and has a smaller dynamic range (177). The field of lable-free quantification will surely grow and holds the chance to open new perspectives of protein analysis in the near future by improving the reproducibility of these methods.

2.4.8. Absolute Quantification

Especially for the investigation of potential biomarkers in clinical research absolute protein quantification attains an increased relevance (178). For this approach stable isotope-labeled standard peptides are synthesized which are unique for a special protein (179). The resulting modified peptides resemble their native counterparts but are substituted with an isotope-coded amino acid. Defined amounts of these isotope-labeled standards are added to the protein sample. After MS analysis of the mixture the ratio of the ion intensity between the native peptide and its respective standard is calculated. Since the total amount of added standard peptide is known, it is feasible to calculate the total amount of the native peptide within the sample (180). In combination with MRM absolute quantification respectably enhances sensitivity because of the preselection process of precursor ions considered for detection (181–183). This leads to very low noise levels paired by a quantification range of up to five orders of magnitude. There are different options for isotope-labeled standard peptide generation (179, 184). In most cases, the synthesis of the so-called AQUA-peptides (AQUA for absolute quantification) is supported by specialized companies. These synthetic peptides on-demand can directly be used by the customer. In the quantification concatamers (QConCat) approach a synthetic gene that expresses concatenated standard peptides is generated (185, 186). This artificial protein is added to the sample and peptides are released upon proteolytic digestion. The protein standards for the absolute quantification (PSAQ) approach uses intact isotope-labeled proteins that can be spiked into the sample at the beginning of sample preparation (187). Due to this most accurate measurements are obtained. Last but not least, the AQUA approach holds the advantage that also PTMs can be quantified.

3. Membrane Proteomics: A Difficult Task

Understanding the physiology of the human brain is one of the most challenging approaches of our century. Within neuronal networks a multitude of diverse cell types are connected to each other encompassing roles in nutrition, growth, electrical information, and cellular plasticity (188). In neuroscience, the outer plasma membrane of nerve cells or glia is of unequivocal interest since most cellular processes are affected due to signals conducted via membrane proteins (188). The highly dynamic architecture of nerve cells and their synaptic boutons is only one example to underscore the significance of in-depth proteomic research of the nervous membrane compartment (189, 190). In addition, membrane proteins claim for an enormous clinical relevance. In detail, about two-thirds of all known drug targets turned out to be membrane proteins (191, 192). This is not surprising taken into account the central roles of membrane proteins in signal transduction or intercellular contact. In contrast to their significance in cellular communication, it still remains a challenge to analyze membrane proteins due to their diverse hydrophobic nature. Therefore, membrane proteins are underrepresented in analyses of cell lysates amenable for most cytosolic proteins (193, 194). Consequently, like no other class of proteins, membrane proteins require a whole set of specialized methods for a robust description. For initial solubilization, the differences in hydrophobicity of membrane proteins can be harnessed early in sample preparation by the combination of different detergents (195, 196). After solubilization, proteins have to be extracted from the solution (197, 198). This can be done by mixtures of ethanol and acetone in order to remove lipids (197). The precipitated proteins are subsequently dissolved by various detergents (195, 197).

As already mentioned, the quality of protein identification is generally hampered by the complexity of protein lysates. For membrane proteins, thus, prior enrichment steps have been proven as promising approaches (for a general overview see (199–201)). Unless not fully eliminated, pre-analytical steps can help to protect membranous preparations from crude contaminations with other proteins or organelles. On the other hand, the major drawback of stringent, high-purity preparations is to eventually lose scarce proteins in the list. Since many membrane proteins are low in abundance, an enrichment step can be decisive for subsequent identification. Different enrichment steps are common in proteomics research of which the main principles are described in the next section.

Density gradient centrifugation after cellular disintegration is a rather old but widely used method for the enrichment of cellular membranes (202). Via aqueous two-phase purification, membrane proteins are separated according to their hydrophobicity from cytosolic proteins (203, 204). In addition, differential centrifugation is capable to isolate membrane fractions due to their specific densities (205–207). In these preparations, the major obstacle is to minimize the contamination by other cellular proteins (208, 209).

A successful method for membrane isolation is the prior labeling of cells with biotin. After subsequent cell lysis, biotinylated membrane particles are affinity-purified by binding to magnetic streptavidin-coupled beads (193, 209–211). Subsequent washing steps with increasing stringency ensure the removal of contaminants from the solution. After a final centrifugation step, only specifically bound membrane compartments are collected. Especially by using magnetic bead-coupled monoclonal antibodies raised against one fundamental protein of the desired membrane compartment, plasma membranes can effectively be enriched (212, 213).

Affinity chromatography-based methods are an alternative to the above-mentioned antibody-driven enrichment. Since most transmembranal proteins show complex N-linked glycosylation, the affinity of glycoproteins to lectins permits the accumulation of especially membrane proteins (214, 215).

In conventional 2D-gel electrophoresis, integral membrane proteins are poorly resolvable. Therefore, special modifications have been proposed already described in Section 2.2 this chapter.

Within LC-techniques, the above-mentioned MudPIT technology is beneficial since a high loading capacity of strong cation exchange is combined with high resolving power of RP-HPLC to effectively resolve proteins including membrane preparations (see Section 2.3 this chapter). This is due to the fact that highly insoluble proteins produce after enzymatic digestion, at least in part, more or less soluble peptides which are amenable to MS analysis.

For all these approaches, one has to take into account that many detergents may interfere with subsequent separation techniques like LC or may hamper MS detection. Therefore, additional preparation steps may be essential prior to further analyzing steps. This implies eventual loss of low-abundance proteins and high variance in sample preparation. In addition, all these methods show drawbacks in identifying mainly membrane proteins bearing more than one transmembrane domain. It turned out to be beneficial to use nonspecific proteases potentially resulting in more peptides to analyze (201).

4. Concluding Remarks

Proteomics hold a huge repertory of diverse methods characterized by specific benefits or drawbacks. Depending on the approach, e.g., global measurement of the protein inventory of a given cell type or tissue in contrast to the investigation of distinct proteins or peptides in more detail, different methods have to be applied. Generally, different methods can be combined in order to ultimately increase the outcome. For the sake of minimizing technical error summation, the number of consecutive methodological steps should be limited to the absolutely necessary. For the solid and reproducible measurement of proteins with good solubility a multitude of methods is already available. But, especially the identification of membrane proteins representing a crucial group of proteins in terms of linking cellular compartments and establishing intercellular communication awaits further improvements.

Acknowledgement

The authors gratefully note Helga Schulze for image editing.

References

1. Wilkins MR, Sanchez JC, Gooley AA et al (1996) Progress with proteome projects: why all proteins expressed by a genome should be identified and how to do it. Biotechnol Genet Eng Rev 13:19–50
2. Patterson SD, Aebersold RH (2003) Proteomics: the first decade and beyond. Nat Genet 33(Suppl):311–323
3. Pandey A, Mann M (2000) Proteomics to study genes and genomes. Nature 405:837–846
4. de Graauw M, Hensbergen P, van de Water B (2006) Phospho-proteomic analysis of cellular signaling. Electrophoresis 27:2676–2686
5. Mann M, Jensen ON (2003) Proteomic analysis of post-translational modifications. Nat Biotechnol 21:255–261
6. Shi Y, Xu P, Qin J (2011) Ubiquitinated proteome: ready for global? Mol Cell Proteomics 10:R110.006882
7. Selby DS, Larsen MR, Calvano CD, Jensen ON (2008) Identification and characterization of N-glycosylated proteins using proteomics. Methods Mol Biol 484:263–276
8. Bodo J, Hsi ED (2011) Phosphoproteins and the dawn of functional phenotyping. Pathobiology 78:115–121
9. Chouchani ET, James AM, Fearnley IM et al (2011) Proteomic approaches to the characterization of protein thiol modification. Curr Opin Chem Biol 15:120–128
10. Thamsen M, Jakob U (2011) The redoxome: proteomic analysis of cellular redox networks. Curr Opin Chem Biol 15:113–119
11. Gygi SP, Rochon Y, Franza BR, Aebersold R (1999) Correlation between protein and mRNA abundance in yeast. Mol Cell Biol 19:1720–1730
12. Anderson NL, Anderson NG (1998) Proteome and proteomics: new technologies, new concepts, and new words. Electrophoresis 19:1853–1861
13. Konopka G (2011) Functional genomics of the brain: uncovering networks in the CNS using a systems approach. Wiley Interdiscip Rev Syst Biol Med 3:628–648
14. Becker M, Schindler J, Nothwang HG (2006) Neuroproteomics—the tasks lying ahead. Electrophoresis 27:2819–2829
15. Bayés A, Grant SGN (2009) Neuroproteomics: understanding the molecular organization and complexity of the brain. Nat Rev Neurosci 10:635–646

16. Sun C, Rosendahl AH, Ansari D, Andersson R (2011) Proteome-based biomarkers in pancreatic cancer. World J Gastroenterol 17:4845–4852
17. Pawa N, Wright JM, Arulampalam THA (2010) Mass spectrometry based proteomic profiling for pancreatic cancer. JOP 11:423–426
18. Park JP, Park MK, Yun JW (2011) Proteomic biomarkers for diagnosis in acute myocardial infarction. Biomarkers 16:1–11
19. Kalinina J, Peng J, Ritchie JC, Van Meir EG (2011) Proteomics of gliomas: initial biomarker discovery and evolution of technology. Neuro Oncol 13:926–942
20. Roti G, Stegmaier K (2012) Genetic and proteomic approaches to identify cancer drug targets. Br J Cancer 106:254–261
21. del Castillo C, Morales L, Alguacil LF et al (2009) Proteomic analysis of the nucleus accumbens of rats with different vulnerability to cocaine addiction. Neuropharmacology 57:41–48
22. Katagiri T, Hatano N, Aihara M et al (2010) Proteomic analysis of proteins expressing in regions of rat brain by a combination of SDS-PAGE with nano-liquid chromatography-quadrupole-time of flight tandem mass spectrometry. Proteome Sci 8:41
23. Tribl F, Marcus K, Meyer HE et al (2006) Subcellular proteomics reveals neuromelanin granules to be a lysosome-related organelle. J Neural Transm 113:741–749
24. Tribl F, Marcus K, Bringmann G et al (2006) Proteomics of the human brain: sub-proteomes might hold the key to handle brain complexity. J Neural Transm 113:1041–1054
25. Dreger M (2003) Proteome analysis at the level of subcellular structures. Eur J Biochem 270:589–599
26. Taylor SW, Fahy E, Ghosh SS (2003) Global organellar proteomics. Trends Biotechnol 21:82–88
27. Schröder BA, Wrocklage C, Hasilik A, Saftig P (2010) The proteome of lysosomes. Proteomics 10:4053–4076
28. Davidsson P, Folkesson S, Christiansson M et al (2002) Identification of proteins in human cerebrospinal fluid using liquid-phase isoelectric focusing as a prefractionation step followed by two-dimensional gel electrophoresis and matrix-assisted laser desorption/ionisation mass spectrometry. Rapid Commun Mass Spectrom 16:2083–2088
29. Olsen JV, Ong S-E, Mann M (2004) Trypsin cleaves exclusively C-terminal to arginine and lysine residues. Mol Cell Proteomics 3: 608–614
30. Aebersold R, Mann M (2003) Mass spectrometry-based proteomics. Nature 422:198–207
31. Biringer RG, Amato H, Harrington MG et al (2006) Enhanced sequence coverage of proteins in human cerebrospinal fluid using multiple enzymatic digestion and linear ion trap LC-MS/MS. Brief Funct Genomic Proteomic 5:144–153
32. Shevchenko A, Wilm M, Vorm O, Mann M (1996) Mass spectrometric sequencing of proteins silver-stained polyacrylamide gels. Anal Chem 68:850–858
33. Winter D, Steen H (2011) Optimization of cell lysis and protein digestion protocols for the analysis of HeLa S3 cells by LC-MS/MS. Proteomics 11:4726–4730
34. Lu B, McClatchy DB, Kim JY, Yates JR (2008) Strategies for shotgun identification of integral membrane proteins by tandem mass spectrometry. Proteomics 8:3947–3955
35. Anderson L (2005) Candidate-based proteomics in the search for biomarkers of cardiovascular disease. J Physiol 563:23–60
36. Laemmli UK (1970) Cleavage of structural proteins during the assembly of the head of bacteriophage T4. Nature 227:680–685
37. Weber K, Osborn M (1969) The reliability of molecular weight determinations by dodecyl sulfate-polyacrylamide gel electrophoresis. J Biol Chem 244:4406–4412
38. Miller I, Crawford J, Gianazza E (2006) Protein stains for proteomic applications: which, when, why? Proteomics 6:5385–5408
39. Switzer RC, Merril CR, Shifrin S (1979) A highly sensitive silver stain for detecting proteins and peptides in polyacrylamide gels. Anal Biochem 98:231–237
40. Westermeier R, Marouga R (2005) Protein detection methods in proteomics research. Biosci Rep 25:19–32
41. Chevallet M, Luche S, Rabilloud T (2006) Silver staining of proteins in polyacrylamide gels. Nat Protoc 1:1852–1858
42. Neuhoff V, Arold N, Taube D, Ehrhardt W (1988) Improved staining of proteins in polyacrylamide gels including isoelectric focusing gels with clear background at nanogram sensitivity using Coomassie brilliant blue G-250 and R-250. Electrophoresis 9:255–262
43. Towbin H, Staehelin T, Gordon J (1979) Electrophoretic transfer of proteins from polyacrylamide gels to nitrocellulose sheets: procedure and some applications. Proc Natl Acad Sci U S A 76:4350–4354
44. Jensen EC (2012) The basics of Western blotting. Anat Rec (Hoboken) 295:369–371
45. Müller T, Loosse C, Schrötter A et al (2011) The AICD interacting protein DAB1 is

up-regulated in Alzheimer frontal cortex brain samples and causes deregulation of proteins involved in gene expression changes. Curr Alzheimer Res 8:573–582

46. Klose J (1975) Protein mapping by combined isoelectric focusing and electrophoresis of mouse tissues. A novel approach to testing for induced point mutations in mammals. Humangenetik 26:231–243
47. O'Farrell PH (1975) High resolution two-dimensional electrophoresis of proteins. J Biol Chem 250:4007–4021
48. Taylor NL, Heazlewood JL, Millar AH (2011) The Arabidopsis thaliana 2-D gel mitochondrial proteome: refining the value of reference maps for assessing protein abundance, contaminants and post-translational modifications. Proteomics 11:1720–1733
49. Klose J, Kobalz U (1995) Two-dimensional electrophoresis of proteins: an updated protocol and implications for a functional analysis of the genome. Electrophoresis 16:1034–1059
50. Bjellqvist B, Ek K, Righetti PG et al (1982) Isoelectric focusing in immobilized pH gradients: principle, methodology and some applications. J Biochem Biophys Methods 6:317–339
51. Görg A, Postel W, Günther S (1988) The current state of two-dimensional electrophoresis with immobilized pH gradients. Electrophoresis 9:531–546
52. Görg A, Drews O, Lück C et al (2009) 2-DE with IPGs. Electrophoresis 30(Suppl 1): S122–S132
53. Luhn S, Berth M, Hecker M, Bernhardt J (2003) Using standard positions and image fusion to create proteome maps from collections of two-dimensional gel electrophoresis images. Proteomics 3:1117–1127
54. Dowsey AW, English JA, Lisacek F et al (2010) Image analysis tools and emerging algorithms for expression proteomics. Proteomics 10:4226–4257
55. Schägger H, von Jagow G (1991) Blue native electrophoresis for isolation of membrane protein complexes in enzymatically active form. Anal Biochem 199:223–231
56. Zahedi R-P, Meisinger C, Sickmann A (2005) Two-dimensional benzyldimethyl-n-hexadecylammonium chloride/SDS-PAGE for membrane proteomics. Proteomics 5:3581–3588
57. Hartinger J, Stenius K, Högemann D, Jahn R (1996) 16-BAC/SDS-PAGE: a two-dimensional gel electrophoresis system suitable for the separation of integral membrane proteins. Anal Biochem 240:126–133
58. Macfarlane DE (1989) Two dimensional benzyldimethyl-n-hexadecylammonium chloride-sodium dodecyl sulfate preparative polyacrylamide gel electrophoresis: a high capacity high resolution technique for the purification of proteins from complex mixtures. Anal Biochem 176:457–463
59. Helling S, Schmitt E, Joppich C et al (2006) 2-D differential membrane proteome analysis of scarce protein samples. Proteomics 6:4506–4513
60. Eley MH, Burns PC, Kannapell CC, Campbell PS (1979) Cetyltrimethylammonium bromide polyacrylamide gel electrophoresis: estimation of protein subunit molecular weights using cationic detergents. Anal Biochem 92:411–419
61. Rais I, Karas M, Schägger H (2004) Two-dimensional electrophoresis for the isolation of integral membrane proteins and mass spectrometric identification. Proteomics 4:2567–2571
62. Unlü M, Morgan ME, Minden JS (1997) Difference gel electrophoresis: a single gel method for detecting changes in protein extracts. Electrophoresis 18:2071–2077
63. Alban A, David SO, Bjorkesten L et al (2003) A novel experimental design for comparative two-dimensional gel analysis: two-dimensional difference gel electrophoresis incorporating a pooled internal standard. Proteomics 3:36–44
64. Mitulovic G, Mechtler K (2006) HPLC techniques for proteomics analysis-a short overview of latest developments. Brief Funct Genomic Proteomic 5:249–260
65. Carr PW, Stoll DR, Wang X (2011) Perspectives on recent advances in the speed of high-performance liquid chromatography. Anal Chem 83:1890–1900
66. Tao D, Zhang L, Shan Y, Liang Z, Zhang Y (2011) Recent advances in micro-scale and nano-scale high-performance liquid-phase chromatography for proteome research. Anal Bioanal Chem 399:229–241
67. Küster B, Wheeler SF, Hunter AP et al (1997) Sequencing of N-linked oligosaccharides directly from protein gels: in-gel deglycosylation followed by matrix-assisted laser desorption/ionization mass spectrometry and normal-phase high-performance liquid chromatography. Anal Biochem 250:82–101
68. Marcus K, Joppich C, May C et al (2009) High-resolution 2DE. Methods Mol Biol 519:221–240
69. Neville B (1998) Reversed-phase HPLC. In: Rapley R, Walker JM (eds) Molecular biomethods handbook, 1st edn. Humana, New York

70. Erni F, Steuer W, Bosshardt H (1987) Automation and validation of HPLC-systems. Chromatographia 24:201–207
71. Köcher T, Swart R, Mechtler K (2011) Ultra-high-pressure RPLC hyphenated to an LTQ-Orbitrap Velos reveals a linear relation between peak capacity and number of identified peptides. Anal Chem 83:2699–2704
72. Whitehouse CM, Dreyer RN, Yamashita M, Fenn JB (1985) Electrospray interface for liquid chromatographs and mass spectrometers. Anal Chem 57:675–679
73. Washburn MP, Wolters D, Yates JR (2001) Large-scale analysis of the yeast proteome by multidimensional protein identification technology. Nat Biotechnol 19:242–247
74. Zhang X, Fang A, Riley CP et al (2010) Multi-dimensional liquid chromatography in proteomics–a review. Anal Chim Acta 664:101–113
75. Wang Z, Hill S, Luther JM et al (2012) Proteomic analysis of urine exosomes by multidimensional protein identification technology (MudPIT). Proteomics 12:329–338
76. Chervet JP, Ursem M, Salzmann JP (1996) Instrumental requirements for nanoscale liquid chromatography. Anal Chem 68:1507–1512
77. Contrepois K, Ezan E, Mann C, Fenaille F (2010) Ultra-high performance liquid chromatography-mass spectrometry for the fast profiling of histone post-translational modifications. J Proteome Res 9:5501–5509
78. Guillarme D, Ruta J, Rudaz S, Veuthey J-L (2010) New trends in fast and high-resolution liquid chromatography: a critical comparison of existing approaches. Anal Bioanal Chem 397:1069–1082
79. Jorgenson JW (2010) Capillary liquid chromatography at ultrahigh pressures. Annu Rev Anal Chem (Palo Alto Calif) 3:129–150
80. Glish GL, Vachet RW (2003) The basics of mass spectrometry in the twenty-first century. Nat Rev Drug Discov 2:140–150
81. Xie F, Liu T, Qian W-J et al (2011) Liquid chromatography-mass spectrometry-based quantitative proteomics. J Biol Chem 286:25443–25449
82. Han J, Datla R, Chan S, Borchers CH (2009) Mass spectrometry-based technologies for high-throughput metabolomics. Bioanalysis 1:1665–1684
83. Karas M, Hillenkamp F (1988) Laser desorption ionization of proteins with molecular masses exceeding 10,000 daltons. Anal Chem 60:2299–2301
84. Nyman TA (2001) The role of mass spectrometry in proteome studies. Biomol Eng 18:221–227
85. Fenn JB, Mann M, Meng CK et al (1989) Electrospray ionization for mass spectrometry of large biomolecules. Science 246:64–71
86. Zaluzec EJ, Gage DA, Watson JT (1995) Matrix-assisted laser desorption ionization mass spectrometry: applications in peptide and protein characterization. Protein Expr Purif 6:109–123
87. Karas M, Glückmann M, Schäfer J (2000) Ionization in matrix-assisted laser desorption/ionization: singly charged molecular ions are the lucky survivors. J Mass Spectrom 35:1–12
88. Nordhoff E, Egelhofer V, Giavalisco P et al (2001) Large-gel two-dimensional electrophoresis-matrix assisted laser desorption/ionization-time of flight-mass spectrometry: an analytical challenge for studying complex protein mixtures. Electrophoresis 22: 2844–2855
89. Stühler K, Meyer HE (2004) MALDI: more than peptide mass fingerprints. Curr Opin Mol Ther 6:239–248
90. Loo JA, Udseth HR, Smith RD (1989) Peptide and protein analysis by electrospray ionization-mass spectrometry and capillary electrophoresis-mass spectrometry. Anal Chem 179:404–412
91. Cech NB, Enke CG (2001) Practical implications of some recent studies in electrospray ionization fundamentals. Mass Spectrom Rev 20:362–387
92. Iribarne JV (1976) On the evaporation of small ions from charged droplets. J Chem Phys 64:2287
93. Dole M (1968) Molecular beams of macroions. J Chem Phys 49:2240
94. Domon B, Aebersold R (2006) Mass spectrometry and protein analysis. Science 312:212–217
95. Wollnik H (1993) Time-of-flight mass analyzers. Mass Spectrom Rev 12:89–114
96. Balogh MP (2004) Debating resolution and mass accuracy in mass spectrometry. Spectroscopy 19:34–40
97. Schwartz JC, Senko MW, Syka JEP (2002) A two-dimensional quadrupole ion trap mass spectrometer. J Am Soc Mass Spectrom 13:659–669
98. March RE (2000) Quadrupole ion trap mass spectrometry: a view at the turn of the century. Int J Mass Spectrom 200:285–312
99. Wilm M, Neubauer G, Mann M (1996) Parent ion scans of unseparated peptide mixtures. Anal Chem 68:527–533
100. Steen H, Küster B, Fernandez M et al (2001) Detection of tyrosine phosphorylated peptides

by precursor ion scanning quadrupole TOF mass spectrometry in positive ion mode. Anal Chem 73:1440–1448
101. Hunter AP, Games DE (1994) Chromatographic and mass spectrometric methods for the identification of phosphorylation sites in phosphoproteins. Rapid Commun Mass Spectrom 8:559–570
102. Schlosser A, Pipkorn R, Bossemeyer D, Lehmann WD (2001) Analysis of protein phosphorylation by a combination of elastase digestion and neutral loss tandem mass spectrometry. Anal Chem 73:170–176
103. Yocum AK, Chinnaiyan AM (2009) Current affairs in quantitative targeted proteomics: multiple reaction monitoring-mass spectrometry. Brief Funct Genomic Proteomic 8:145–157
104. Busch F, Paul W (1961) Isotopentrennung mit dem elektrischen Massenfilter. Zeitschrift für Physik 164:581–587
105. Douglas DJ, Frank AJ, Mao D (2005) Linear ion traps in mass spectrometry. Mass Spectrom Rev 24:1–29
106. Mikesh LM, Ueberheide B, Chi A et al (2006) The utility of ETD mass spectrometry in proteomic analysis. Biochim Biophys Acta 1764:1811–1822
107. Chi A, Huttenhower C, Geer LY et al (2007) Analysis of phosphorylation sites on proteins from Saccharomyces cerevisiae by electron transfer dissociation (ETD) mass spectrometry. Proc Natl Acad Sci U S A 104: 2193–2198
108. Perdivara I, Petrovich R, Allinquant B et al (2009) Elucidation of O-glycosylation structures of the beta-amyloid precursor protein by liquid chromatography-mass spectrometry using electron transfer dissociation and collision induced dissociation. J Proteome Res 8:631–642
109. Alley WR, Mechref Y, Novotny MV (2009) Characterization of glycopeptides by combining collision-induced dissociation and electron-transfer dissociation mass spectrometry data. Rapid Commun Mass Spectrom 23:161–170
110. Wiesner J, Premsler T, Sickmann A (2008) Application of electron transfer dissociation (ETD) for the analysis of posttranslational modifications. Proteomics 8:4466–4483
111. Wang Y, Franzen J (1992) The non-linear resonance QUISTOR Part 1. Potential distribution in hyperboloidal QUISTORs. Int J Mass Spectrom Ion Process 112:167–178
112. Wang Y, Franzen J, Wanczek KP (1993) The non-linear resonance ion trap. Part 2. A general theoretical analysis. Int J Mass Spectrom Ion Process 124:125–144
113. Wang Y, Franzen J (1994) The non-linear ion trap. Part 3. Multipole components in three types of practical ion trap. Int J Mass Spectrom Ion Process 132:155–172
114. Franzen J (1993) The non-linear ion trap. Part 4. Mass selective instability scan with multipole superposition. Int J Mass Spectrom Ion Process 125:165–170
115. Franzen J (1994) The non-linear ion trap. Part 5. Nature of non-linear resonances and resonant ion ejection. Int J Mass Spectrom Ion Process 130:15–40
116. Marshall AG, Hendrickson CL, Jackson GS (1998) Fourier transform ion cyclotron resonance mass spectrometry: a primer. Mass Spectrom Rev 17:1–35
117. Comisarow MB, Marshall AG (1974) Fourier transform ion cyclotron resonance spectroscopy. Chem Phys Lett 25:282–283
118. Goodlett DR, Bruce JE, Anderson GA et al (2000) Protein identification with a single accurate mass of a cysteine-containing peptide and constrained database searching. Anal Chem 72:1112–1118
119. Hu Q, Noll RJ, Li H et al (2005) The orbitrap: a new mass spectrometer. J Mass Spectrom 40:430–443
120. Perry RH, Cooks RG, Noll RJ (2008) Orbitrap mass spectrometry: instrumentation, ion motion and applications. Mass Spectrom Rev 27:661–699
121. Makarov A (2000) Electrostatic axially harmonic orbital trapping: a high-performance technique of mass analysis. Anal Chem 72:1156–1162
122. Scigelova M, Makarov A (2006) Orbitrap mass analyzer–overview and applications in proteomics. Proteomics 6(Suppl 2):16–21
123. Aebersold R, Goodlett DR (2001) Mass spectrometry in proteomics. Chem Rev 101:269–295
124. Spengler B, Kirsch D, Kaufmann R, Jaeger E (1992) Peptide sequencing by matrix-assisted laser-desorption mass spectrometry. Rapid Commun Mass Spectrom 6:105–108
125. de Hoffmann E (1996) Tandem mass spectrometry: a primer. J Mass Spectrom 31:129–137
126. Steen H, Küster B, Mann M (2001) Quadrupole time-of-flight versus triple-quadrupole mass spectrometry for the determination of phosphopeptides by precursor ion scanning. J Mass Spectrom 36:782–790
127. Aldini G, Regazzoni L, Orioli M et al (2008) A tandem MS precursor-ion scan approach to

identify variable covalent modification of albumin Cys34: a new tool for studying vascular carbonylation. J Mass Spectrom 43:1470–1481
128. Hopfgartner G, Varesio E, Tschäppät V et al (2004) Triple quadrupole linear ion trap mass spectrometer for the analysis of small molecules and macromolecules. J Mass Spectrom 39:845–855
129. Yates JR, Speicher S, Griffin PR, Hunkapiller T (1993) Peptide mass maps: a highly informative approach to protein identification. Anal Biochem 214:397–408
130. Boyd RK (1994) Linked-scan techniques for MS/MS using tandem-in-space instruments. Mass Spectrom Rev 13:359–410
131. Carr SA, Huddleston MJ, Annan RS (1996) Selective detection and sequencing of phosphopeptides at the femtomole level by mass spectrometry. Anal Biochem 239:180–192
132. Huddleston MJ, Bean MF, Carr SA (1993) Collisional fragmentation of glycopeptides by electrospray ionization LC/MS and LC/MS/MS: methods for selective detection of glycopeptides in protein digests. Anal Biochem 65:877–884
133. Gadgil HS, Bondarenko PV, Treuheit MJ, Ren D (2007) Screening and sequencing of glycated proteins by neutral loss scan LC/MS/MS method. Anal Biochem 79:5991–5999
134. Langenfeld E, Zanger UM, Jung K et al (2009) Mass spectrometry-based absolute quantification of microsomal cytochrome P450 2D6 in human liver. Proteomics 9:2313–2323
135. Unwin RD, Griffiths JR, Leverentz MK et al (2005) Multiple reaction monitoring to identify sites of protein phosphorylation with high sensitivity. Mol Cell Proteomics 4:1134–1144
136. Annan RS, Carr SA (1997) The essential role of mass spectrometry in characterizing protein structure: mapping posttranslational modifications. J Protein Chem 16:391–402
137. Williamson BL, Marchese J, Morrice NA (2006) Automated identification and quantification of protein phosphorylation sites by LC/MS on a hybrid triple quadrupole linear ion trap mass spectrometer. Mol Cell Proteomics 5:337–346
138. Perkins DN, Pappin DJ, Creasy DM, Cottrell JS (1999) Probability-based protein identification by searching sequence databases using mass spectrometry data. Electrophoresis 20:3551–3567
139. Eng JK, McCormack AL, Yates JR (1994) An approach to correlate tandem mass spectral data of peptides with amino acid sequences in a protein database. J Am Soc Mass Spectrom 5:976–989
140. Biemann K (1990) Appendix 5. Nomenclature for peptide fragment ions (positive ions). Methods Enzymol 193:886–887
141. Zhang W, Chait BT (2000) ProFound: an expert system for protein identification using mass spectrometric peptide mapping information. Anal Chem 72:2482–2489
142. McLafferty FW, Tureek F (eds) (1993) Interpretation of mass spectra, 4th edn. University Science, California
143. Zhu W, Smith JW, Huang C-M (2010) Mass spectrometry-based label-free quantitative proteomics. J Biomed Biotechnol 2010:840518
144. Ong S-E, Mann M (2005) Mass spectrometry-based proteomics turns quantitative. Nat Chem Biol 1:252–262
145. Bantscheff M, Schirle M, Sweetman G et al (2007) Quantitative mass spectrometry in proteomics: a critical review. Anal Bioanal Chem 389:1017–1031
146. Bettmer J (2010) Application of isotope dilution ICP-MS techniques to quantitative proteomics. Anal Bioanal Chem 397:3495–3502
147. Gygi SP, Rist B, Gerber SA et al (1999) Quantitative analysis of complex protein mixtures using isotope-coded affinity tags. Nat Biotechnol 17:994–999
148. Ross PL, Huang YN, Marchese JN et al (2004) Multiplexed protein quantitation in Saccharomyces cerevisiae using amine-reactive isobaric tagging reagents. Mol Cell Proteomics 3:1154–1169
149. Schmidt A, Kellermann J, Lottspeich F (2005) A novel strategy for quantitative proteomics using isotope-coded protein labels. Proteomics 5:4–15
150. Yao X, Freas A, Ramirez J et al (2001) Proteolytic 18O labeling for comparative proteomics: model studies with two serotypes of adenovirus. Anal Chem 73:2836–2842
151. Staes A, Demol H, Van Damme J et al (2004) Global differential non-gel proteomics by quantitative and stable labeling of tryptic peptides with oxygen-18. J Proteome Res 3:786–791
152. Shiio Y, Aebersold R (2006) Quantitative proteome analysis using isotope-coded affinity tags and mass spectrometry. Nat Protoc 1:139–145
153. Thompson A, Schäfer J, Kuhn K et al (2003) Tandem mass tags: a novel quantification strategy for comparative analysis of complex protein mixtures by MS/MS. Anal Chem 75:1895–1904

154. Boehm AM, Pütz S, Altenhöfer D (2007) Precise protein quantification based on peptide quantification using iTRAQ. BMC Bioinformatics 8:214
155. Aggarwal K, Choe LH, Lee KH (2006) Shotgun proteomics using the iTRAQ isobaric tags. Brief Funct Genomics Proteomics 5:112–120
156. Bantscheff M, Boesche M, Eberhard D et al (2008) Robust and sensitive iTRAQ quantification on an LTQ Orbitrap mass spectrometer. Mol Cell Proteomics 7:1702–1713
157. Dayon L, Turck N, Scherl A et al (2010) From relative to absolute quantification of tryptic peptides with tandem mass tags: application to cerebrospinal fluid. Chimia (Aarau) 64:132–135
158. Brunner A, Keidel E-M, Dosch D et al (2010) ICPLQuant—a software for non-isobaric isotopic labeling proteomics. Proteomics 10:315–326
159. Ong S-E, Blagoev B, Kratchmarova I et al (2002) Stable isotope labeling by amino acids in cell culture, SILAC, as a simple and accurate approach to expression proteomics. Mol Cell Proteomics 1:376–386
160. Amanchy R, Kalume DE, Iwahori A et al (2005) Phosphoproteome analysis of HeLa cells using stable isotope labeling with amino acids in cell culture (SILAC). J Proteome Res 4:1661–1671
161. Meierhofer D, Wang X, Huang L, Kaiser P (2008) Quantitative analysis of global ubiquitination in HeLa cells by mass spectrometry. J Proteome Res 7:4566–4576
162. Oeljeklaus S, Reinartz BS, Wolf J et al (2012) Identification of core components and transient interactors of the peroxisomal importomer by dual-track stable isotope labeling with amino acids in cell culture analysis. J Proteome Res 11:2567–2580
163. Krijgsveld J, Ketting RF, Mahmoudi T et al (2003) Metabolic labeling of *C. elegans* and *D. melanogaster* for quantitative proteomics. Nat Biotechnol 21:927–931
164. Larance M, Bailly AP, Pourkarimi E et al (2011) Stable-isotope labeling with amino acids in nematodes. Nat Methods 8:849–851
165. Sury MD, Chen J-X, Selbach M (2010) The SILAC fly allows for accurate protein quantification in vivo. Mol Cell Proteomics 9:2173–2183
166. Krüger M, Moser M, Ussar S et al (2008) SILAC mouse for quantitative proteomics uncovers kindlin-3 as an essential factor for red blood cell function. Cell 134:353–364
167. Walther DM, Mann M (2011) Accurate quantification of more than 4000 mouse tissue proteins reveals minimal proteome changes during aging. Mol Cell Proteomics 10:M110.004523
168. Oda Y, Huang K, Cross FR et al (1999) Accurate quantitation of protein expression and site-specific phosphorylation. Proc Natl Acad Sci U S A 96:6591–6596
169. Old WM, Meyer-Arendt K, Aveline-Wolf L et al (2005) Comparison of label-free methods for quantifying human proteins by shotgun proteomics. Mol Cell Proteomics 4:1487–1502
170. Neilson KA, Ali N, Muralidharan S et al (2011) Less label, more free: approaches in label-free quantitative mass spectrometry. Proteomics 11:535–553
171. Carvalho PC, Hewel J, Barbosa VC, Yates JR (2008) Identifying differences in protein expression levels by spectral counting and feature selection. Genet Mol Res 7:342–356
172. Liu H, Sadygov RG, Yates JR (2004) A model for random sampling and estimation of relative protein abundance in shotgun proteomics. Anal Chem 76:4193–4201
173. Bondarenko PV, Chelius D, Shaler TA (2002) Identification and relative quantitation of protein mixtures by enzymatic digestion followed by capillary reversed-phase liquid chromatography-tandem mass spectrometry. Anal Chem 74:4741–4749
174. Chelius D, Bondarenko PV (2002) Quantitative profiling of proteins in complex mixtures using liquid chromatography and mass spectrometry. J Proteome Res 1:317–323
175. Higgs RE, Knierman MD, Gelfanova V et al (2005) Comprehensive label-free method for the relative quantification of proteins from biological samples. J Proteome Res 4:1442–1450
176. Wang G, Wu WW, Zeng W et al (2006) Label-free protein quantification using LC-coupled ion trap or FT mass spectrometry: reproducibility, linearity, and application with complex proteomes. J Proteome Res 5:1214–1223
177. Zybailov B, Coleman MK, Florens L, Washburn MP (2005) Correlation of relative abundance ratios derived from peptide ion chromatograms and spectrum counting for quantitative proteomic analysis using stable isotope labeling. Anal Chem 77:6218–6224
178. Pan S, Aebersold R, Chen R et al (2009) Mass spectrometry based targeted protein quantification: methods and applications. J Proteome Res 8:787–797
179. Brun V, Masselon C, Garin J, Dupuis A (2009) Isotope dilution strategies for absolute quantitative proteomics. J Proteomics 72:740–749

180. Gerber SA, Rush J, Stemman O et al (2003) Absolute quantification of proteins and phosphoproteins from cell lysates by tandem MS. Proc Natl Acad Sci U S A 100:6940–6945
181. Stahl-Zeng J, Lange V, Ossola R et al (2007) High sensitivity detection of plasma proteins by multiple reaction monitoring of N-glycosites. Mol Cell Proteomics 6:1809–1817
182. Anderson L, Hunter CL (2006) Quantitative mass spectrometric multiple reaction monitoring assays for major plasma proteins. Mol Cell Proteomics 5:573–588
183. Keshishian H, Addona T, Burgess M et al (2007) Quantitative, multiplexed assays for low abundance proteins in plasma by targeted mass spectrometry and stable isotope dilution. Mol Cell Proteomics 6:2212–2229
184. Brun V, Dupuis A, Adrait A et al (2007) Isotope-labeled protein standards: toward absolute quantitative proteomics. Mol Cell Proteomics 6:2139–2149
185. Brownridge P, Holman SW, Gaskell SJ et al (2011) Global absolute quantification of a proteome: challenges in the deployment of a QconCAT strategy. Proteomics 11: 2957–2970
186. Pratt JM, Simpson DM, Doherty MK et al (2006) Multiplexed absolute quantification for proteomics using concatenated signature peptides encoded by QconCAT genes. Nat Protoc 1:1029–1043
187. Dupuis A, Hennekinne J-A, Garin J, Brun V (2008) Protein standard absolute quantification (PSAQ) for improved investigation of staphylococcal food poisoning outbreaks. Proteomics 8:4633–4636
188. Kandel ER, Schwartz JH, Jessell TM (eds) (2000) Principles of neural science, 4th edn. McGraw-Hill, New York
189. McEwen BS (2012) The ever-changing brain: cellular and molecular mechanisms for the effects of stressful experiences. Dev Neurobiol 72:878–890
190. Gogolla N, Galimberti I, Caroni P (2007) Structural plasticity of axon terminals in the adult. Curr Opin Neurobiol 17:516–524
191. Hopkins AL, Groom CR (2002) The druggable genome. Nat Rev Drug Discov 1:727–730
192. Rabilloud T (2003) Membrane proteins ride shotgun. Nat Biotechnol 21:508–510
193. Macher BA, Yen T-Y (2007) Proteins at membrane surfaces-a review of approaches. Mol Biosyst 3:705–713
194. Santoni V, Molloy M, Rabilloud T (2000) Membrane proteins and proteomics: un amour impossible? Electrophoresis 21:1054–1070
195. Josic D, Zeilinger K (1996) Membrane proteins. Methods Enzymol 271:113–134
196. Roy I, Mondal K, Gupta MN (2007) Leveraging protein purification strategies in proteomics. J Chromatogr B Analyt Technol Biomed Life Sci 849:32–42
197. Clifton JG, Li X, Reutter W et al (2007) Comparative proteomics of rat liver and Morris hepatoma 7777 plasma membranes. J Chromatogr B Analyt Technol Biomed Life Sci 849:293–301
198. Clifton JG, Brown MK, Huang F et al (2006) Identification of members of the annexin family in the detergent-insoluble fraction of rat Morris hepatoma plasma membranes. J Chromatogr A 1123:205–211
199. Rabilloud T (2009) Membrane proteins and proteomics: love is possible, but so difficult. Electrophoresis 30(Suppl 1):S174–S180
200. Josic D, Clifton JG (2007) Mammalian plasma membrane proteomics. Proteomics 7:3010–3029
201. Cordwell SJ, Thingholm TE (2010) Technologies for plasma membrane proteomics. Proteomics 10:611–627
202. Tauber R, Reutter W (1978) Protein degradation in the plasma membrane of regenerating liver and Morris hepatomas. Eur J Biochem 83:37–45
203. Cao R, Li X, Liu Z et al (2006) Integration of a two-phase partition method into proteomics research on rat liver plasma membrane proteins. J Proteome Res 5:634–642
204. Schindler J, Lewandrowski U, Sickmann A et al (2006) Proteomic analysis of brain plasma membranes isolated by affinity two-phase partitioning. Mol Cell Proteomics 5:390–400
205. Blonder J, Terunuma A, Conrads TP et al (2004) A proteomic characterization of the plasma membrane of human epidermis by high-throughput mass spectrometry. J Invest Dermatol 123:691–699
206. Navarre C, Degand H, Bennett KL et al (2002) Subproteomics: identification of plasma membrane proteins from the yeast Saccharomyces cerevisiae. Proteomics 2:1706–1714
207. Zhang L, Xie J, Wang X et al (2005) Proteomic analysis of mouse liver plasma membrane: use of differential extraction to enrich hydrophobic membrane proteins. Proteomics 5:4510–4524
208. Chang PS, Absood A, Linderman JJ, Omann GM (2004) Magnetic bead isolation of neutrophil plasma membranes and quantification of membrane-associated guanine nucleotide binding proteins. Anal Biochem 325:175–184

209. Zhang W, Zhou G, Zhao Y et al (2003) Affinity enrichment of plasma membrane for proteomics analysis. Electrophoresis 24:2855–2863
210. Elia G (2008) Biotinylation reagents for the study of cell surface proteins. Proteomics 8:4012–4024
211. Zhao Y, Zhang W, Kho Y, Zhao Y (2004) Proteomic analysis of integral plasma membrane proteins. Anal Chem 76:1817–1823
212. Zhang W, Wang H, Wang J et al (2006) Multiresidue determination of zeranol and related compounds in bovine muscle by gas chromatography/mass spectrometry with immunoaffinity cleanup. J AOAC Int 89:1677–1681
213. Lawson EL, Clifton JG, Huang F et al (2006) Use of magnetic beads with immobilized monoclonal antibodies for isolation of highly pure plasma membranes. Electrophoresis 27:2747–2758
214. Ghosh D, Krokhin O, Antonovici M et al (2004) Lectin affinity as an approach to the proteomic analysis of membrane glycoproteins. J Proteome Res 3:841–850
215. Kullolli M, Hancock WS, Hincapie M (2008) Preparation of a high-performance multi-lectin affinity chromatography (HP-M-LAC) adsorbent for the analysis of human plasma glycoproteins. J Sep Sci 31:2733–2739

Part III

The Neuronal Cytoskeleton

Chapter 6

Analyzing the Cytoskeletal Transcriptome: Sex Differences in Rat Hypothalamus

Sanda Iacobas, Genevieve Neal-Perry, and Dumitru Andrei Iacobas

Abstract

This chapter presents the protocol established in our laboratory to determine the cytoskeletal genomic fabric (CGF), defined as the most inter-coordinately and stably expressed gene web whose encoding proteins form the intracellular scaffolding. In addition to detailed experimental procedures and conventional bioinformatics, novel analytical tools are introduced to reveal hidden features of the cytoskeletal transcriptome. Thus, the Prominent Gene Analysis ranks CGF genes according to their contribution to fabric synchronous expression and robustness against variable conditions, Pair-Wise Relevance determines the fabric topology, and the "transcriptomic distance" quantifies fabric differences between conditions. The new tools use expanded transcriptomic quantifiers, considering not only the expression level but also the control and intercoordination of the fabric genes.

This method is here used to analyze the sex differences in rat hypothalamic CGF and the transcriptomic networks by which CGF and the myelination genomic fabric (MGF) modulate each other. The analysis revealed profound differences between the adult males and estrus-stage females in the CGF topology and CGF–MGF interaction that can provide a basis for the substantial sex dichotomy in neurodegenerative diseases.

Key words: Gene expression, Genomic fabric, Ctnnal1, Coro1b, Nckap1, Spna2, Tmsb4x

1. Introduction

The intracellular scaffolding (cytoskeleton) is an elaborated cytoplasmic protein structure (1) central in determining cell and organ size and morphology (2), intracellular transport (3), and cell division (4). Whether involved in wrapping the axons with myelin sheath (5, 6), trafficking vesicles to special locations (7), or forming mitotic spindles (8) to name just a few functions, the cytoskeleton continuously adapts its molecular composition and topology to cellular demands (9). Such multiple roles require contribution of a

Rolf Dermietzel (ed.), *The Cytoskeleton: Imaging, Isolation, and Interaction*, Neuromethods, vol. 79,
DOI 10.1007/978-1-62703-266-7_6, © Springer Science+Business Media, LLC 2013

large number of proteins whose synthesis, assembly, and interactions are dynamically regulated by a wide diversity of internal and external stimuli (10, 11). Therefore, high-throughput experimental methods and advanced computational procedures are necessary to capture the complexity of cytoskeleton structure and functions. Among these gene expression microarrays (12) offer outstanding capabilities when empowered by adequate bioinformatics.

Our laboratory has extensive experience in profiling the transcriptomes of mouse, rat, and human tissues, and cell cultures subjected to various conditions with both commercial (Illumina, San Diego, CA, USA; Agilent, Santa Clara, CA, USA; Affimetrix, Santa Clara, CA, USA) and university (Einstein, Rockefeller, Yale, Duke) gene expression platforms. Here we present the protocol to analyze the cytoskeletal transcriptome in hypothalami collected from gonad intact adult male and female (at estrus) rats using Agilent 60 mer4 × 44k whole genome rat arrays and a variety of computational tools. Agilent arrays are used because of their high quality, complete genome coverage, competitive price, and commercial availability. The experimental protocol is composed of the following steps: (1) extraction of total RNA; (2) reverse transcription in the presence of fluorescent labels; (3) microarray hybridization; (4) washing; (5) scanning; (6) feature extraction and primary data processing (quality control, background extraction, normalization); and (7) conventional and advanced analyses.

Conventional microarray data analysis identifies the up-/down-regulated and turned on/off genes when comparing two conditions and clusters these genes according to the gene ontology (GO) categories (as biological process, molecular function, and cellular location) of the encoded proteins. However, although conventional analysis provides reliable, genome-wide information with regard to expression of individual genes, this is only a very small part of the information that can be obtained by simultaneously quantifying thousands of genes in the same sample. The reason is that genes are not independently but rather coordinately expressed to ensure a kind of "transcriptomic stoichiometry" (13) of the functional pathways. Therefore, manipulation of a single gene has ripple effects on numerous others, located on all chromosomes and involved in a wide diversity of processes (14–17). Expression levels of individual genes depend on local conditions, giving rise to the so-called genomic noise when comparing biological replicas. However, expression variability is not uniform among genes within biological replicas, suggesting differential control of transcript abundance by the homeostatic mechanisms. Thus, expression of key genes is confined into narrow intervals while that of others may more readily adapt to changes in the environmental conditions. Moreover, expression variability allows calculation of Pearson correlation coefficients between expression levels of gene pairs in biological replicas and identification of significantly synergistically, antagonistically, or independently expressed genes (13).

Because of expression coordination, any gene is directly or indirectly related to every functional pathway. However, genes are not of equal importance in controlling the pathways. Therefore, we have developed the Prominent Gene Analysis (PGA) (18–20) as an alternative to the Principal Component Analysis (PCA) (21–23) to select the most relevant genes for a given biological process. The difference between these procedures is that while PCA considers as "centroids" the most alterable (up-/down-regulated) genes in a set of conditions, PGA selects those forming the most interconnected and stably expressed functional web in each condition. Since expression of key genes is most likely kept under a stricter control by the homeostatic mechanisms to preserve the function, they exhibit less alteration than other genes and consequently they are less likely to be selected by PCA. In contrast, PGA selects genes whose interconnection synchronizes the functional pathway components, while their expression stability renders the fabric resistant to change in environmental conditions. In previous papers, we provided evidence that a fabric exhibits characteristic gene composition depending on tissue and sex (19) and remodels its topology during development (18) and in response to various stimuli (20), disease (24), and stressful conditions (18).

Fabric remodeling is controlled by both intra- and inter-fabric modulating transcriptomic networks, such as the control of the myelination genomic fabric (MGF) in immortalized precursor oligodendrocytes by Ca^{2+}-signaling (20). The modulating networks may also extend beyond cell borders through intercellular communication (20, 25, 26). We have recently introduced the Pair-Wise Relevance (PWR) Analysis to visualize and compare how genes couple to model fabric topology (27). The PWR of a gene-pair is represented by a "mountain" whose elevation is proportional to the contribution of that pair to the fabric expression level, stability, and coordination. PWR analysis can be also used to quantify how gene fabrics of different pathways modulate each other by identifying the gene-pairs across functional networks that are both most influential and least affected by the variability in local conditions. In this chapter, we use PWR analysis not only to determine the topology of cytoskeletal genomic fabric (CGF) but also to quantify the transcriptomic interactions between CGF and MGF. Although several authors reported that cytoskeleton plays important roles in myelination (e.g., (28, 29)) the involved molecular mechanisms are still uncertain. As illustrated below, PWR analysis of CGF–MGF interaction identifies key players of the two fabrics.

2. Materials

Table 1 lists the necessary materials to determine the transcriptomic differences between two conditions with four biological replicas of each for statistical significance when using Agilent rat chips (platform described in http://www.ncbi.nlm.nih.gov/geo/query/acc.cgi?acc=GPL14746). Since the vendors pack the reagents in larger amounts than necessary for an 8-sample experiment, we added how many RNA samples can be processed and how many experiments comparing two conditions each in four biological replicas can be performed with the listed amounts.

The required equipment includes −20°C and −80°C freezers, refrigerated and vacuum centrifuges, NanoDrop 2000 Spectrophotometer (Thermo Fisher Scientific Inc, New York, NY, USA), 2100 Electrophoresis Bioanalyzer (Agilent, Santa Clara, CA, USA), Thermal Cycler (GMI, Bellevue WA, USA), Agilent hybridization chamber, water bath, shaker, (Agilent) hybridization oven, Agilent G2565CA microarray scanner system with SureScan high resolution technology (http://www.chem.agilent.com/Library/usermanuals/Public/G2505-90019_ScannerC_User.pdf), and a computer workstation (>2.0 GHz, >8 GB RAM, >100 GB available disk space).

Table 1
Materials for a gene expression experiment to compare two conditions with four biological replicas each

#	Vendor	Catalog#	Item	Qty	Smpls	Exps
1	Agilent	G2519F	Rat 4×44K, 60mer, 4×44K features, 4 microarrays per slide	2	16	2
2	Agilent	5188-5242	Gene Expression Hybridization Kit	1	40	5
3	Agilent	G2534-60011	Package 5 backings 4 array per slide	1	40	5
4	Qiagen	74104	RNeasy mini kit (50 rxns=25 extraction + 25 washes)	1	25	3
5	Agilent	5190-2306	Low Input Quick Amp Labeling Kit, Two-Color	1	48	6
6	Agilent	5188-5327	Gene Expression Wash Pack	1	80	10
7	Agilent	5185-5979	Stabilization and drying solution 500 ml	1	16	2
8	Agilent	5188-5279	RNA Spike In Kit for Two color v4.0	1	48	6
9	Agilent	5067-1511	RNA 6000 Nano Kit	1	300	37.5

Smpls = number of samples that can be analyzed, *Exps* = number of 8-sample experiments that can be performed with the listed materials. To order, contact: Agilent (Santa Clara, CA, USA, http://www.home.agilent.com/agilent/home.jspx?cc=US&lc=eng), Qiagen (Valencia, CA, USA, http://www.qiagen.com/default.aspx)

3. Methods

3.1. Sample Preparation

The experimental protocol was approved by the AECOM Institutional Animal Care and Use Committee (IACUC). We have followed the IACUC protocols for prevention of disease, daily observation, and surveillance for assessment of animal health, and the methods of animal handling, restraint, anesthesia, and analgesia. Animal euthanasia respected the guidelines of the Panel on Euthanasia of the American Veterinary Medical Association. Four male and four female (estrus stage) 30-week-old rats were anesthetized with isoflurane, sacrificed, brains quickly removed, and hypothalami dissected. Total RNA was extracted from each sample with Qiagen kit (#4 in Table 1); RNA concentration was determined using a NanoDrop ND-2000 UV–VIS Spectrophotometer (http://www.nanodrop.com/Library/nd-1000-v3.7-users-manual-8.5x11.pdf) and its purity with Agilent RNA 6000 Nano kit (#9) in an Agilent 2100 Bioanalyzer (http://gcf.pbrc.edu/docs/Agilent/Agilent%20Manual.pdf).

We used an in-house improved version of the standard Agilent protocol (described in http://www.chem.agilent.com/Library/usermanuals/Public/G4140-90050_GeneExpression _Two_Color_v6.5.pdf) starting with 1–2 μg total RNA instead of 25–200 ng that reduces the amplification to less than 5×. Low Input Quick Amp labeling Kit (#5) that contains T7 RNA polymerase was used to simultaneously amplify total RNA and the positive controls (RNA Spike in Kit for Two Color, #8), and incorporate cyanine 3- or cyanine 5-labeled CTP (Cy3 or Cy5). After purification in Qiagen (Valencia, CA, USA) RNeasy mini kit (#4) columns the concentration, yield, and specific activity of the resulting fluorescent cRNA are determined at the NanoDrop. If the yield is <825 ng and the specific activity is <6 pmol Cy3 or Cy5 per μg cRNA, data can be unreliable and it is necessary to repeat the cRNA preparation.

3.2. Microarray Hybridization and Washing

825 ng from each of the Cy3- or Cy5-labeled samples of two biological replicas were mixed with Agilent Gene Expression Hybridization Kit (#2) and the final volume adjusted to 100 μL. The solution was slowly dispensed onto the gasket (#3) placed in the hybridization chamber, the chip (#1) with the active face downward is arranged over the gasket, and the "sandwich" firmly closed. The samples were hybridized to the array at 65°C for 17 h in the oven while vertically rotating (10 rot/min) to wet the gasket and asses the mobility of the bubbles.

The hybridized chip is washed at room temperature using the Gene Expression Wash Pack (#6) and Stabilization and drying solution (#7) then immediately scanned.

3.3. Scanning, Feature Extraction, and Primary Data Processing

The arrays were scanned with an Agilent G2565A dual laser scanner at 5 μm resolution (pixel size) and 20-bit scan mode and the resulted Tagged Image File Format (tiffs) obtained through use of (Agilent) Scan Control Software 8.3 analyzed with (Agilent) Feature Extraction 10.7.3.1 software (http://www.chem.agilent.com/Library/usermanuals/Public/G4460-90017_FE_10.5_Installation.pdf). All spots affected by local corruption or with the median foreground fluorescence less than twice the median background fluorescence are removed from the analysis. The background-subtracted signals were normalized iteratively to the medians of the valid spots, alternating intra-array and inter-array normalization until the fluctuation of valid spots becomes less than 5% between successive iteration steps. The gene expression data complying with the "Minimum Information About Microarray Experiments" (MIAME) were deposited and are accessible at http://www.ncbi.nlm.nih.gov/geo/query/acc.cgi?acc=GSE38450. 10,558 well-annotated unigenes were quantified in the hypothalami of all male and female rats used in this experiment. Out of these genes, 166 were assigned to gene ontology categories related to cytoskeleton by the software S.O.U.R.C.E. (http://smd.stanford.edu/cgi-bin/source/sourceSearch).

3.4. Analysis of Individual Gene Expression Level and Control

Detection of significantly differentially expressed genes relied on both absolute (>1.5×) fold-change and statistical significance ($p<0.05$) of the heteroscedastic t-test for equality of two means. The p-value is computed with a Bonferroni type adjustment applied to the groups of spots redundantly probing the same gene (30). In the illustrated experiment, levels of 14 (8.4%) CGF genes were significantly higher in males than females, while no gene exhibited significantly higher expression in female (listed in Table 2).

For each gene, we have also estimated the control of transcript abundance by determining the Gene Expression Stability (GES) as the percentile locating the middle of the chi-square interval estimate of the expression variability within biological replicas (13). In this measure, GES = 100 indicates the most controlled (i.e., the least variably expressed) and GES = 1 the least controlled (i.e., the most variably expressed) genes. On average, male cytoskeletal genes exhibited significantly higher variability than female counterparts (34.8 vs. 29.0), indicating while female cytoskeleton is more robust against change in external conditions, male cytoskeleton adapts easier to such changes. The most controlled gene in males and females was Ctnnal1 and Coro1b, respectively. Arc was the least controlled gene expressed in males and Pfn1 in females, while Actn2, Cfl1, and Cttnbp2 were very stably expressed in one sex and very unstably expressed in the other.

3.5. Analysis of Expression Coordination

Genes are considered as synergistically, antagonistically, or independently expressed if the Pearson correlation coefficient between their expression levels within biological replicas is, respectively,

Table 2
Differentially expressed (U = up-regulated, *red* background) and controlled (S = switch between very stably and very unstably expressed, *blue*) cytoskeletal genes together with the highest (H, *purple*) and the lowest (L, *yellow*) stringency of the homeostatic control of transcript abundance

Gene	Symbol	R	M/F	*p*-val	GES-M	GES-F
Actin, gamma 1	Actg1	U	1.53	0.029	67.2	33.4
ADP-ribosylation-like factor 6 interacting protein 5	Arl6ip5	U	2.61	0.015	34.0	16.6
CAP, adenylate cyclase-associated protein 1 (yeast)	Cap1	U	2.38	0.013	40.1	12.3
Capping protein (actin filament) muscle Z-line, alpha 1	Capza1	U	1.55	0.009	79.2	45.5
Capping protein (actin filament) muscle Z-line, alpha 2	Capza2	U	1.95	0.016	60.5	7.5
Cytoplasmic linker-associated protein 2	Clasp2	U	1.59	0.037	53.9	24.4
Ermin, ERM-like protein	Ermn	U	1.82	0.016	90.7	16.0
Erythrocyte protein band 4.9	Epb4.9	U	1.52	0.036	50.9	49.7
Intraflagellar transport 80 homolog (Chlamydomonas)	Ift80	U	1.51	0.019	64.8	55.7
Mkl protein	Mkl	U	2.17	0.024	44.3	0.9
Protein tyrosine phosphatase 4a1	Ptp4a1	U	2.06	0.003	77.4	19.6
Replication protein A1	Rpa1	U	1.51	0.037	78.0	23.8
Rho/rac guanine nucleotide exchange factor(GEF)2	Arhgef2	U	1.62	0.047	41.9	39.5
Thy-1 cell surface antigen	Thy1	U	1.63	0.021	57.5	84.6
Actinin alpha 2	Actn2	S	1.94	0.358	1.5	86.4
Cofilin 1, non-muscle	Cfl1	S	1.58	0.099	85.2	3.3
Cortactin-binding protein 2	Cttnbp2	S	2.01	0.354	0.9	81.0
Activity-regulated cytoskeleton-associated protein	Arc	L	4.26	0.408	0.3	27.4
Proflin 1	Pfn1	L	1.82	0.141	44.9	0.3
Catenin (cadherin-associated protein), alpha-like 1	Ctnnal1	H	1.08	0.585	99.7	35.8
Coronin, actin-binding protein, 1B	Coro1b	H	1.32	0.048	68.4	99.7

M/F = expression ratio male/female (negative for larger expression in female), *p-val* = *p*-value of the heteroscedastic *t*-test of equality of two means with Bonferroni correction, *GES-M*/*GES-F* = Gene Expression Stability in male/female hypothalamus

$\rho > 0.9$, $\rho < -0.9$, or $|\rho| < 0.05$. While expression coordination indicates whether the genes are synergistically or antagonistically involved in a biological process, molecular function, or cellular structure, their independence defines the nonoverlapping regions of functional pathways. Figure 1 presents the statistically significant ($p < 0.05$) synergistic and antagonistic expression coordination of a subset of 56 cytoskeletal genes in male and female hypothalami. Note that

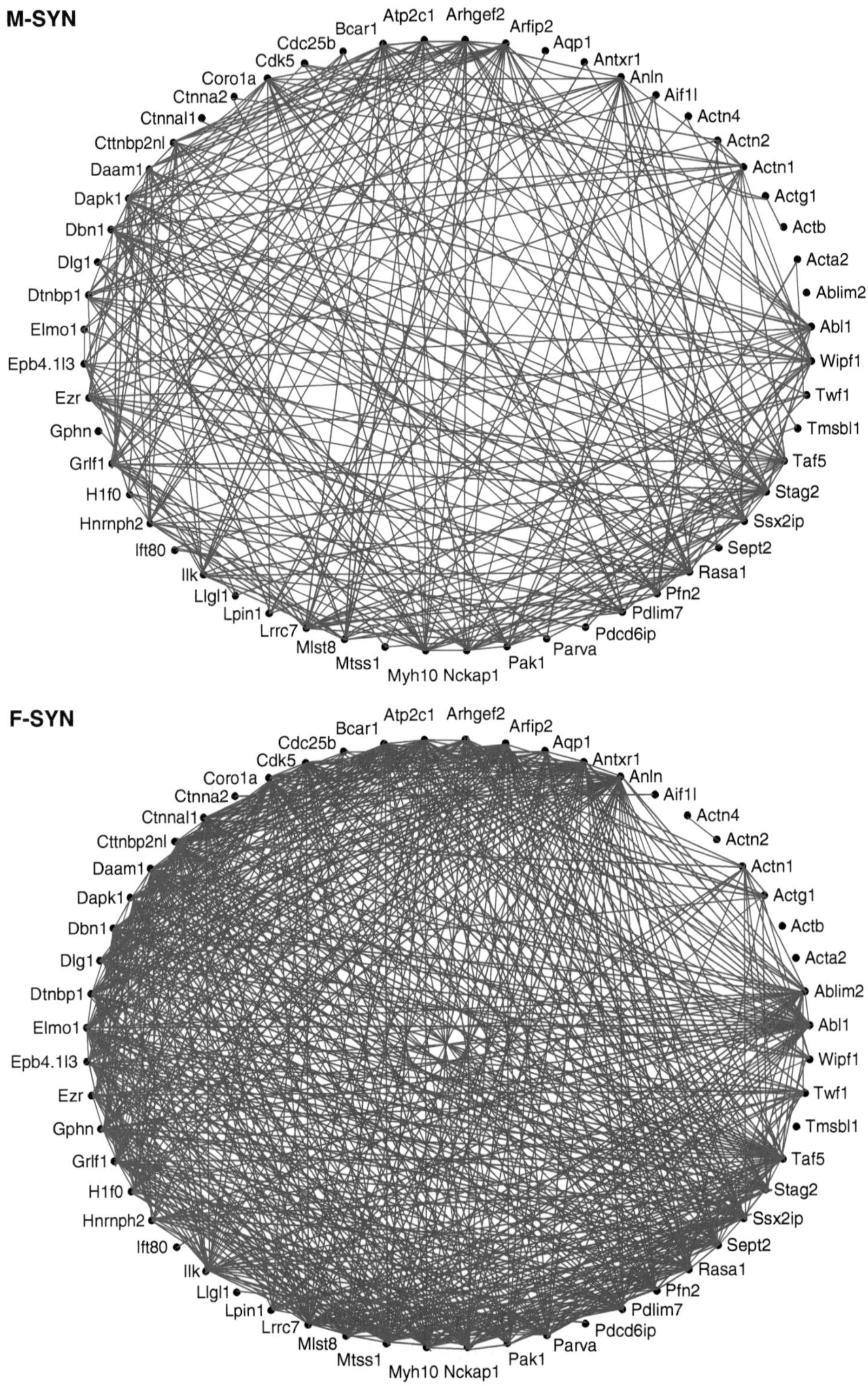

Fig. 1. Networks of 56 cytoskeletal genes in hypothalami of male (M) and estrus-stage female (F) rats. *Red/blue lines* indicate that the linked genes are synergistically (SYN) or antagonistically (ANT) expressed.

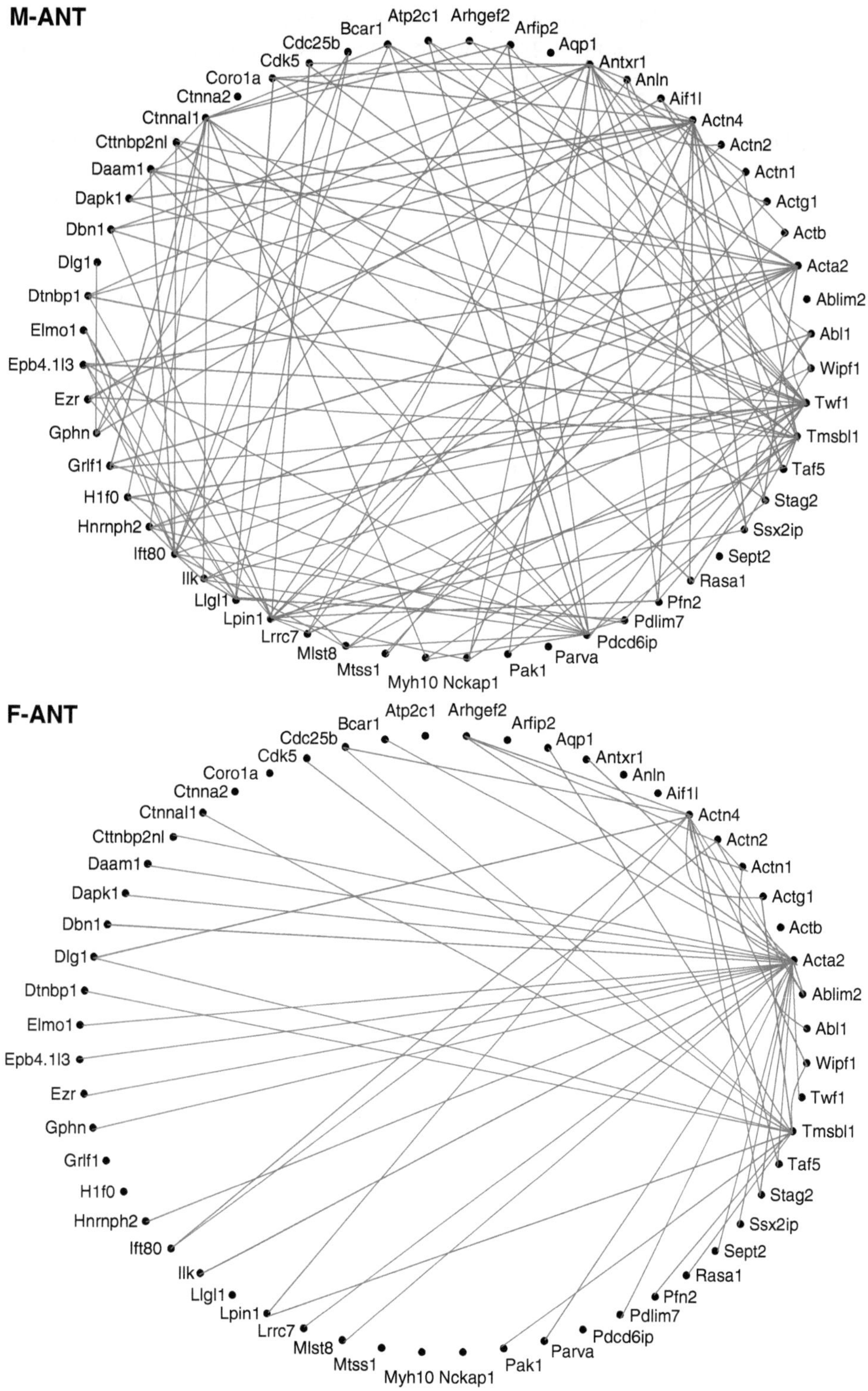

Fig. 1. (continued)

the synergistic coordination is significantly higher in female (449 vs. 308; 1.46×), while antagonistic coordination is significantly higher in male (141 vs. 52; 2.71×).

3.6. Transcriptomic Distance

Up to this point, we found that certain cytoskeletal genes are differentially expressed, controlled, and coordinated in the hipothalami of adult males and females but what is the significance of each of these differences on the hypothalamic cytoskeletal transcriptome?

In order to address this question, we here introduce the transcriptomic distance as the most comprehensive measure of the differences between the genomic fabrics of two conditions for it encompasses differences in expression level, control, and networking:

$$\left(D_{tot}^{(M,F)}\right)^2 = \frac{1}{\{CGF\}} \sum_{i \in CGF} \underbrace{\left(\left(1 - p_i^{(M,F)}\right)\left(\left|x_i^{(M,F)}\right| - 1\right)\right)^2}_{\text{differential expression level}}$$
$$+ \frac{1}{\{CGF\}} \sum_{i \in CGF} \underbrace{\left(\frac{\langle X \rangle_{CGF}^{(M)}}{X_i^{(M)}} - \frac{\langle X \rangle_{CGF}^{(F)}}{X_i^{(F)}}\right)^2}_{\text{differential control}} + \underbrace{2/\{CGF\} \sum_{j \in CGF} \frac{\left(r_{i,j}^{(M)} - r_{i,j}^{(F)}\right)^2}{\{CGF\} - 1}}_{\text{differential networking}}. \quad (1)$$

The distance is normalized for the average gene ({CGF} = 166 = number of CGF genes in this study). Interestingly, when applied to the rat hypothalamic CGF, the analysis indicates that the sexual dimorphism mainly results from differential expression control and networking rather than differential expression level of genes (Fig. 2). This result suggests the introduction of a major paradigm shift in how genomic studies are approached. Thus, instead of looking for changes in individual genes (biomarkers), we should consider *genomic fabric remodeling* that includes but is not limited to expression regulation of individual genes.

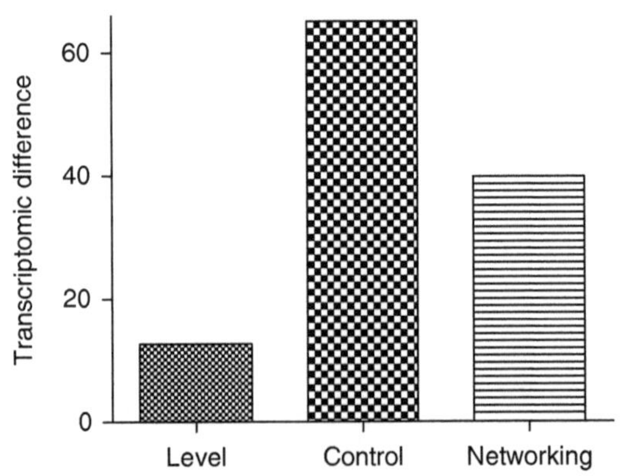

Fig. 2. Contribution of differential gene expression level, control, and networking to the transcriptomic differences between the hypothalamic CGFs of the two sexes.

3.7. Prominence Gene Analysis

A major problem in genomic studies is gene assignment to various GO categories, including selection of genes responsible for forming the cytoskeleton. Conventional methods use software such as S.O.U.R.C.E., GenMapp (www.genmapp.org), KEGG (http://www.genome.jp/kegg/pathway.html), or IPA (http://ingenuity.com). Many of the pathway genes are selected by using the PCA that, as presented in the Introduction, chooses the most alterable genes that are not necessarily the most important ones. Although these pathway assignments are very popular because they are inclusive (containing all reported relationships between genes/proteins without tissue, sex, or age specificity) and rigid (topology not being affected by variability in local conditions), they may not accurately reflect the biological reality. Nevertheless, they are excellent text-miners of the scientific literature and starting platforms for further refinements.

PGA (17–19) was introduced by us to rank the genes according to their contribution to the expression intercoordination and stability of a genomic fabric. The prominence (GP) is the percentile of the prominence score *GPS* of gene i within a gene fabric (here, CGF), the first percentile indicating the most relevant and the 100th percentile the least relevant fabric gene. GPSs of individual genes are used to determine the Fabric Prominence Score (FPS):

$$GPS_i^{(s)} \equiv \frac{X_i^{(s)}}{\sum\limits_{j \in CGF, j \neq i} \left(\rho_{i,j}^{(s)}\right)^2}, \quad FPS^{(s)} \equiv \frac{\sum\limits_{i \in CGF} GPS_i^{(s)}}{\sum\limits_{i \in CGF} a_i^{(s)} X_i^{(s)}}$$

$$s = M, F\ (sex)\ a_i^{(s)} = average\ relative\ expression\ level\ of\ gene$$

$$X_\gamma^{(s)} = mid\ \chi^2\ interval\ estimate\ of\ expression\ variability \qquad (2)$$

With PGA one can iteratively refine fabric gene composition by replacing the least prominent gene with another from outside the initial selection whose coordination profile is highly similar or opposite to that of a gene from the top percentiles. If the prominence of the fabric formed by this gene replacement is higher (i.e., smaller FPS) then a new hierarchy is established and the process continues until no further gene can be replaced. Thus, almost all expressed genes are tested and ranked by PGA for their relevance to the fabric so that new key players may be identified and the "established" pathways redesigned for each tissue, sex, age, and set of conditions (pathophysiological and environmental).

Figure 3 presents the prominences of the S.O.U.R.C.E. selected cytoskeletal genes in male and female hypothalami. Of note is that while Cdc25b, Ran, and Sirpa exhibited high prominence (GP < 10) and Aqp1, Arc, and Fmnl1 low prominence (GP > 90) in both sexes, the rest of high and low prominent genes were different. Moreover, Actn2 and Ermn exhibited high prominence in one sex and low prominence in the other. This analysis suggests that CGF genes may play different roles in the two sexes.

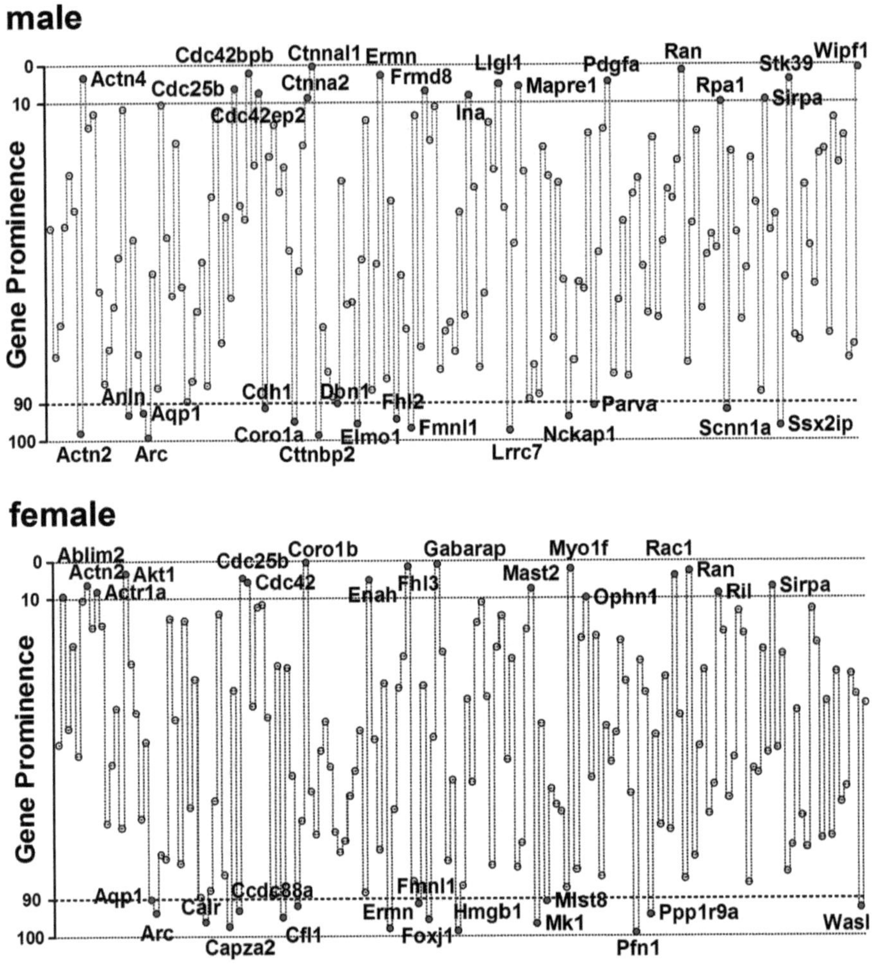

Fig. 3. The prominences of the selected 166 cytoskeletal genes in male and female hypothalami. Note that excepting Cdc25b, Ran, and Sirpa (GP < 10) and Aqp1, Arc, and Fmnl1 (GP > 90) different genes populate the higher (GP < 10) and lower (GP > 90) prominent gene regions.

3.8. Pair-Wise Relevance Analysis

This analysis reveals the topology of genomic fabrics by quantifying the contributions of all pairs that can be formed with fabric genes. Moreover, it can be used to quantify and rank the inter-fabric modulatory networks. The PWR increases with the average expression levels (a) and Pearson correlation coefficient (ρ) between expression levels of the paired genes in biological replicas, and decreases with their relative expression variability X:

$$\begin{gathered}
PWR\ subarrayli \\
\forall i \in CGF; i \neq j \in \Gamma = CGF \vee MGF \;\&\; s = M, F \\
PWR_{ij}^{(s)} = \frac{a_i^{(s)} a_j^{(s)}}{\langle a \rangle_{CGF}^{(s)} \langle a \rangle_{\Gamma}^{(s)}} \left|\rho_{ij}^{(s)}\right|^2 \frac{\langle X \rangle_{CGF}^{(s)} \langle X \rangle^{(s)}}{X_i^{(s)} X_j^{(s)}} \\
Where: \langle ? \rangle_{fabric}^{(sex)} = average\ "?"\ within\ indicated\ fabric\ for\ that\ sex
\end{gathered} \tag{3}$$

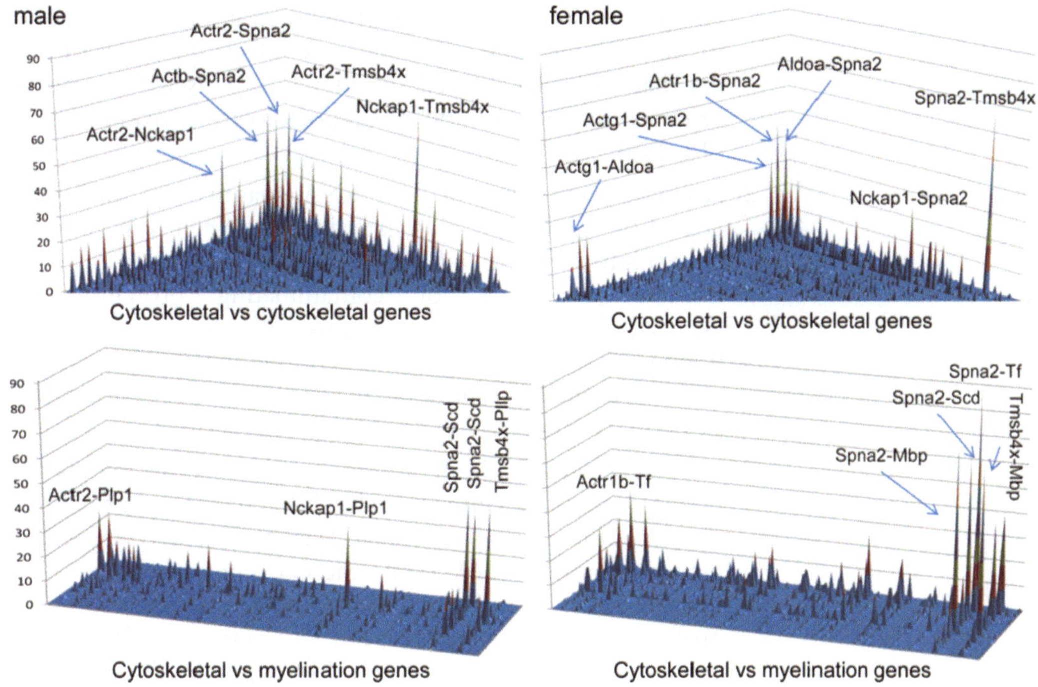

Fig. 4. The most relevant gene pairs within CGF and between CGF and MGF in male and female rat hypothalami.

Figure 4 presents the most relevant gene-pairs within CGF and between cytoskeletal and myelination genes.

This analysis also revealed substantial sex differences in the topology of CGF and CGF–MGF interaction. It confirmed the dominance of pairs formed by actins (Actb, Actg1, Actr1b, Actr2) with actin cytoskeleton regulator (Tmsb4x) and Spna2 (known for restricting cAMP from disrupting microtubules (31)) in the CGF. However, the importance of pairs formed by NCK-associated protein1 (Nckap1) with actins, Tmsb4x, and Spna2 was an unexpected result that requires further investigation. PWR analysis also confirmed the interaction between cytoskeletal genes and myelination biomarkers such as Mbp (32, 33), TF (34), and Scd (35).

4. Notes

1. A high-throughput method not only measures thousands of features at a time in the same sample but, when properly processed, it can also allow determination of how these features are related to each other in functional pathways. There is no single molecule investigation suitable for understanding the

complexity of the cytoskeleton biological systems since manipulation of one molecular type affects hundreds of others that are not explored simultaneously (as we and others reported in various knockout/knockdown/knock-in studies (13–17)).

2. Accurate assessments of total RNA quantity and quality are crucial for gene expression experiments. When using a nanodrop, high-quality total RNA should have $1.8 < A_{260}/A_{280} < 2.0$ (indicating absence of contaminating proteins) and $A_{260}/A_{230} > 2.0$ (indicating absence of other organic compounds). The resulting electropherogram obtained when analyzing the quality using a bioanalyzer should report an RNA Integrity Number (RIN) >8.0 (ideal RIN = 10).
3. Although mathematically similar, our gene expression coordination analysis uses an essentially different criterion to cluster the genes than the conventional "weighted gene co-expression network analysis" (36). While the conventional analysis clusters the genes according to their similar/opposite expression profiles across experiments (as in (37)), ours identifies the synergistically, antagonistically, and independently expressed partners in a series of biological replicas. Gene interactions can differ substantially under different environmental and nutritional conditions.
4. Most conventional gene analysis methods only consider the expression of the most responsive/alterable genes, ignoring the contributions of gene expression control, gene–gene expression (synergistic or antagonist), coordination, and interaction among genomic fabrics. In contrast, our analysis of gene expression takes into account also the impact of gene environment, providing more accurate information about coordination of gene networks.
5. The novel "transcriptomic distance" is the most comprehensive measure of fabric differences between conditions one can imagine that can be used (for instance) to quantify the transcriptomic alterations during the progression of a disease. It can be also a better alternative to the "transcriptomic recovery efficiency" that we have introduced to evaluate treatments with bone marrow stem cells (38, 39).

Acknowledgement

Dr. David C. Spray's (Einstein) comments on biological interpretation of experimental data are gratefully acknowledged.

References

1. Wickstead B, Gull K (2011) The evolution of the cytoskeleton. J Cell Biol 194(4):513–525, Review
2. Bond J, Woods CG (2006) Cytoskeletal genes regulating brain size. Curr Opin Cell Biol 18(1):95–101
3. Niewiadomska G, Baksalerska-Pazera M, Riedel G (2006) Cytoskeletal transport in the aging brain: focus on the cholinergic system. Rev Neurosci 17(6):581–618, Review
4. Evans JP, Robinson DN (2011) The spatial and mechanical challenges of female meiosis. Mol Reprod Dev 78(10–11):769–777
5. Richter-Landsberg C (2008) The cytoskeleton in oligodendrocytes. Microtubule dynamics in health and disease. J Mol Neurosci 35(1):55–63, Review
6. Susuki K, Raphael AR, Ogawa Y et al (2011) Schwann cell spectrins modulate peripheral nerve myelination. Proc Natl Acad Sci U S A 108(19):8009–8014
7. Kreft M, Potokar M, Stenovec M et al (2009) Regulated exocytosis and vesicle trafficking in astrocytes. Ann N Y Acad Sci 1152:30–42, Review
8. Morin X, Bellaïche Y (2011) Mitotic spindle orientation in asymmetric and symmetric cell divisions during animal development. Dev Cell 21(1):102–119, Review
9. Quintá HR, Galigniana NM, Erlejman AG et al (2011) Management of cytoskeleton architecture by molecular chaperones and immunophilins. Cell Signal 23(12):1907–1920
10. Sandbo N, Dulin N (2011) Actin cytoskeleton in myofibroblast differentiation: ultrastructure defining form and driving function. Transl Res 158(4):181–196
11. Dupin I, Etienne-Manneville S (2011) Nuclear positioning: mechanisms and functions. Int J Biochem Cell Biol 43:1698–1707
12. Lau CL, PerreauVM CMJ et al (2011) Transcriptomic profiling of astrocytes treated with the Rho kinase inhibitor fasudil reveals cytoskeletal and pro-survival responses. J Cell Physiol 227:1199–1211. doi:10.1002/jcp.22838
13. Iacobas DA, Iacobas S, Spray DC (2007) Connexin43 and the brain transcriptome of the newborn mice. Genomics 89(1):113–123
14. Iacobas DA, Urban M, Iacobas S et al (2003) Array analysis of gene expression in connexin43 null astrocytes. Physiol Genomics 15(3):177–190
15. Iacobas DA, Iacobas S, Li WEI et al (2005) Genes controlling multiple functional pathways are transcriptionally regulated in connexin43 null mouse heart. Physiol Genomics 20:211–223
16. Spray DC, Iacobas DA (2007) Organizational principles of the connexin-related brain transcriptome. J Membr Biol 218(1–3):39–47
17. Iacobas DA, Iacobas S, Urban-Maldonado M et al (2008) Similar transcriptomic alterations in Cx43 knock-down and knock-out astrocytes. Cell Commun Adhes 15(1):195–206
18. Iacobas DA, Iacobas S, Haddad GG (2010) Heart rhythm genomic fabric in hypoxia. Biochem Biophys Res Commun 391(4):1769–1774
19. Iacobas DA, Iacobas S, Thomas N et al (2010) Sex-dependent gene regulatory networks of the heart rhythm. Funct Integr Genomics 10(1):73–86
20. Iacobas S, Iacobas DA (2011) Astrocyte proximity modulates the myelination gene fabric of oligodendrocytes. Neuron Glia Biol 6(3):157–169
21. Draghici S (2003) Data analysis tools for DNA microarrays. Chapman & Hall, Boca Raton, pp 223–224
22. Stekel D (2003) Microarray bioinformatics. Cambridge University Press, Cambridge, pp 133–134
23. Wu MC, Lin X (2009) Prior biological knowledge-based approaches for the analysis of genome-wide expression profiles using gene sets and pathways. Stat Methods Med Res 18(6):577–593
24. Iacobas DA, Iacobas S, Werner P et al (2007) Alteration of transcriptomic networks in adoptive-transfer experimental autoimmune encephalomyelitis. Front Integr Neurosci 1:10. doi:10.3389/neuro.07/010.2007
25. Iacobas DA, Suadicani SO, Spray DC et al (2006) A stochastic 2D model of intercellular Ca2+ wave spread in glia. Biophys J 90(1):24–41
26. Iacobas DA, Iacobas S, Spray DC (2007) Connexin-dependent transcellular transcriptomic networks in mouse brain. Prog Biophys Mol Biol 94(1–2):168–184
27. Iacobas S, Thomas NM, Iacobas DA (2012) Plasticity of the myelination genomic fabric. Mol Genet Genomics 287(3):237–246
28. Bauer NG, Richter-Landsberg C, Ffrench-Constant C (2009) Role of the oligodendroglial cytoskeleton in differentiation and myelination. Glia 57(16):1691–1705, Review
29. Boggs JM, Gao W, Zhao J et al (2010) Participation of galactosylceramide and sulfatide

in glycosynapses between oligodendrocyte or myelin membranes. FEBS Lett 584(9): 1771–1778

30. Iacobas DA, Iacobas S, Urban-Maldonado M et al (2005) Sensitivity of the brain transcriptome to connexin ablation. Biochim Biophys Acta 1711:183–196

31. Creighton J, Zhu B, Alexeyev M et al (2008) Spectrin-anchored phosphodiesterase4D4 restricts cAMP from disrupting microtubules and inducing endothelial cell gap formation. J Cell Sci 121(Pt 1):110–119

32. Boggs JM (2006) Myelin basic protein: a multifunctional protein. Cell Mol Life Sci 63(17):1945–1961, Review

33. Harauz G, Ladizhansky V, Boggs JM (2009) Structural polymorphism and multifunctionality of myelin basic protein. Biochemistry 48(34):8094–8104, Review

34. Ortiz EH, Pasquini LA, Soto EF et al (2005) Apotransferrin and the cytoskeleton of oligodendroglial cells. J Neurosci Res 82(6): 822–830

35. Rodríguez-Rodríguez RA, Tabernero A, Velasco A et al (2004) Theneurotrophic effect of oleic acid includes dendritic differentiation and the expression of the neuronal basic helix-loop-helix transcription factor NeuroD2. J Neurochem 88(5):1041–1051

36. Zhang B, Horvath S (2005) A general framework for weighted gene co-expression network analysis. Stat Appl Genet Mol Biol 4:article17

37. Kang HJ, Kawasawa YI, Cheng F et al (2011) Spatio-temporal transcriptome of the human brain. Nature 478(7370):483–489

38. Soares MB, Lima RS, Souza BS, Vasconcelos JF, Rocha LL, Dos Santos RR, Iacobas S, Goldenberg RC, Lisanti MP, Iacobas DA, Tanowitz HB, Spray DC, Campos de Carvalho AC (2011) Reversion of gene expression alterations in hearts of mice with chronic chagasic cardiomyopathy after transplantation of bone marrow cells. Cell Cycle 10:1448–1455

39. Lachtermacher S, Esporcatte BL, Fortes Fda S, Rocha NN, Montalvão F, Costa PC, Belem L, Rabischoffisky A, Faria Neto HC, Vasconcellos R, Iacobas DA, Iacobas S, Spray DC, Thomas NM, Goldenberg RC, de Carvalho AC (2012) Functional and transcriptomic recovery of infarcted mouse myocardium treated with bone marrow mononuclear cells. Stem Cell Rev 8(1):251–261

Chapter 7

Proteomic Approaches for the Study of Electrical Synapses and Associated Protein-Interaction Complexes

Cantas Alev, Georg Zoidl, and Rolf Dermietzel

Abstract

Recent advances in the identification and analysis of protein–protein interaction complexes associated with synapses and synaptic proteins deepened not only our insights into the molecular composition and dynamic structural makeup of interneuronal connections but contributed also significantly to our understanding of the molecular and mechanistic aspects underlying functional plasticity in neuronal networks. In particular proteome analytical tools, combining traditional isolation protocols with modern mass spectrometric approaches, were utilized successfully for the molecular analysis of chemical synapses and other neuronal subcellular structures revealing new and exciting insights into the temporal and spatial changes of the proteins composing or associated with for example synaptic vesicles, synaptic membranes, or postsynaptic densities (PSDs). Proteomic approaches may thus offer also a chance to gain valuable insights into the so far elusive molecular composition of electrical synapses, the Cinderella fated and long neglected little brethren of "classical" chemical synapses. In this chapter we provide an experimental basis of how such an analysis can be designed, with a major focus on the most abundant electrical synapse protein, connexin36.

Key words: Electrical synapses, Connexin36, Shotgun proteomics, Phosphorylation analysis, Nano-LC-ESI-MS/MS, CaMKII, Tubulin

1. Introduction

In contrast to the wealth of information on the molecular and functional properties of chemical synapses, our understanding of the molecular aspects involved in the formation, regulation, and function of electrical synapses, despite their first description more than half a century ago (1, 2), is very limited. Many questions especially regarding their exact composition and structural makeup as well as the functional relationship to and molecular interactions with other synaptic components and proteins await clarification.

Numerous successful attempts to unravel the molecular composition of the chemical synapse, its associated pre- and postsynaptic

Rolf Dermietzel (ed.), *The Cytoskeleton: Imaging, Isolation, and Interaction*, Neuromethods, vol. 79,
DOI 10.1007/978-1-62703-266-7_7, © Springer Science+Business Media, LLC 2013

protein complexes, and protein interaction-networks centered around synapse-embedded neuronal receptors, e.g., AMPA or NMDA receptors, have dramatically increased our understanding of the dynamic molecular processes involved in the mediation and regulation of synaptic transmission and neuronal activity, with proteomic approaches being an integral part of several of these studies (3–7).

One of the major aims of the here presented protocols is therefore to offer a platform for the analogous identification and characterization of novel interaction partners of neuronal connexins at sites of electrical synapses, focusing on and exemplified by the most abundant gap junction molecule connexin36 (Cx36). Very few proteins, all of which initially detected in non-neuronal contexts and shown to bind to the cytoplasmic domains of other gap junction proteins alike, have so far been shown to interact with Cx36 (8–11). The first and central objective of the here presented protocols is thus to provide tools that may help to identify novel and specific interaction partners of Cx36. In order to overcome the limitations of classical gel-based proteomic strategies, we utilized a powerful shotgun proteomic approach based on nano-liquid chromatography electrospray ionization tandem mass spectrometry (nano-LC-ESI-MS/MS), permitting the direct analysis of very complex protein mixtures with a high dynamic range. This approach led to the reliable identification of Cx36 interaction partners from in vitro affinity purified samples and immunoprecipitated in vivo protein complexes isolated from murine brain lysates of wild-type and $Cx36^{-/-}$ knockout control mice.

Our applied mass-spectrometric approach permits not only the identification of peptide- and corresponding protein-sequences but also the detection of posttranslational modifications (PTMs) such as phosphorylated amino acid residues present on either Cx36 or its associated binding partners. Two of the most promising protein-interaction candidates of Cx36, tubulin, and CaMKII will be described in more detail enabling the reader to convey further in-depth validation and functional characterization experiments.

Recently several reports suggested that electrical synapses can undergo activity-dependent long-lasting modulatory changes and are involved in processes formerly believed to be exclusively associated with chemical synapses such as synaptic plasticity, learning, and memory formation (12–18). The identification of neuronal interaction partners of Cx36 could thus help to understand the currently enigmatic molecular processes underlying these proposed activity-dependent long-lasting modulatory changes of electrical synapses containing Cx36.

A dynamic protein-interaction network centered around the cytoplasmic domains of postsynaptically embedded glutamate receptors binding to CaMKII was shown to be essential for long-lasting modulatory processes (e.g., long-term potentiation, LTP, and long-term depression, LTD) (19–21). Another question which we would like to address using the here presented protocols is whether an analogous protein interaction network associated with electrical synapses and centered around the cytoplasmic domains of neuronal connexins (e.g., Cx36) exists and whether such a complex might show similarities or even share components with the "interactome" of chemical synapses.

2. Methods

2.1. Protein Interaction Assays

2.1.1. Immunoprecipitation of Connexin36 Protein-Interaction Complexes

In order to assess in vivo protein-interactions of Cx36, co-immunoprecipitation (Co-IP) experiments utilizing murine brain lysates of wild-type and Cx36-deficient mice are performed. Cx36-protein interaction complexes can be generated for example by using anti-Cx36 antibodies (i.e., #51-6300, Zymed, San Francisco, CA) in combination with the ProFound® Co-IP Kit (Pierce, Rockford, IL) following the manufacturer's instructions. Lysates must be precleared with non-antibody-bound agarose beads (Pierce) for 2 h at 4°C prior to use. Co-IPs generated via the use of Cx36 knockout animal brain lysates (transgenic animals with a deletion of the respective protein provide the best controls for pull-down assays) are defined as controls and used together with "real" wild-type samples for subsequent analytical assessment. As part of the gel-based proteomic approach, putative protein complexes must be eluted from the used binding-matrix with Laemmli-sample buffer by incubation at 65°C for 10 min. After brief centrifugation, samples can be loaded on 10–15% acrylamide gels, fractionated via SDS-PAGE, and visualized by Coomassie and/or silver staining. Selected bands, clearly different between wild-type and knockout IPs, will then be cut out and used for mass spectrometry analysis. Alternatively, Co-IP samples are applicable for a non-gel-based proteomic approach (see Section 2.2.4), with the respective protein complexes being directly brought into solution by 30-min incubation at 50°C in 8 M Urea containing proteomic lysis buffer and used for in-solution digestion without prior gel separation. A minimum of three controls and three wild-type Cx36-Co-IPs should be treated in this way and used for nano-liquid chromatography tandem mass spectrometry analysis (nano-LC-ESI-MS/MS). Co-IPs can be also used for Western blot analysis validating the presence or the absence of identified interaction partners (e.g., tubulin) in the isolated in vivo complexes.

2.1.2. Affinity Capture Experiments

Purified GST-fusion proteins covering the entire length of the cytoplasmic loop of Cx36 (aa100-201) and carboxy-terminal tail (aa272-321) including mutated fusion proteins missing certain aspects of putative connexin-binding sites extracted from in silico screens (see Section 2.2.1) as well as peptides corresponding to predicted binding domains can be chosen for further in vitro binding assays.

2.1.3. "Pull-Down" with Connexin36 GST-Fusion Proteins

GST-fusion proteins covering the cytoplasmic loop (CL) or tail (CT) of Cx36 are used as baits to capture possible in vitro interaction partners of this gap junction molecule from brain lysates of adult wild-type mice. The ProFound® Pull-Down GST Protein: Protein Interaction Kit (Pierce, Rockford, IL) may be utilized for this purpose, with binding and washing conditions performed as instructed by the manufacturer. Murine brain lysates of adult C57BL/6 mice are generated and precleared with Glutathione Sepharose-4B beads (Pharmacia) at 4°C with rotation for a minimum of 2 h prior to use. GST-pull-downs using GST only as bait were defined as controls and used accordingly. Affinity captured proteins in analogy to Co-IP samples are suitable either directly for "shotgun" proteomic analysis (non-gel-based approach) or can be subject to SDS-PAGE and used for Western Blot or "classical" gel-based proteomic analysis and evaluation. A minimum of three samples for GST-Cx36-CT and GST-Cx36-CL and six samples for GST alone should be included, eluted directly from the binding matrix with Laemmli-sample buffer or proteomic lysis buffer, and used for gel-based or shotgun proteomic analysis, respectively (see Sections 2.2.2 and 2.2.4 this chapter).

2.1.4. Affinity Capture of Tubulin by Connexin36-Derived Peptides

For the analysis of interacting cytoskeletal proteins with electrical synapses we will describe tubulin at first hand as it was identified as a putative interaction partner of Cx36 by both proteomic approaches. Tubulin binding can be further validated by Western Blot analysis of GST-pull-down samples. As part of an additional validation step, equal amounts of biotinylated peptides covering different parts of the Cx36-binding domains are immobilized onto NeutrAvidin beads (Pierce, Rockford, IL) and incubated with wild-type adult and P8 (postnatal day 8) brain lysates for 4 h at 4°C in rotation. Following extensive washing steps, bound proteins need to be eluted by addition of Laemmli-sample buffer and incubation at 65°C for 10 min. Supernatants are then resolved by SDS-PAGE followed by Western Blot analysis with anti-tubulin antibody (#T-3526, Sigma).

2.1.5. Microtubule Co-sedimentation Assay

This assay is performed with the Microtubule-Associated protein (MAP) Spin-Down Assay Kit (Cytoskeleton, Denver) according to the instructions of the manufacturer. Tubulin is polymerized in vitro to microtubules using Taxol as instructed by the manufacturer. Equal amounts of GST-Cx36-CT and GST alone were incubated either in the presence or the absence of stabilized microtubules for 20 min at RT. After ultracentrifugation (Beckman) at 100,000 × *g* for 40 min at RT aliquots of the supernatant and pellet fractions are taken and analyzed via SDS-PAGE. Western Blot analysis with anti-GST-HRP (Pierce, Rockford, IL) allows to identify the presence or the absence of fusion proteins co-sedimented together with stabilized microtubules into the pellet fraction.

2.1.6. Mirror Resonance Analysis

To verify in situ interaction between baits and interacting partners mirror resonance spectroscopy offers a reliable tool to study binding efficiency under different physiological conditions. An IAsys apparatus as described previously (22) has been proven to be a suitable device for this purpose. In brief, the bait (in our case CaMKII) is bound covalently to carboxymethyl dextran-coated cuvettes using 1-ethyl-3-(3-dimethylaminopropyl) carbodiimide-HCl (EDC) and N-hydroxysuccinimide (NHS) according to the instructions of the manufacturer (Affinity Sensors). Peptides corresponding to different binding regions of the synaptic protein (Cx36) are added to CaMKII-covered cuvettes in PBS containing 0.05% Tween 20 and removed by a 3-min wash with 10 mM HCl, with dose–response curves being generated for each peptide on two separate bait cuvettes and a minimum of three comparisons of rank order of binding at 25 μM peptide concentrations being performed (23) (Fig. 1).

2.2. Mass Spectrometry and Proteomics

2.2.1. In Silico Prediction of Protein-Binding Motifs

A straightforward approach to start for screening of putative protein-binding domains and motifs in a given protein is the use of open-source software. In our case ScanProsite (http://prosite.expasy.org/scanprosite/) (24) and Motif Scan (http://myhits.isb-sib.ch/cgi-bin/motif_scan) were applied with a focus on putative evolutionary conserved motifs in the cytoplasmic loop and carboxy-terminal tail of the electrical synapse protein Cx36. Evolutionary conservation of domains can be evaluated manually via alignment and comparison of selected motifs using amino acid

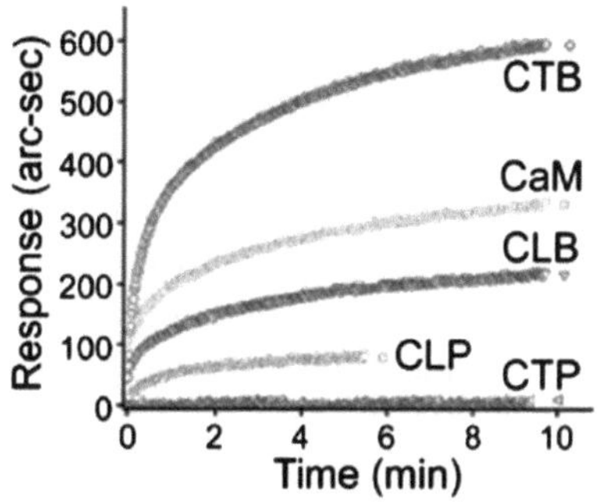

Fig. 1. Quantitative analysis of the binding between connexin36 and CaMKII. The binding between the peptides corresponding to four different cytoplasmic domains of connexin36 (CTB, CLB and CTP, CLP) and CaMKII was measured using mirror resonance spectroscopy. The binding of calmodulin was also measured (positive control). Strongest binding could be observed for CTB, followed by calmodulin and the peptide covering the binding domain in the loop (CLB). Peptides covering the putative phosphorylation sites (CLP and CTP) showed almost no binding to the kinase (23).

sequence information of the corresponding mammalian, reptilian, and amphibian orthologs/homologs. If one looks for specific binding sites, for instance for putative calmodulin-binding sites, more specified databases need to be exploited, such as the Calmodulin Target Database (http://calcium.uhnres.utoronto.ca/ctdb/ctdb/sequence.html) (25). For comparison and alignment of multiple specific sequences alignment software such as ClustalW (http://www.genome.jp/tools/clustalw/) can be applied.

2.2.2. Gel-Based "Classical" Proteomic Analysis of Samples

In case of positive hits one should start with a gel-based "classical" proteomic approach utilizing GST-pull-down and/or Co-IP samples (see Sections 2.1.1 and 2.1.3). After separation of the protein samples by SDS-PAGE and staining with Coomassie blue and/or silver staining, evaluation for the presence of specific bands present only in the "true" samples but absent from the corresponding controls (in our case Cx36 knockout (Cx36 $^{-/-}$ C57 black mice) animals) needs to be done. Cut out protein bands of interest using a light box and one-time-use scalpels, transfer them into 1.5 ml microcentrifuge tubes, and subject probes to "in-gel" trypsin digestion (see the following paragraph). A similar approach is applicable for mass spectrometric identification of in vitro phosphorylation sites, with in vitro phosphorylated fusion proteins being cut out from corresponding SDS-PAGE gels and subjected to in-gel digestion (see Section 2.3.3).

2.2.3. In-Gel Digestion of Excised Protein Bands

Processing of the gel slabs: Wash the gel pieces twice with 400 μl of water for 15 min followed by incubation with 200 μl of 50% acetonitrile (Sigma) in 100 mM ammonium bicarbonate (Sigma) for 15 min at RT. After rehydration with 100 mM ammonium bicarbonate, incubate the gel pieces with 100 μl acetonitrile for 15 min at RT, and remove acetonitrile from shrunk gel pieces with 100 μl of 10 mM dithiothreitol (DTT) (Sigma) in 100 mM ammonium bicarbonate. Incubate the gel pieces for 30 min at 56°C.

Add 55 mM iodoacetamide (IAA) (Sigma) in 100 mM ammonium bicarbonate and incubate the gel pieces at RT for 30 min in the dark. Wash the gel pieces twice with 50% acetonitrile in 100 mM ammonium bicarbonate and dry them in a vacuum chamber.

Perform tryptic digestion with 20–30 μl of 10 ng/μl sequencing grade trypsin (Promega, Madison, WI) in 100 mM ammonium bicarbonate with 1 mM $CaCl_2$ on each dried gel piece. Check 30 min later whether the amount of trypsin solution is enough to swell and cover the entire gel piece. Add more trypsin solution, if necessary, and incubate the gel pieces at 37°C overnight. Spin down the samples in a table centrifuge and transfer the tryptic peptides containing supernatants into new microcentrifuge tubes. For the complete extraction of peptides incubate the gel pieces first with 50 μl of ammonium bicarbonate and then with 50 μl of acetonitrile for 15 min at RT. After a final incubation with 0.5% formic acid/50% acetonitrile for 15 min at RT, all recovered supernatants must be vacuum dried and reconstituted in 0.1% formic acid/2% acetonitrile. Samples can then be used for nano-LC-ESI-MS/MS analysis as described (see Section 2.2.6 and "Analysis of phosphorylated peptides by nano-LC-ESI-MS/MS" this chapter).

2.2.4. Non-gel-Based "Shotgun" Proteomic Analysis of Samples

For non-gel-based proteomic analysis GST-pull-downs and Co-IP samples (wild type and control) can directly be utilized from the carrier matrices. In this case all proteins bound to the used beads need to be eluted using 8 M Urea with 1 mM DTT in 100 mM ammonium bicarbonate followed by in-solution digestion (see the following paragraph) and analyzing via nano-LC-ESI-MS/MS mass spectrometry which allows identification of the proteins present in each sample.

2.2.5. In-Solution Digestion of Pull-Down and Co-IP Samples

Beads containing the interacting protein complexes are re-suspended in an equal amount of proteomic lysis buffer containing 8 M Urea (Sigma) and 1 mM DTT (Sigma) in 100 mM ammonium bicarbonate (usually 20 μl buffer to 20 μl of beads), carefully mixed by pipetting, and incubated for 30 min at 50°C. Following the elution of the protein interaction complexes (and all other proteins) from the beads an equal amount (20 μl) of an alkylation buffer containing 20 mM iodoacetamide (Sigma) in 100 mM ammonium bicarbonate (end concentration of IAA 10 mM) is added and the samples are incubated for 1 h at RT in the dark. Each sample is mixed with sequencing grade trypsin (Promega) diluted in 100 mM ammonium bicarbonate containing 2 mM $CaCl_2$ to a concentration of 20 ng/μl and an amount equal to the liquid volume already present in each sample vial (40 μl). Samples are then incubated overnight at 37°C. The next day digestion is stopped and the samples acidified by addition of 2 μl trifluoroacetic acid (TFA). Samples are centrifuged at maximum speed and separated from the insoluble beads. Samples are loaded with the help of a pressurization device onto fused-silica capillaries (100 μm I.D. × 365 μm O.D.) filled with a C_{18} reverse-phase (RP) packing material and use for nano-LC-ESI-MS/MS.

2.2.6. Nano-LC-ESI-MS/MS Analysis of In-Solution Digested Samples

The mass spectrometric analysis of in-solution digested samples via nano-LC-ESI-MS/MS is performed with a Finnigan® LTQ® ion trap mass spectrometer (Thermo Electron Corporation, San Jose, CA)-based proteomic workstation. The nano-LC-ESI-MS/MS workstation used for this purpose consists of a Dual-Gradient Rapid Separation HPLC pump (Dionex, Amsterdam, The Netherlands), a FAMOS® autosampler (Dionex, Amsterdam, The Netherlands), a Switchos (LC Packings) column switching unit, and a Finnigan® LTQ® ion trap mass spectrometer (provider see above). The LTQ® operates via Xcalibur to acquire a full MS scan between 350 and 2,000 m/z followed by full MS/MS scans of the three most intensive ions from the preceding MS scan. The heated desolvation capillary is set to 180°C. The relative collision energy for collision-induced dissociation is set to 35%. Dynamic exclusion is enabled with a repeat count of 2, a repeat duration of 0.5 min, and a 3-min exclusion duration window. Samples are loaded onto 15 cm fused silica columns. Fritless 100 μm capillaries are packed using a pressure cell in house with Eclipse XDB C_{18} (Agilent). The column flow is set to 0.15–0.25 μl/min and the spray voltage to 1.8 kV. The buffer solutions used for liquid chromatography are 5% acetonitrile, 0.012% heptafluorobutyric acid, 0.5% acetic acid (buffer A) and 80% acetonitrile, 0.012% heptafluorobutyric acid, 0.5% acetic acid (buffer B). After equilibration for 5 min with buffer A, a linear gradient (A→B) is generated within 60 min. In order to prevent the possible carryover of tryptic peptides from the previous sample to the next, two wash steps need to be interspersed between each mass spectrometric run.

2.2.7. Database Analysis of Nano-LC-ESI-MS/MS Results

The SEQUEST® (Thermo Finnigan, San Jose, CA) algorithm provides a reliable tool to interpret MS/MS spectra generated during the nano-LC-ESI-MS/MS analysis of shotgun proteomic samples (26–28). Only results with cross-correlation scores (XCorr) of single-charged, double-charged, and triple-charged ions greater than 1.5, 2.5, or 3.5 should be accepted. XCALIBUR results fulfilling the minimal inclusion criteria and containing information on the peptides and proteins identified in each sample can be exported into Excel and analyzed manually (Fig. 2). Proteins identified from and suggested to be present in the GST-pull-down and Co-IP samples should be compared with the ones found for the corresponding control samples (pull-down using GST only or Co-IP using control brain lysate). The MS/MS spectra obtained for in-gel trypsin digestion samples are analyzed using MASCOT with SwisProt as a default database. Only proteins with single peptide MASCOT scores greater than 25 and a total score of more than 100 for the corresponding protein should be considered to be present in the excised and digested SDS-PAGE band.

2.2.8. Application of Post-filtering Criteria for the Prediction of "Likely" Candidates

In order to minimize the inclusion of possible false positive candidates a set of stringent analytical rules need to be applied. Peptides and corresponding proteins are only considered to be "present" and thus included into the first step of analysis if they fulfilled the minimal output criteria of, e.g., SEQUEST with a minimal cross-correlation score for charged peptides of XCorr-Scores: 1+>1.5, 2+>2.5, and 3+>3.5. All proteins, which are also identified and present in the control samples, are subsequently eliminated from the initial list, even if the corresponding proteins are identified in all Co-IPs or GST pull-down samples but are found in only one of the analyzed controls. These harsh criteria are likely to increase the number of false negatives, but were nevertheless implemented in

Fig. 2. Representative screen-shot of a "shotgun" proteomic analysis result page (generated via SEQUEST software Xcalibur) of a GST-Cx36-CT pull-down sample. Note that several peptides with high probability can be assigned to "true" candidates such as (#7) Neuronal tropomodulin and (#9) Drebrin as well as unspecific "false" binding partners such as (#10) alpha 1 actin precursor or (#8) ATP synthase H+ transporting protein.

order to keep the number of false positive candidates as low as possible. Before any candidates can be considered for further functional evaluation, additional criteria need to be fulfilled. Candidates not only should be absent from any of the analyzed control samples but are also expected to be present in both of the applied strategies, Cx36-Co-IP and GST-Cx36-CT/CL pull-down alike. In order to further narrow down the list of potential candidates, only proteins, which are identified from at least four different samples, Co-IPs (wild type), and/or Cx36-CT/CL GST-pull-downs alike, are considered.

2.3. Phosphorylation Analysis

2.3.1. Prediction of Phosphorylation Sites

In case that interaction of the electrical synapse protein (Cx36) with kinases is also in the focus of the study, we provide here protocols which can be utilized for the prediction, identification, and validation of phosphorylation sites present in synaptic proteins (23, 29, 30).

For prediction of phosphorylation sites of, e.g., Cx36 (gi|2828580|) open-access tools such as NetPhos (http://www.cbs.dtu.dk/services/NetPhos/) (31), Group-based prediction system GPS (http://gps.biocuckoo.org/) (32), Disorder enhanced phosphorylation sites predictor DISPHOS (http://www.dabi.temple.edu/disphos/) (33), or KinasePhos (http://kinasephos.mbc.nctu.edu.tw/) (34) are suitable choices. Highest stringency levels should be applied for all used software with scores expected to be significantly higher than the default cutoff values in all cases. Phosphorylation sites fulfilling the inclusion criteria and predicted by multiple programs can be considered for further analysis. Predicted phosphorylation sites should be validated via mass spectrometric analysis of corresponding in vitro phosphorylated Cx36 fusion proteins and peptides. In the following we provide protocols for two kinases (CaMKII and PKA) which have been identified to interact with Cx36 (23, 29).

2.3.2. In Vitro Phosphorylation Assays

In Vitro Phosphorylation of Connexin36 by CaMKII

Cx36 fusion proteins (GST-Cx36-CL, GST-Cx36-CT) are in vitro phosphorylated by CaMKII and used either for mass spectrometric analysis of phosphorylation sites or binding/phosphorylation studies. Fusion proteins (GST-Cx36-CL, GST-Cx36-CT) are immobilized onto GST-agarose beads (Pierce), washed three times with PBS, and incubated with either non-phosphorylated or autophosphorylated CaMKII in the presence (hot) or the absence (cold) of [γ-^{32}P]-ATP for 30 min at 30°C. For auto-activation recombinant CaMKII (250 U) is incubated at 30°C in the presence of 150 μM ATP in CaMKII-buffer containing Ca^{2+} or Ca^{2+}/calmodulin (New England Biolabs, Beverly, MA). Beads are washed five times with PBS, bound proteins are eluted using Laemmli-sample buffer by incubation at 65°C for 5 min, supernatants are fractionated via SDS-PAGE, and [^{32}P]-phosphorylated fusion proteins are visualized via autoradiography. In addition to full-length fusion proteins, mutated versions lacking the predicted binding or phosphorylation domains for CaMKII can also be utilized. In vitro phosphorylated fusion proteins (cold) are excised from Coomassie Brilliant Blue®-stained gels and used for nano-liquid chromatography tandem mass spectrometric identification of phosphorylation sites (see this chapter "Analysis of Phosphorylated Peptides by Nano-LC-ESI-MS/MS").

In Vitro Phosphorylation of Connexin36 by PKA

Fusion proteins of Cx36 are bound to the appropriate beads, washed, and pre-incubated with recombinant PKA and the corresponding phosphorylation buffer (Promega, Madison, USA) according to the manufacturer's instructions followed by an incubation for 30 min at 30°C either with (hot) or without (cold) [γ^{32}P]-ATP. [^{32}P]-phosphorylated and therefore radioactively labeled fusion proteins can be visualized via autoradiographic analysis of the corresponding dried SDS-PAGE gels. For mass spectrometric identification of phosphorylation sites in vitro phosphorylated fusion proteins (cold) are excised from Coomassie Brilliant Blue®-stained gels and used for nano-LC-ESI-MS/MS (see "Analysis of Phosphorylated Peptides by Nano-LC-ESI-MS/MS" this section).

2.3.3. Identification of Connexin36 Phosphorylation Sites via Mass Spectrometry

Cx36 peptides are phosphorylated by in vitro treatment with CaMKII or PKA in the appropriate assay buffers (New England Biolabs, Beverly, MA) at 30°C for 60 min according to the manufacturer's instructions. The reactions are terminated by addition of 0.1% TFA solution. The phosphorylated peptides are desalted and concentrated using ZipTip® (Millipore, Schwalbach, Germany) and analyzed subsequently using the 4000QTrap and LTQ® systems as described below. Protein bands of in vitro phosphorylated fusion proteins covering the amino acid sequences of the carboxy-terminus and the cytoplasmic loop of Cx36 will be excised from Coomassie Brilliant Blue®-stained SDS-PAGE gels, cut into 1 mm^3 pieces, and washed 3 times with 50% acetonitrile/25 mM ammonium bicarbonate for 10 min with shaking. Incubate gel pieces with 25 mM ammonium bicarbonate and 10 mM DTT for 45 min at 56°C, wash with acetonitrile, and then incubate with 25 mM ammonium bicarbonate and 55 mM iodoacetamide for 30 min in the dark. Subsequently wash the gel pieces with acetonitrile, air-dry, rehydrate them with 12.5 ng/μl trypsin (Promega, Madison, WI, sequencing grade) in 50 mM ammonium bicarbonate, and then incubate the pieces at 37°C overnight. Peptides are extracted from the gel pieces using 5% formic acid/50% acetonitrile for 30 min. The eluted peptides are vacuum dried and dissolved in 0.1% formic acid/2% acetonitrile.

Analysis of Phosphorylated Peptides by Nano-LC-ESI-MS/MS

The mass spectrometric setup consists of a Chorus220 nanoflow LC system connected to a 4000QTrap mass spectrometer. One microliter of each sample is directly loaded onto a 200 mm × 5 cm MICHROM Magic C_{18} column at 1,000 nl/min. The 4000QTrap is operated in the multiple reaction monitoring (MRM) mode. Resulting spectra are searched against a nonredundant NCBI database using MASCOT.

Phosphosite mapping is also performed using a combined setup consisting of an HTS-PAL® autosampler (CTC Analytics), a Paradigm MS4 nanoHPLC (Michrom BioResources), and an LTQ® (Thermo Finnigan, San Jose, CA) linear ion trap mass spectrometer, operated in the data-dependent neutral loss MS^3 scanning mode. Four successive scans are performed with an initial full MS scan over a range of 450–2,000 *m/z*, followed by second and third data-dependent scans of the top two most abundant ions obtained in the first scan and an MS^3 analysis in order to determine MS^2 spectra with neutral losses corresponding to phosphate groups, obtained in the second and third scans. Automatic MS/MS spectra should be assessed from the highest peak in each scan by setting relative collision energy of 35% and exclusion time of 15 min for molecules of the same m/z value range. The MS^3 data obtained for each tryptic digest analyzed (phosphorylated as well as non-phosphorylated proteins) are searched using TurboSEQUEST (XCorr-Scores: 1+>1.9, 2+>2.5, 3+>3.7) within Bioworks as well as against a nonredundant NCBI database using MASCOT. Predicted phosphorylation sites can be validated manually via SEQUEST using MS^2 and MS^3 raw data.

3. Perspectives

3.1. Advantages and Limitations of the Applied Proteomic Strategies

Proteomic approaches to identify interaction partners of specific proteins or to unravel the composition of entire protein–protein interaction complexes have been used successfully for quite some time (35–38) (see also Chap. 3.3 of this volume). Mass spectrometry-based technological approaches were also successfully applied in the context of neuroscience research, i.e., the identification of postsynaptic protein interaction complexes (39–42). Classical "gel-based" proteomic approaches also helped to identify binding partners of several (non-neuronal) connexins (43–46), albeit a comprehensive and large-scale "non-gel"-based approach to unravel the putative interaction partners of (neuronal) connexins has been missing. The here applied protocols combining nano-LC-ESI-MS/MS-based mass spectrometry with two independent methods of protein interaction partner enrichment and isolation (affinity capture and Co-IP) might be helpful to foster further evaluation of, e.g., cytoskeletal elements anchoring at gap junction sites. Detailed protocols of two of these candidates, tubulin and CaMKII, were chosen for further in-depth description which may allow subsequent studies to enrich the list of interacting partners at electrical synapse sites. We initially faced the same obstacles as most of the large-scale proteomic approaches aiming for the analysis of protein interaction complexes, mainly, the high number of identified "false positive" unspecific binding partners: a problem that can be circumvented by, e.g., the use of a two-step sample "purification" method (36, 47–49). By analyzing a large number of control samples, thereby creating a "library" of unspecific interaction partners, for both of the applied purification methods, we could exclude most of the identified proteins as likely to be "false-positive" resulting in the generation of a rather short list of highly probable "true-positive" candidates. The fact that our inclusion criteria are very harsh led to the exclusion of some known interaction partners of Cx36, such as ZO1 (9, 10) and calmodulin (11). Both proteins were identified from several of the affinity capture and Co-IP samples but were also present in one of the nine control samples each, turning them into false negatives. This indicates that the generated "raw" lists of Cx36 interaction partners might include several more "real" Cx36-binding proteins, which are currently excluded as false negatives. Several improvements could be applied to the current approach possibly resulting in the increase of the overall number of "true candidates": for instance an increase in the number of samples analyzed from the here described 3–6 or more, or the use of fusion proteins with different fusion tags on either end in order to perform a two-step affinity purification (TAP) protocol. TAP may possibly reduce unspecific binding, permitting the detection of low-abundance or weakly binding interaction partners. The use of commercially available liquid chromatography columns, instead of the "self-packed" ones, would likely also increase the level of sample-to-sample reproducibility of the (secondary) mass spectrometric analytical step. Further optimization of the initial fusion protein production and purification steps as well as an improvement of the affinity capture conditions might also lead to an increase in the number of candidates fulfilling our harsh selection criteria. In this regard it should be pointed out that the GST-fusion proteins of the Cx36 cytoplasmic loop tend to form inclusion bodies and require harsher solubilization conditions when compared to the fusion proteins of the Cx36 carboxy-terminal tail. Despite the possibility for further improvement, the here presented protocols do represent a first promising start for further detailed analyses of the pre- and postsynaptic nexus associated with electrical synapses.

Acknowledgment

We would like to thank Drs. Mamoru Matsubara, Stephanie Urschel, and Dirk Wolters for their kind and constructive support during mass spectrometric analysis and identification of protein-interaction partners and phosphorylation sites of Cx36.

References

1. Furshpan EJ, Potter DD (1959) Transmission at the giant motor synapses of the crayfish. J Physiol 145(2):289–325
2. Furshpan EJ, Potter DD (1957) Mechanism of nerve-impulse transmission at a crayfish synapse. Nature 180(4581):342–343
3. Costain WJ et al (2010) Proteomic analysis of synaptosomal protein expression reveals that cerebral ischemia alters lysosomal Psap processing. Proteomics 10(18):3272–3291
4. Bai F, Witzmann FA (2007) Synaptosome proteomics. Subcell Biochem 43:77–98
5. Witzmann FA et al (2005) A proteomic survey of rat cerebral cortical synaptosomes. Proteomics 5(8):2177–2201
6. Grant SG et al (2005) Synapse proteomics of multiprotein complexes: en route from genes to nervous system diseases. Hum Mol Genet 14 Spec No. 2:R225–R234
7. Kittler JT, Moss SJ, editors. The Dynamic Synapse: Molecular Methods in Ionotropic Receptor Biology. Boca Raton (FL): CRC Press; 2006
8. Li X, Lu S, Nagy JI (2009) Direct association of connexin36 with zonula occludens-2 and zonula occludens-3. Neurochem Int 54(5–6): 393–402
9. Flores CE et al (2008) Interaction between connexin35 and zonula occludens-1 and its potential role in the regulation of electrical synapses. Proc Natl Acad Sci U S A 105(34):12545–12550
10. Li X et al (2004) Neuronal connexin36 association with zonula occludens-1 protein (ZO-1) in mouse brain and interaction with the first PDZ domain of ZO-1. Eur J Neurosci 19(8):2132–2146
11. Burr GS et al (2005) Calcium-dependent binding of calmodulin to neuronal gap junction proteins. Biochem Biophys Res Commun 335(4):1191–1198
12. Bissiere S et al (2011) Electrical synapses control hippocampal contributions to fear learning and memory. Science 331(6013):87–91
13. LeBeau FE et al (2003) The role of electrical signaling via gap junctions in the generation of fast network oscillations. Brain Res Bull 62(1):3–13
14. Rozental R, Giaume C, Spray DC (2000) Gap junctions in the nervous system. Brain Res Brain Res Rev 32(1):11–15
15. Haas JS, Zavala B, Landisman CE (2011) Activity-dependent long-term depression of

electrical synapses. Science 334(6054): 389–393

16. Hayut I et al (2011) LTS and FS inhibitory interneurons, short-term synaptic plasticity, and cortical circuit dynamics. PLoS Comput Biol 7(10):e1002248
17. Wang Y, Belousov AB (2011) Deletion of neuronal gap junction protein connexin 36 impairs hippocampal LTP. Neurosci Lett 502(1):30–32
18. Wu CL et al (2011) Heterotypic gap junctions between two neurons in the drosophila brain are critical for memory. Curr Biol 21(10):848–854
19. Lisman JE, McIntyre CC (2001) Synaptic plasticity: a molecular memory switch. Curr Biol 11(19):R788–R791
20. Barria A, Malinow R (2005) NMDA receptor subunit composition controls synaptic plasticity by regulating binding to CaMKII. Neuron 48(2):289–301
21. Sanhueza M et al (2011) Role of the CaMKII/NMDA Receptor Complex in the Maintenance of Synaptic Strength. J Neurosci 31(25): 9170–9178
22. Duffy HS et al (2002) pH-dependent intramolecular binding and structure involving Cx43 cytoplasmic domains. J Biol Chem 277(39):36706–36714
23. Alev C et al (2008) The neuronal connexin36 interacts with and is phosphorylated by CaMKII in a way similar to CaMKII interaction with glutamate receptors. Proc Natl Acad Sci U S A 105(52):20964–20969
24. Gattiker A, Gasteiger E, Bairoch A (2002) ScanProsite: a reference implementation of a PROSITE scanning tool. Appl Bioinformatics 1(2):107–108
25. Yap KL et al (2000) Calmodulin target database. J Struct Funct Genomics 1(1):8–14
26. Washburn MP, Wolters D, Yates JR 3rd (2001) Large-scale analysis of the yeast proteome by multidimensional protein identification technology. Nat Biotechnol 19(3):242–247
27. Link AJ et al (1999) Direct analysis of protein complexes using mass spectrometry. Nat Biotechnol 17(7):676–682
28. Elias JE et al (2005) Comparative evaluation of mass spectrometry platforms used in large-scale proteomics investigations. Nat Methods 2(9):667–675
29. Urschel S et al (2006) Protein kinase A-mediated phosphorylation of connexin36 in mouse retina results in decreased gap junctional communication between AII amacrine cells. J Biol Chem 281(44):33163–33171
30. Kothmann WW et al (2007) Connexin 35/36 is phosphorylated at regulatory sites in the retina. Vis Neurosci 24(3):363–375
31. Blom N, Gammeltoft S, Brunak S (1999) Sequence and structure-based prediction of eukaryotic protein phosphorylation sites. J Mol Biol 294(5):1351–1362
32. Xue Y et al (2011) GPS 2.1: enhanced prediction of kinase-specific phosphorylation sites with an algorithm of motif length selection. Protein Eng Des Sel 24(3):255–260
33. Iakoucheva LM et al (2004) The importance of intrinsic disorder for protein phosphorylation. Nucleic Acids Res 32(3):1037–1049
34. Huang HD et al (2005) KinasePhos: a web tool for identifying protein kinase-specific phosphorylation sites. Nucleic Acids Res 33(Web Server issue):W226–W229
35. Hilder TL et al (2007) Proteomic identification of the cerebral cavernous malformation signaling complex. J Proteome Res 6(11):4343–4355
36. Chang IF (2006) Mass spectrometry-based proteomic analysis of the epitope-tag affinity purified protein complexes in eukaryotes. Proteomics 6(23):6158–6166
37. Barnouin K (2004) Two-dimensional gel electrophoresis for analysis of protein complexes. Methods Mol Biol 261:479–498
38. Graumann J et al (2004) Applicability of tandem affinity purification MudPIT to pathway proteomics in yeast. Mol Cell Proteomics 3(3):226–237
39. Li X et al (2009) Proteomic screen for multiprotein complexes in synaptic plasma membrane from rat hippocampus by blue native gel electrophoresis and tandem mass spectrometry. J Proteome Res 8(7):3475–3486
40. Jordan BA et al (2004) Identification and verification of novel rodent postsynaptic density proteins. Mol Cell Proteomics 3(9):857–871
41. Satoh K et al (2002) Identification of activity-regulated proteins in the postsynaptic density fraction. Genes Cells 7(2):187–197
42. Klemmer P, Smit AB, Li KW (2009) Proteomics analysis of immuno-precipitated synaptic protein complexes. J Proteomics 72(1):82–90
43. Butkevich E et al (2004) Drebrin is a novel connexin-43 binding partner that links gap junctions to the submembrane cytoskeleton. Curr Biol 14(8):650–658
44. Singh D et al (2005) Connexin 43 interacts with zona occludens-1 and -2 proteins in a cell cycle stage-specific manner. J Biol Chem 280(34):30416–30421
45. Singh D, Lampe PD (2003) Identification of connexin-43 interacting proteins. Cell Commun Adhes 10(4–6):215–220
46. Giepmans BN et al (2001) Gap junction protein connexin-43 interacts directly with microtubules. Curr Biol 11(17):1364–1368

47. Kitao H, Takata M (2006) Purification of TAP-tagged proteins by two-step pull down from DT40 cells. Subcell Biochem 40:409–413
48. Tagwerker C et al (2006) A tandem affinity tag for two-step purification under fully denaturing conditions: application in ubiquitin profiling and protein complex identification combined with in vivocross-linking. Mol Cell Proteomics 5(4):737–748
49. Kaneko A et al (2004) Tandem affinity purification of the Candida albicans septin protein complex. Yeast 21(12):1025–1033

Chapter 8

The Cytoskeleton as a Pharmacological Target in Neurodegenerative Diseases

Saar Oz and Iliana Gozes

Abstract

The cytoskeleton plays a key role in maintaining the highly asymmetrical shape and structural polarity of neurons that are essential for neuritogenesis and neuronal physiology. In neurodegenerative diseases, the cytoskeleton is abnormally assembled and impairment of neurotransmission occurs. Cumulative evidence suggests that neurodegenerative diseases and psychiatric illnesses are associated with cytoskeletal alterations in neurons that, in turn, lose synaptic connectivity and the ability to transmit incoming axonal information to the somatodendritic domain. Major components of the neuronal cytoskeleton are the microtubules (MT). MTs, composed of the tubulin heterodimer backbone and decorated/regulated by MT-associated proteins (MAPs), provide a dynamic skeleton for cellular structures as well as trails for motor protein movement of important cargo.

The following suggested set of selected assays are intended to evaluate MT changes and effects by candidate drugs. These methods assess different characteristics of MTs in versatile assays as follows. (1) MT polymerization assays. (2) Quantifying polymerized vs. soluble tubulin by fractionation of tubulin pools and evaluating percent of polymerized tubulin as well as MAPs, by size fractionations immunoassays. (3) Tubulin isoform expression, for example, beta 3 tubulin expression at the level of RNA and protein as a marker for neuronal differentiation and neurite outgrowth using quantitative reverse transcription polymerase chain reaction (RT-PCR), immunoblotting, and immunofluorescence. (4) The microtubule-associated protein 2, MAP2 expression as a marker for neurite outgrowth and neuronal survival. (5) The microtubule-associated protein tau expression, phosphorylation, and aggregation as a marker for tau pathology (tauopathy), a major hallmark of Alzheimer's disease (AD). (6) MT dynamics by distinguishing stable and dynamic microtubules through their content of posttranslationally modified α-tubulin. MTs that are rich in tyrosinated/detyrosinated tubulin are visualized using specific antibodies and immunofluorescence. (7) Measurements of cellular/axoplasmic transport.

Key words: Methods to measure microtubules (MTs), Microtubule (MT) dynamics, Tau, MAP2, Beta 3 tubulin, Tubulin pools, Tauopathy

Rolf Dermietzel (ed.), *The Cytoskeleton: Imaging, Isolation, and Interaction*, Neuromethods, vol. 79,
DOI 10.1007/978-1-62703-266-7_8, © Springer Science+Business Media, LLC 2013

1. Introduction: Microtubules (MTs) and Neurodegeneration

Emerging evidence suggests that mutations contributing to neurodegenerative diseases are associated with dysfunction of cytoskeletal components that influence vesicular biogenesis, vesicle/organelle trafficking, and synaptic signaling. The data suggest a mechanism by which cytoskeletal disruption initiates a cascade of events including mitochondrial dysfunction and oxidative stress that, ultimately, activates the DNA damage response. Neurodegeneration in a number of distinct diseases is linked to dysfunction of the cytoskeleton as an underlying cause of cell death. Progressive loss of neuronal polarity is a major histopathological event in neurodegenerative diseases and schizophrenia that precedes neuronal death and disappearance of synaptic connectivity (1–10). The first reports on cytoskeletal alterations in neurodegenerative diseases described deposition of cytoskeletal debris in neuronal tissue of Parkinson's and Alzheimer's disease (AD) patients. Based on this observation, it was suggested that these diseases can be collectively named cytoskeletal disorders (11, 12). Defects of the cytoskeleton may be a common feature contributing to neurodegeneration. Stoichiometric ratios obtained by quantification of cytoskeletal proteins were aberrant in AD and Pick's disease and in Down syndrome. These data indicated diminished cytoskeletal proteins content in neurons of dementia patients and suggested diminished neuronal density and neuronal cytoskeletal disassembly (13). Drugs that prevent the loss of polarity and cytoskeleton retraction intrinsic to these diseases can be extremely useful in treatment of neurodegenerative diseases. The ideal drug would strengthen MTs against loss and also correct MT distortions that may have already occurred, while still allowing the MTs to participate in their normal dynamics. This short review emphasizes selected methodology associated with MT studies at the cell culture/in vitro level.

1.1. Alzheimer's Disease (AD)

AD the most common types of dementia, is a brain disorder affecting the elderly. AD is characterized by the formation of two main protein aggregates, β-amyloid plaques and tau neurofibrillary tangles (NFTs), which are involved in a process leading to progressive neuronal degeneration and death. The neuropathological hallmarks of AD encompass neuronal damage and death in brain regions critical for learning and memory including the neocortex, hippocampus, amygdala, anterior thalamus, basal forebrain (14–16).

Calcium homeostasis and activation of NMDA receptors have been proposed as mediators of β-amyloid-induced cell death. Nitric oxide (NO) must also be considered as a mediator of the cytotoxic effects of β-amyloid. Activation of the NMDA subtype of the glutamate receptor induces the production of nitric oxide by a calcium and calcium-dependent stimulation of neuronal NO synthase.

NO synthesized endogenously after NMDA activation can lead to neurotoxicity, possibly resulting from the reaction of nitric oxide with superoxide anions (17).

The main protein component of NFTs is the MT-associated protein (MAP), tau. Tau regulates assembly and stabilization of MT and MT dynamics. Mutations in tau cause a decrease in MT stabilization (18). Tau is a substrate for hyperphosphorylation leading to the formation of the pathogenic NFTs (19, 20). Tau is normally located in axons but accumulates in various fibrillar morphologies within the perikarya (as neurofibrillary tangles, NFTs) or within dendrites (as neuropil threads) (21). This redistribution of the intracellular compartmentalization of tau is associated with a change in tau conformation and the appearance of multiple phosphoepitopes. NFTs are aggregates of filamentous tau polymers that comprise a portion of the fibrillar pathologies in AD.

1.2. Tauopathies

Tauopathies refer to a group of disorders that present accumulation of insoluble hyperphosphorylated tau protein in the brain. More than 20 different degenerative disorders are characterized by some degree of neurofibrillary degeneration (see also other Chapters 9 this volume) and can be classified as tauopathies (22). *Tau* gene mutations are pathogenic for frontotemporal dementia (FTD) with Parkinsonism linked to chromosome 17 (FTDP-17), and tau polymorphisms are genetic risk factors for sporadic progressive supranuclear palsy (PSP) and corticobasal degeneration (CBD). In CBD and PSP, only 4-tau isoforms aggregate into twisted and straight filaments, respectively; they appear as a major tau doublet. In Pick's disease, 3-tau isoforms aggregate into random coiled filaments. These differences in tau isoforms may be related to either degeneration of a particular cell population in a given disorder or to aberrant cell trafficking of a particular tau isoform (23).

1.3. Amyotrophic Lateral Sclerosis

Modifications of tubulin by nitration are also associated with motor neuron loss in amyotrophic lateral sclerosis (ALS). Mutations in the copper–zinc superoxide dismutase (SOD1) are known to underlie 2% of familial cases of ALS (24). Evidence indicates that SOD mutations may be directly linked to defects in both cytoskeleton components and vesicular transport motors. Aggregates containing both neurofilaments and kinesin are hallmarks of ALS, suggesting that these targets may be sequestered in disease. Kinesin and dynein facilitate transport of organelles along microtubules in an anterograde and retrograde direction, respectively (25). In ALS, there is not only selective loss of kinesin motors, but there is also measurable slowing of axonal transport in motor neurons (26). The data indicate that impairment of slow axonal transport may be an early event in ALS pathophysiology (26). The role of SOD in this process is not entirely clear. However, recent evidence indicates there may be disruption of tubulin by nitration near the dynactin binding site in motor neurons (27). ALS is a cytoskeletal disease where neurofilaments and peripherin are abnormally accumulated in axonal spheroid bodies (10).

1.4. Schizophrenia

Cytoarchitectural abnormalities and cytoskeletal dysfunction that arise during development have been observed in the schizophrenia brain, including abnormal expression of two microtubule-associated proteins (MAP2 and MAP5) in specific subfields of the hippocampal formation in schizophrenia (28, 29). Further studies have shown alterations in phosphorylation and distribution of MAP1B in specific brain areas of schizophrenia patients (10). The MAP6 protein, also known as stable tubule-only polypeptide (STOP) stabilizes cold-stable microtubules that are important for brain development and function. A linkage between MAP6 gene and schizophrenia has been also reported (30). The gene Disrupted-In-Schizophrenia-1 (DISC1), originally identified at the breakpoint of a chromosomal translocation that is linked to a rare familial schizophrenia, has been genetically implicated in schizophrenia in other populations (31).

The DISC1 protein is localized at the centrosome, the cell principal MT organizing center, regulating the assembly of the α and β tubulin heterodimer into MTs. Centrosome function, specifically the cytoplasmic dynein regulator protein Nuclear Distribution Element-Like (NUDEL), is important to neuronal migration during brain development. NUDEL binding to the lissencephaly (LIS1) protein is required for cerebral cortical histogenesis (32).

Also, a family-based association study of kinesin heavy chain member 2 gene (KIF2) and schizophrenia suggested that the KIF2 gene is a potential schizophrenia susceptibility gene (33).

Thus, potential disrupted axoplasmic transport is associated with schizophrenia.

1.5. MT Stability and Dynamics: α-Tubulin Tyrosination Cycle and MT Dynamics

The tubulin tyrosination cycle involves the removal of the carboxy-terminal tyrosine residue of the α-tubulin chain by an uncharacterized tubulin carboxypeptidase (TCP) and the re-addition of a tyrosine residue at the same location by tubulin tyrosine ligase (TTL), in an ATP-dependent reaction (34–36).

Although TCP may still be considered uncharacterized, recent literature suggests that the cytoplasmic ATP/GTP binding protein-like 2 (AGBL2) is TCP (37). TTL, the enzyme that catalyzes the tubulin tyrosination reaction, adds tyrosine to soluble, detyrosinated tubulin dimmers and does not react efficiently with tubulin once it is assembled into MTs (34, 36, 38, 39). Indirect evidence indicates that TCP is likely to prefer polymerized substrate and that binding to MTs may be required for its function (38, 40–42). TCP acts slowly on MTs, whereas TTL recharges Glu-tubulin quickly upon release from MTs (39). As a consequence of substrate specificity of TTL and of kinetic differences of TTL and TCP, tyrosinated tubulin (Tyr-tubulin) is the main component of dynamic MTs and has a short half-life ($t_{1/2} < 5–10$ min), being the dominant tubulin variant in cycling cells in vitro, whereas Glu-tubulin has a long half-life ($t_{1/2} > 1–2$ h) and is a marker of long-lived stable MTs (34, 38, 43–46). Although detyrosination itself does not stabilize MTs (47), it can be used as a marker for how long an MT has been assembled. The amount of Tyr-tubulin in the MTs was highest in the cell body and at the growth cone. These observations indicate that newly assembled MTs are concentrated in the proximal and distal regions of growing axons and therefore that the cell body and growth cone are the most active sites of MT assembly dynamics in neurons that are actively extending axons (48–50). A few Tyr-Tubulin containing MTs penetrate into the peripheral regions of growth cones, which generally do not contain an MT network (51–53). These observations suggest that MT dynamics differ throughout the axon and growth cone. In dendrites where MTs display mixed polarity, Tyr-Tubulin exists at higher level than Glu-Tubulin (54, 55). Spatial variations in the relative content of tyrosinated and Glu-tubulin along individual MTs reflect differences in their dynamic properties. Studies further showed that tubulin PTMs are involved in interactions between MTs and certain MAPs or motor proteins. These observations have suggested that modified MTs play a role in neurite outgrowth and maturation as well as in maintaining neuronal morphology. Results suggest that reduced stable Glu-MTs are a primary consequence of tau accumulation that initiates mechanisms underlying astrocyte dysfunction and death in human neurodegenerative glial tauopathies (56).

1.6. Beta 3 Tubulin

Original studies identified tubulin isotypes specific to the brain that increase in number with brain development, expressed in a single neuron and show neuronal specificity (57–59). The beta 3 isotype (β3-tubulin) is neuron-specific (60–63) and appears to play a role in early neuritogenesis, either concomitantly or in coordination with neuronal MAPs (64). β3-Tubulin expression is increased during periods of axon guidance and maturation (65). Furthermore, β3-tubulin is associated with protection against oxidative stress. It has been suggested that β3-tubulin plays a role in neurite extension by enhancing MT polymerization during early neuritogenesis (66). Recent studies suggest that β3-tubulin renders the MT less sensitive to oxidative damage (67, 68). Another recent study associated mutations in human β3-tubulin in perturbation of MT dynamics (increases in de-tyrosinated tubulin), kinesin interactions, and axon guidance (69), which are linked to a spectrum of human nervous system disorders that are now called the TUBB3 syndromes. Each of the mutation causes ocular motility disorder, whereas some also result in intellectual and behavioral impairments, facial paralysis, and/or later-onset axonal sensorimotor polyneuropathy.

Tubulin isotype expression, and tubulin posttranslation modifications, like α-tubulin detyrosination is regulated by specific enzymes, and each can profoundly impact the capacity of the MT to interact with other proteins (70–76).

2. Materials and Methods

2.1. Selected Assays Intended to Evaluate MT Changes and Effects by Candidate Drugs

1. MT polymerization assays.

In detail:

2. Quantifying polymerized vs. soluble tubulin by fractionation of tubulin pools and evaluating percent of polymerized tubulin as well as MAPs, by size fractionations and immunoassays.
3. Evaluate MT dynamics by distinguishing stable and dynamic MTs through their content of posttranslationally modified α-tubulin. MTs that are rich in tyrosinated/detyrosinated tubulin are visualized using specific antibodies and immunofluorescence.

In short:

4. Tubulin isoform expression, for example, β3-tubulin expression in the level of RNA and protein as a marker for neuronal differentiation and neurite outgrowth using quantitative reverse transcription polymerase chain reaction (RT-PCR), immunoblotting, and immunofluorescence.
5. The MT-associated protein 2, MAP2 expression as a marker for neurite outgrowth and neuronal survival.
6. The MT-associated protein tau expression, phosphorylation and aggregation as a marker for tau pathology (tauopathy), a major hallmark of Alzheimer's disease (AD).
7. Measurements of cellular/axoplasmic transport.

2.2. Selected Methods in Detail

1. MT polymerization assays.

In vitro MT polymerization assays are carried out on purified tubulin and there are commercially available kits for these experiments as described: (http://www.cytoskeleton.com/tubulins) and see also (77).

In cell polymerization assays can be carried out after nocodazole depolymerization of microtubules as described (78).

2.2.1. Cell Culture Methods (Please See Also (79))

2. Quantifying polymerized vs. soluble tubulin by fractionation of tubulin pools and evaluating percent of polymerized tubulin as well as MAPs, by fractionations and immunoassays.

Cell Culture Methods (Please See Also (79))

Rat Cerebral Cortical Astrocyte Cell Cultures

Newborn rats (Harlan, Jerusalem, Israel) are sacrificed by decapitation, and the brain is removed (80). The cortex is dissected, and the meninges are removed. The tissue is minced with scissors and placed in Hank's balanced salts solution 1 (S1), containing HBSS (Biological Industries, Beit Haemek, Israel), 15 mM HEPES buffer, pH 7.3 (Biological Industries, Beit Haemek, Israel), and 0.25% trypsin (Biological Industries) in an incubator at 37°C 5% CO_2 for 20 min. The cells are then placed in 5 mL of solution 2 (S2) containing 10% heat inactivated fetalserum (Biological Industries), 0.1% gentamycin sulfate solution (Biological Industries), and 0.1% penicillin–streptomycin–nystatin solution (Biological Industries) in Dulbecco's modified Eagle's medium (DMEM, Sigma, Rehovot, Israel). The cells are allowed to settle and are then transferred to a new tube containing 2.5 mL of S2 and triturated using a Pasteur pipette. The process is repeated twice more. Once all the cells are suspended, cell density is determined using a hemocytometer (Neubauer Improved, Germany) and 15×10^6 cells/15 mL S2 are inoculated into each 75-cm^2 flask (Corning, Corning, NY, USA). Cells are incubated at 37°C 10% CO_2. The medium is changed after 24 h, and cells are grown until confluent.

Rat Cerebral Cortical Astrocyte Cell Subcultures

The flasks are shaken to dislodge residual neurons and oligodendrocytes that may be present. Flasks are then washed with 10 mL cold HBSS×1, HEPES 15 mM. 5 mL of versene-trypsin solution (BioLab, Jerusalem, Israel) are added to each flask, and the flasks are incubated at room temperature for 5 min to remove astrocytes. The flasks are shaken to dislodge the cells. The versene-trypsin solution is neutralized with 5 mL of S2. The cell suspension is collected and centrifuged at $100 \times g$ for 10 min. The supernatant is removed and the cells resuspended in S2. Cells are inoculated into 75-cm^2 flasks or plated in 35-mm dishes (Corning, Corning, NY, USA) and incubated until confluent at 37°C 10% CO_2.

Enriched Neuronal Cultures

The cells are prepared as described for rat cerebral cortical astrocytes cell cultures. After suspending the cells in S2, they are centrifuged at $100 \times g$ for 5 min and the supernatant is discarded. The cell pellet is resuspended in solution *S3* containing 5% heat inactivated horse serum (Biological Industries), 0.1% gentamycin, 0.1% penicillin–streptomycin–nystatin, 1% N3 (defined medium components essential for neuronal development in culture) (81), 15 μg/mL 5′-fluoro-2-deoxyuridine (FUDR, Sigma, Rehovot, Israel), and 3 μg/mL uridine (Sigma, Rehovot, Israel) in DMEM. Cells are counted in a hemocytometer, diluted in S3, and seeded in poly-L-lisine-coated 35-mm dishes (3×10^6 cells/dish). The medium is changed the next day to S3 without FUDR and uridine. Cells are allowed to grow for 1 week at 37°C 5% CO_2 before experiments are performed.

Primary Neuronal Cultures

Primary cultures of neurons are prepared as follows. Newborn rats are sacrificed at postnatal day 1. Cerebral cortex tissue is then dissected and dissociated individually from each pup with the Papain Dissociation System (PDS), (Worthington Biochemical Corporation) according to the manufacturer's instructions. Cortical neurons are maintained in Neurobasal medium (NB) (Gibco) supplemented with NeuroCult B27-SM1 (STEMCELL), 1% Glutamax (Gibco), and grown on poly-D-lysine-coated cell culture glass cover slips. The cells are incubated in 5% CO_2 in a humidified incubator at 37°C.

Mixed Neuroglial Cultures

Neurons are prepared as described for enriched neuronal cultures. 3×10^5 cells per 35-mm culture plate are seeded on 8-day-old astrocytes prepared as described for rat cerebral cortical astrocyte cell subculture. Cells are allowed to grow for 1 week at 37°C 5% CO_2 before experiments are performed (81).

Cell Lines

PC12 cells (Pheochromocytoma cells) (ATCC, Bethesda, MD, USA) are seeded at 3×10^4 cells/cm^2 on poly-L-Lysine-coated plastic tissue culture dishes (Corning, Lowell, MA, USA) to form an adherent monolayer. Cells are maintained in Dulbecco's modified Eagle's medium (DMEM) supplemented with 10% horse serum (HS), 5% fetal calf serum (FCS), 2 mM glutamine, 100 U/mL penicillin, and 100 mg/mL streptomycin (Biological Industries, Beit Haemek, Israel). PC12 differentiation is induced by nerve growth factor (NGF, Sigma) at concentrations of 50 ng/mL by replacing half of the medium every other day until the cells acquire a differentiated morphology. We defined differentiated cells as bearing two or more neurites with lengths equal to or longer than twice the diameter of the cell body (82). The cells are incubated in 5% CO_2 in a humidified incubator at 37°C. The cells are subcultured every 5 days at a 4:1 split ratio by pipetting. The medium is changed every 2 or 3 days after adhesion.

NIH3T3 (mouse fibroblasts) (ATCC, Bethesda, MD, USA) are cultured in DMEM containing 10% heat inactivated fetal calf serum, 2 mM L-Glutamine, 100 U/mL penicillin, and 0.1 mg/mL streptomycin (Biological Industries) in 5% CO_2 at 37°C (growth conditions). Every 3–4 days cells are split using trypsin–EDTA solution B (Biological Industries).

Poly-L-Lysine Coating of Dishes

One milliliter of 10 μg/mL poly-L-lysine hydrobromide (Sigma, Rehovot, Israel) in sterile double-distilled water (sDDW) is added to 35-mm dishes. After 1 h, dishes are washed three times with sDDW and kept at room temperature (78).

2.2.2. Protein Extraction

Whole Cell Lysate Preparation

Cells are incubated for various durations in the absence or presence of drug treatment. Cytoplasmic fraction extraction is performed as follows. Cell cultures, grown in 6-well plates, are incubated for 10 min in ice cold modified RIPA lysis buffer (50 mM Tris–HCl pH 7.4, 150 mM NaCl, 2 mM EGTA, 1% Triton X-100, 0.1% SDS, 0.1% sodium deoxycholate) supplemented with proteinase inhibitor cocktail diluted 1:100 (Sigma, Saint Louis, MO, USA). Cells are scraped using rubber policeman, transferred to collection tubes and incubated on ice for 30 min, followed by centrifugation of 12,000 × *g* for 10 min at 4°C. The resulting supernatants are then collected and stored at −80°C. Aliquots are taken to determine protein concentration.

Cytoplasmic and nuclear fractions are prepared using NE-PER Nuclear and Cytoplasmic Extraction Reagents (Pierce, Rockford, IL, USA). Cells or tissue are harvested and homogenized according to manufacturer's protocol supplemented with anti-proteases cocktail diluted 1:100 (Sigma, Rehovot, Israel). NE-PER Reagents Kit maintains nuclear and cytoplasmic cell compartments by lysing the cell membrane, recovering the nucleus intact, and then extracts nuclear proteins separately.

Proteins are quantified as described below. Samples are each mixed with sample buffer (SB) (SDS-polyacrylamide gel electrophoresis sample buffer 45% glycerol, 20% β-mercaptoethanol, 9.2% SDS, 0.04% bromophenol blue, and 0.3 M Tris–HCl, pH 6.8) and heated at 95°C for 5 min. Samples are analyzed by immunoblotting see below.

Polymerized vs. Soluble Tubulin (and Related MAPs) Separation Assay

To quantify tubulin polymerization, a simple assay was developed by modifying a method originally described (74, 75). Cells grown to confluence in 6-well plates are washed with MT-buffer (80 mM PIPES pH 6.8, mM $MgCl_2$, 2 mM EGTA, 5% Glycerol) w/o Triton X-100 and lysed at 37°C for 5 min, with 150 μl of MT-buffer with 0.5% Triton X-100 supplemented with proteinase inhibitor cocktail diluted 1:100 (Sigma, Saint Louis, MO, USA) in order to extract soluble (cytosolic) tubulin (S); Pelleted cells are rinsed once again with equal volume of Modified RIPA lysis buffer (50 mM Tris–HCl pH 7.4, 150 mM NaCl, 2 mM EGTA, 1% Triton X-100, 0.1% SDS, 0.1% sodium deoxycholate) supplemented with proteinase inhibitor cocktail diluted 1:100 (Sigma, Saint Louis, MO, USA) in order to collect the polymerized (cytoskeletal) tubulin (P). The cytosolic and cytoskeletal fractions are each mixed with sample buffer (SB) (SDS-polyacrylamide gel electrophoresis sample buffer 45% glycerol, 20% β-mercaptoethanol, 9.2% SDS, 0.04% bromophenol blue, and 0.3 M Tris–HCl, pH 6.8) and heated at 95°C for 5 min. Equal volume of each fraction is analyzed by immunoblotting (see below). The percentage of polymerized tubulin is determined by dividing the densitometry value of polymerized tubulin by the total tubulin content (the sum of the densitometry values of soluble and polymerized tubulin). The distribution of tubulin between the soluble and polymerized fractions could be altered to some extent by experimental conditions, so care must be taken to maintain consistency, and each experiment should be performed with appropriate controls.

2.2.3. Protein Quantification

BCA

Proteins are quantified using BCA™ Protein Assay Kit (Pierce, Rockford, IL, USA). This method is based on the reduction of Cu^{2+} to Cu^{1+} by protein in an alkaline medium (the Biuret reaction) that is coupled to the colorimetric detection of the cuprous cation (Cu^{+1}) by using a reagent containing bicinchoninic acid (BCA). The purple-colored reaction product of this assay is formed by the chelation of two molecules of BCA with one cuprous ion. The resulting water-soluble complex exhibits a strong absorbance at 562 nm that is linear with increasing protein concentrations over a working range of 20–2,000 μg/mL (83, 84).

Bradford

Some reagents are not compatible with the BCA™ Protein Assay but they are compatible with the Bio-Rad Protein Assay. Therefore, the later is used once the buffer that had been used contained the incompatible reagents. The Bio-Rad Protein Assay, based on the method of Bradford, is an accurate procedure for determining concentration of solubilized protein. It involves the addition of an acidic dye to protein solution, and subsequent measurement at 595 nm with a spectrophotometer or microplate reader. In this assay, differential color change of a dye occurs in response to various concentrations of protein. The absorbance maximum for an acidic solution of Coomassie blue dye (Pierce, Rockford, IL, USA) shifts from 465 to 595 nm when binding to protein occurs. The Coomassie blue dye binds to primarily basic and aromatic amino acid residues, especially arginine. Comparison to a standard curve provides a relative measurement of protein concentration.

Immunoblotting Analysis

This procedure is used to separate proteins on an SDS-polyacrylamide gel electrophoresis according to their size, together with known molecular weight protein markers, transfer the separated proteins to a membrane, and identify the desired protein using specific antibodies. Sample buffer X5 is added to protein samples that are further denatured by boiling at 95 °C for 5 min. 15–25 μg protein extracts per lane are separated by electrophoresis on a 10% (w/v) polyacrylamide gel (Bio-Rad, Hercules, CA, USA) containing 0.1% SDS or 4–20% precast iGels (NuSep, Bogart, GA, USA). Molecular weight is determined using Wide Range (6–250 kDa) Multicolored Protein Markers (Invitrogen, Carlsbad, CA, USA). Following electrophoresis, the protein bands are either visualized using the Coomassie blue staining (Pierce, Rockford, IL, USA) according to manufacturer's instructions, or they transferred to nitrocellulose membranes. Nonspecific antigen sites are blocked using a solution containing 5% nonfat dried milk (w/v) in TBS-T (20 mM Tris pH 7.5, 136.8 mM NaCl, and 0.05% v/v Tween 20) for 1 h at room temperature.

Antigen detection is performed using antibodies directed against the studied protein. Antibody–antigen complexes are detected using Horseradish peroxidase (HRP) conjugated secondary antibodies. Reactive bands are visualized by chemiluminescence kit

(Pierce) on Kodak Biomax ML Scientific imaging film (Kodak, Chalon-sur-Saône, France). The densitometric analysis of western blots is performed with MiniBIS Pro Gel imaging system and software (DNR, Maale Hachamisha, Israel).

3. Evaluate MT dynamics by distinguishing stable and dynamic microtubules through their content of posttranslationally modified α-tubulin. MTs that are rich in tyrosinated/detyrosinated tubulin are visualized using specific antibodies and immunofluorescence.

Microplate Reader: "In Cell Western"

The assay is performed as previously described (85). This assay aims to measure and evaluate MT-Destabilizing and MT-Stabilizing agents by utilizing the α-tubulin tyrosination cycle. Briefly, cells are permeabilized for 5 min prior to fixation in order to remove soluble tubulin, fixed and incubated with tubulin antibodies (anti-tyrosinated α-tubulin (clone YL1/2) (Abcam) for newly formed, dynamic MT and anti-detyrosinated (Glu) α-tubulin (clone L4) (Abcys) for stable long-lasting MT, at a 1:4,000 dilution, followed by secondary antibodies and Hoechst. Fluorescence is measured using the Tecan Microplate fluorescent reader (Neoteck scientific instrumentation, Switzerland) infinite F200 model, Magellan software version 6.3. Fluorescence signal is normalized to the corresponding Hoechst signal in each well, which gives an indication of the relative amount of cells per well.

NOTE: To minimize false signal from scattered light and background fluorescence, use black-walled multi-well plates.

36,000 cells per well are seeded in 96-well microplate in 90 μl of medium. Cells are allowed to grow for 24 h.

NOTE: Cells should be grown, treated, fixed, and stained directly in multi-well plates.

Each well is then supplemented with 10 μl medium containing test compounds and control compounds (random locations). Plates are incubated 2 h at 5% CO_2 at 37°C. After incubation cells are permeabilized for 5 min with MT-buffer containing Triton X-100 at 37°C in order to extract soluble (cytosolic) tubulin, and then fixed with either cold methanol or 3.7% PFA.

NOTE: This general fixation protocol will work with most antibodies and cell lines. However, we recommend trying different fixation methods to identify the optimal fixation protocol for each antibody and/or cell line.

Wash the cells twice with TBS-T (20 mM Tris pH 7.5, 136.8 mM NaCl, and 0.05% v/v Tween 20). Block cells in 5% normal serum from same species as secondary antibody (e.g., normal goat serum, normal donkey serum) or 2% Bovine serum albumin (BSA), in TBS-T 1 h at room temp.

NOTE: For double-labeling, prepare a cocktail of mouse and rabbit primary antibodies at their appropriate dilutions in TBST.

MT and nuclear staining: 50 μl per well of anti-tyr-Tubulin (YL1/2) and anti-glu-tubulin (L4) primary antibodies at 1:4,000 dilution. Incubate 2 h at 37°C or overnight at 4°C. Wash the cells at least twice with TBS-T.

NOTE: For double-labeling, prepare a cocktail of fluorochrome-conjugated rat and rabbit secondary antibodies.

Secondary antibodies: add 50 μl per well rabbit and rat fluorochrome-conjugated secondary antibodies and Hoechst (1 μg/mL), prepare the dilution in 5% normal serum in 2% BSA/TBS-T and incubate 1 h at room temperature in the dark. Wash the cells at least three times with TBS-T. Add 50 μl per well of 50% glycerol-PBS solution. Measure fluorescence using a microplate reader.

NOTE: When using any primary or fluorochrome-conjugated secondary antibody for the first time, titrate the antibody to determine which dilution allows for the strongest specific signal with the least background for your sample.

Immunostaining and Confocal Microscopy Image Analysis

50,000 cells per well are seeded on glass coverslips in 24-well plates. Cells are allowed to grow for 24 h. Each well is then supplemented with test compounds and control compounds (random locations). Plates are incubated 2 h at 5% CO_2 at 37°C. After incubation cells are fixed and permeabilized simultaneously, with 3% paraformaldehyde, 0.075% glutaraldehyde (Fluka Biochemika) in MT-buffer with Triton X-100, for 10 min, quenched with 1 mg/mL $NaBH_4$ in PBS.

NOTE: This general fixation protocol will work with most antibodies and cell lines. However, we recommend trying different fixation methods to identify the optimal fixation protocol for each antibody and/or cell line.

Wash the cells twice with TBS-T. Block cells in 5% normal serum from same species as secondary antibody (e.g., normal goat serum, normal donkey serum) or 2% BSA, in TBS-T 1 h at room temp.

NOTE: For double-labeling, prepare a cocktail of mouse and rabbit primary antibodies at their appropriate dilutions in TBS-T.

Microtubule and nuclear staining: Use anti-tyr-Tubulin (YL1/2) and anti-glu-tubulin (L4) primary antibodies at 1:4,000 dilution. Incubate 2 h at 37°C or overnight at 4°C. Wash the cells at least twice with TBS-T.

NOTE: For double-labeling, prepare a cocktail of fluorochrome-conjugated anti-rat and anti-rabbit secondary antibodies.

Secondary antibodies: add rabbit and rat fluorochrome-conjugated secondary antibodies and Hoechst (1 μg/mL), prepare the dilution in 5% normal serum in 2% BSA/TBS-T and incubate 1 h at room temperature in the dark. Wash the cells at least three times with TBS-T. Mount the glasses with stained cells on slides.

NOTE: When using any primary or fluorochrome-conjugated secondary antibody for the first time titrate the antibody to determine which dilution allows for the strongest specific signal with the least background for your sample.

Images are collected with a confocal laser scanning microscope with 63X oil immersion optics, laser lines at 488, 561 nm. When comparing fluorescence intensities, identical CLSM parameters (e.g., scanning line, laser light, gain, offset, etc.) should be used. All of the fluorescent signals acquired should be above the autofluorescent background as measured from a control slide stained with the secondary antibody without a primary antibody. To compare the relative levels of protein expression, we used the average integrated density (AID) image analysis procedure for cell immunostains. In brief, integrated density is defined by the sum of the values of the pixels in the selected region of interest (ROI). This is equivalent to the product of area and mean gray value. AID for the positive stained area was determined by measuring the fluorescent intensity of the ROI, which is above the positive cut-off intensity. Positive cut-off intensities were determined based on the fluorescence intensities histogram for each antibody staining. For measurements of the MTs in a given cell, the focal plane in the confocal microscope was chosen as the one showing the maximal area on the *z*-axis (focal axis). Analysis can be performed using the MICA software (Cytoview, Petach Tikva, Israel), ImageJ (NIH, Bethesda, MD, USA), or any other commercially available software.

2.3. Selected Methods in Short

4. Tubulin isoform expression, for example, β3-tubulin expression in the level of RNA and protein as a marker for neuronal differentiation and neurite outgrowth using quantitative reverse transcription polymerase chain reaction (RT-PCR), immunoblotting and immunofluorescence.

For protein quantification, methods described above are applicable. For quantification at the level of RNA, this is the subject of many other manuscripts. Please see our primary manuscripts detailing the technology for other proteins (86, 87).

5. The microtubule-associated protein 2, MAP2 expression as a marker for neurite outgrowth and neuronal survival.

We have used MAP2-FITC immunofluorescent labeling to measure neurite outgrowth (88).

Our further studies used MAP2 immunolabeling to measure neuronal survival (89).

6. The microtubule-associated protein tau expression, phosphorylation, and aggregation as a marker for tau pathology (tauopathy), a major hallmark of Alzheimer's disease (AD).

In a number of manuscripts we have looked at tau levels of expression, tau phosphorylation, and tau aggregation (90–95).

Tau aggregation can also be measured in vitro and inhibitors of tau aggregation are being developed as we reviewed before (96).

Live cell imaging, tau distribution in cells can be studied with advanced imaging technologies (97).

7. *Measurements of cellular/axoplasmic transport.*

We have suggested a method to assess cellular microtubule-related mitochondrial transport as delineated in our paper (98). Methods for assessing axoplasmic transport deserve a manuscript of their own. In general following labeled amino acids (99) or video microscopy (100) or MRI imaging (101).

3. In Conclusion

This method manuscript touches the tip of the iceberg in terms of MT assessments. There are a number of animal models with MT deficiencies such as tauopathy models as described above as well as models of schizophrenia (also mentioned above) that can be used for testing MT interacting neuroprotecting drugs (102).

There are also many more MAPs that interact with the MT lattice and additional methods to study MT dynamics which all can contribute to drug development against neurodegenerative disease (103–106). This review concentrates on MT dynamics at the level of tubulin tyrosination and tubulin in cell polymerization. Selected other methods are mentioned in brief with reference to appropriate literature. There are classical methods and outcomes assessing MT dynamics in vitro and these are summarized in recent reviews (107, 108)

The methods described or referenced should pave the path to better understanding of neurodegeneration leading to the development of cytoskeletal-protecting drugs.

Acknowledgements

This work is in partial fulfillment of the Ph.D. thesis requirements of Saar Oz. Support was provided by AMN Foundation, Canadian Friends of Tel Aviv University - Montreal Circle of Friends, Joe & Grace Alter, Barbara & Don Seal, the Oberfeld family, the Adams family and Allon Therapeutics Inc. Professor Illana Gozes is the incumbent of the Lily and Avraham Gildor Chair for the Investigation of Growth Factors, and the Director of the Adams

Super Center for Brain Studies, the Levie-Edersheim-Gitter Institute for Functional Brain Imaging and the Dr. Diana and Zelman Elton (Elbaum) Laboratory for Molecular Neuroendocrinology at Tel Aviv University.

References

1. Flament-Durand J, Couck AM (1979) Spongiform alterations in brain biopsies of presenile dementia. Acta Neuropathol 46:159–162
2. Ellisman M, Ranganathan R, Deerinck T, Young S, Terry R, Mirra S (1987) Neuronal fibrillar cytoskeleton and endomembrane system organization in Alzheimer's disease. Adv Behav Biol. 34:61–75; Plenum Press, New York, NY
3. Terry RD (1998) The cytoskeleton in Alzheimer disease. J Neural Transm Suppl 53:141–145
4. Cash AD, Aliev G, Siedlak SL, Nunomura A, Fujioka H, Zhu X, Raina AK, Vinters HV, Tabaton M, Johnson AB, Paula-Barbosa M, Avila J, Jones PK, Castellani RJ, Smith MA, Perry G (2003) Microtubule reduction in Alzheimer's disease and aging is independent of tau filament formation. Am J Pathol 162:1623–1627
5. Tanemura K, Murayama M, Akagi T, Hashikawa T, Tominaga T, Ichikawa M, Yamaguchi H, Takashima A (2002) Neurodegeneration with tau accumulation in a transgenic mouse expressing V337M human tau. J Neurosci 22:133–141
6. Schaefer MK, Schmalbruch H, Buhler E, Lopez C, Martin N, Guenet JL, Haase G (2007) Progressive motor neuronopathy: a critical role of the tubulin chaperone TBCE in axonal tubulin routing from the Golgi apparatus. J Neurosci 27:8779–8789
7. Tang-Schomer MD, Patel AR, Baas PW, Smith DH (2010) Mechanical breaking of microtubules in axons during dynamic stretch injury underlies delayed elasticity, microtubule disassembly, and axon degeneration. FASEB J 24:1401–1410
8. Hall GF, Chu B, Lee G, Yao J (2000) Human tau filaments induce microtubule and synapse loss in an in vivo model of neurofibrillary degenerative disease. J Cell Sci 113(Pt 8):1373–1387
9. Doran B, Gherbesi N, Hendricks G, Flavell RA, Davis RJ, Gangwani L (2006) Deficiency of the zinc finger protein ZPR1 causes neurodegeneration. Proc Natl Acad Sci USA 103:7471–7475
10. Benitez-King G, Ramirez-Rodriguez G, Ortiz L, Meza I (2004) The neuronal cytoskeleton as a potential therapeutical target in neurodegenerative diseases and schizophrenia. Curr Drug Targets CNS Neurol Disord 3:515–533
11. Calne DB, Eisen A (1989) The relationship between Alzheimer's disease, Parkinson's disease and motor neuron disease. Can J Neurol Sci 16:547–550
12. McMurray CT (2000) Neurodegeneration: diseases of the cytoskeleton? Cell Death Differ 7:861–865
13. Pollak D, Cairns N, Lubec G (2003) Cytoskeleton derangement in brain of patients with Down syndrome, Alzheimer's disease and Pick's disease. J Neural Transm Suppl 67:149–158
14. Morrison JH, Hof PR (1997) Life and death of neurons in the aging brain. Science 278:412–419
15. Arnold SE, Hyman BT, Flory J, Damasio AR, Van Hoesen GW (1991) The topographical and neuroanatomical distribution of neurofibrillary tangles and neuritic plaques in the cerebral cortex of patients with Alzheimer's disease. Cereb Cortex 1:103–116
16. Samuel W, Terry RD, DeTeresa R, Butters N, Masliah E (1994) Clinical correlates of cortical and nucleus basalis pathology in Alzheimer dementia. Arch Neurol 51:772–778
17. Le WD, Colom LV, Xie WJ, Smith RG, Alexianu M, Appel SH (1995) Cell death induced by beta-amyloid 1–40 in MES 23.5 hybrid clone: the role of nitric oxide and NMDA-gated channel activation leading to apoptosis. Brain Res 686:49–60
18. Mukrasch MD, Biernat J, von Bergen M, Griesinger C, Mandelkow E, Zweckstetter M (2005) Sites of tau important for aggregation populate {beta}-structure and bind to microtubules and polyanions. J Biol Chem 280:24978–24986
19. Hyman BT, Augustinack JC, Ingelsson M (2005) Transcriptional and conformational changes of the tau molecule in Alzheimer's disease. Biochim Biophys Acta 1739:150–157
20. Higuchi M, Lee VM, Trojanowski JQ (2002) Tau and axonopathy in neurodegenerative disorders. Neuromolecular Med 2:131–150

21. LaFerla FM, Oddo S (2005) Alzheimer's disease: Abeta, tau and synaptic dysfunction. Trends Mol Med 11:170–176
22. Williams DR (2006) Tauopathies: classification and clinical update on neurodegenerative diseases associated with microtubule-associated protein tau. Intern Med J 36:652–660
23. Buee L, Delacourte A (1999) Comparative biochemistry of tau in progressive supranuclear palsy, corticobasal degeneration, FTDP-17 and Pick's disease. Brain Pathol 9:681–693
24. Martin LJ, Price AC, Kaiser A, Shaikh AY, Liu Z (2000) Mechanisms for neuronal degeneration in amyotrophic lateral sclerosis and in models of motor neuron death (Review). Int J Mol Med 5:3–13
25. Hirokawa N (1998) Kinesin and dynein superfamily proteins and the mechanism of organelle transport. Science 279:519–526
26. Williamson TL, Cleveland DW (1999) Slowing of axonal transport is a very early event in the toxicity of ALS-linked SOD1 mutants to motor neurons. Nat Neurosci 2:50–56
27. Eiserich JP, Estevez AG, Bamberg TV, Ye YZ, Chumley PH, Beckman JS, Freeman BA (1999) Microtubule dysfunction by post-translational nitrotyrosination of alpha-tubulin: a nitric oxide-dependent mechanism of cellular injury. Proc Natl Acad Sci USA 96:6365–6370
28. Arnold SE, Hyman BT, Van Hoesen GW, Damasio AR (1991) Some cytoarchitectural abnormalities of the entorhinal cortex in schizophrenia. Arch Gen Psychiatry 48:625–632
29. Arnold SE, Lee VM, Gur RE, Trojanowski JQ (1991) Abnormal expression of two microtubule-associated proteins (MAP2 and MAP5) in specific subfields of the hippocampal formation in schizophrenia. Proc Natl Acad Sci USA 88:10850–10854
30. Shimizu H, Iwayama Y, Yamada K, Toyota T, Minabe Y, Nakamura K, Nakajima M, Hattori E, Mori N, Osumi N, Yoshikawa T (2006) Genetic and expression analyses of the STOP (MAP6) gene in schizophrenia. Schizophr Res 84:244–252
31. Chubb JE, Bradshaw NJ, Soares DC, Porteous DJ, Millar JK (2008) The DISC locus in psychiatric illness. Mol Psychiatry 13:36–64
32. Cannon TD, Hennah W, van Erp TG, Thompson PM, Lonnqvist J, Huttunen M, Gasperoni T, Tuulio-Henriksson A, Pirkola T, Toga AW, Kaprio J, Mazziotta J, Peltonen L (2005) Association of DISC1/TRAX haplotypes with schizophrenia, reduced prefrontal gray matter, and impaired short- and long-term memory. Arch Gen Psychiatry 62:1205–1213
33. Kaibuchi K, Tsuboi D (2010) Current perspective on the pathogenesis of schizophrenia from the viewpoint of risk factors such as DISC1 (corrected). Nihon Shinkei Seishin Yakurigaku Zasshi 30:149–152
34. Lafanechere L, Job D (2000) The third tubulin pool. Neurochem Res 25:11–18
35. Barra HS, Arce CA, Argarana CE (1988) Posttranslational tyrosination/detyrosination of tubulin. Mol Neurobiol 2:133–153
36. MacRae TH (1997) Tubulin post-translational modifications—enzymes and their mechanisms of action. Eur J Biochem 244:265–278
37. Sahab ZJ, Hall MD, Me Sung Y, Dakshanamurthy S, Ji Y, Kumar D, Byers SW (2011) Tumor suppressor RARRES1 interacts with cytoplasmic carboxypeptidase AGBL2 to regulate the alpha-tubulin tyrosination cycle. Cancer Res 71:1219–1228
38. Wehland J, Weber K (1987) Turnover of the carboxy-terminal tyrosine of alpha-tubulin and means of reaching elevated levels of detyrosination in living cells. J Cell Sci 88(Pt 2):185–203
39. Wehland J, Weber K (1987) Tubulin-tyrosine ligase has a binding site on beta-tubulin: a two-domain structure of the enzyme. J Cell Biol 104:1059–1067
40. Webster DR, Gundersen GG, Bulinski JC, Borisy GG (1987) Assembly and turnover of detyrosinated tubulin in vivo. J Cell Biol 105:265–276
41. Gundersen GG, Khawaja S, Bulinski JC (1987) Postpolymerization detyrosination of alpha-tubulin: a mechanism for subcellular differentiation of microtubules. J Cell Biol 105:251–264
42. Kreis TE (1987) Microtubules containing detyrosinated tubulin are less dynamic. EMBO J 6:2597–2606
43. Gundersen GG, Kalnoski MH, Bulinski JC (1984) Distinct populations of microtubules: tyrosinated and nontyrosinated alpha tubulin are distributed differently in vivo. Cell 38: 779–789
44. Schulze E, Asai DJ, Bulinski JC, Kirschner M (1987) Posttranslational modification and microtubule stability. J Cell Biol 105:2167–2177
45. Bosc C, Cronk JD, Pirollet F, Watterson DM, Haiech J, Job D, Margolis RL (1996) Cloning, expression, and properties of the microtubule-stabilizing protein STOP. Proc Natl Acad Sci USA 93:2125–2130

46. Westermann S, Weber K (2003) Post-translational modifications regulate microtubule function. Nat Rev Mol Cell Biol 4:938–947
47. Webster DR, Wehland J, Weber K, Borisy GG (1990) Detyrosination of alpha tubulin does not stabilize microtubules in vivo. J Cell Biol 111:113–122
48. Brown A, Slaughter T, Black MM (1992) Newly assembled microtubules are concentrated in the proximal and distal regions of growing axons. J Cell Biol 119:867–882
49. Baas PW, Black MM (1990) Individual microtubules in the axon consist of domains that differ in both composition and stability. J Cell Biol 111:495–509
50. Baas PW, Ahmad FJ, Pienkowski TP, Brown A, Black MM (1993) Sites of microtubule stabilization for the axon. J Neurosci 13: 2177–2185
51. Shea TB (1999) Selective stabilization of microtubules within the proximal region of developing axonal neurites. Brain Res Bull 48:255–261
52. Dent EW, Gertler FB (2003) Cytoskeletal dynamics and transport in growth cone motility and axon guidance. Neuron 40:209–227
53. Fukushima N, Morita Y (2006) Actomyosin-dependent microtubule rearrangement in lysophosphatidic acid-induced neurite remodeling of young cortical neurons. Brain Res 1094:65–75
54. Mansfield SG, Gordon-Weeks PR (1991) Dynamic post-translational modification of tubulin in rat cerebral cortical neurons extending neurites in culture: effects of taxol. J Neurocytol 20:654–666
55. Witte H, Neukirchen D, Bradke F (2008) Microtubule stabilization specifies initial neuronal polarization. J Cell Biol 180:619–632
56. Yoshiyama Y, Zhang B, Bruce J, Trojanowski JQ, Lee VM (2003) Reduction of detyrosinated microtubules and Golgi fragmentation are linked to tau-induced degeneration in astrocytes. J Neurosci 23:10662–10671
57. Gozes I, Littauer UZ (1978) Tubulin microheterogeneity increases with rat brain maturation. Nature 276:411–413
58. Gozes I, Sweadner KJ (1981) Multiple tubulin forms are expressed by a single neurone. Nature 294:477–480
59. Gozes I, Saya D, Littauer UZ (1979) Tubulin microheterogeneity in neuroblastoma and glioma cell lines differs from that of the brain. Brain Res 171:171–175
60. Schwarz PM, Liggins JR, Luduena RF (1998) Beta-tubulin isotypes purified from bovine brain have different relative stabilities. Biochemistry 37:4687–4692
61. Luduena RF (1998) Multiple forms of tubulin: different gene products and covalent modifications. Int Rev Cytol 178:207–275
62. Katsetos CD, Legido A, Perentes E, Mork SJ (2003) Class III beta-tubulin isotype: a key cytoskeletal protein at the crossroads of developmental neurobiology and tumor neuropathology. J Child Neurol 18:851–866, discussion 867
63. Katsetos CD, Herman MM, Mork SJ (2003) Class III beta-tubulin in human development and cancer. Cell Motil Cytoskeleton 55:77–96
64. Ferreira A, Caceres A (1992) Expression of the class III beta-tubulin isotype in developing neurons in culture. J Neurosci Res 32:516–529
65. Jiang YQ, Oblinger MM (1992) Differential regulation of beta III and other tubulin genes during peripheral and central neuron development. J Cell Sci 103(Pt 3):643–651
66. Gard DL, Kirschner MW (1985) A polymer-dependent increase in phosphorylation of beta-tubulin accompanies differentiation of a mouse neuroblastoma cell line. J Cell Biol 100:764–774
67. Guo J, Qiang M, Luduena RF (2011) The distribution of beta-tubulin isotypes in cultured neurons from embryonic, newborn, and adult mouse brains. Brain Res 1420:8–18
68. Guo J, Walss-Bass C, Luduena RF (2010) The beta isotypes of tubulin in neuronal differentiation. Cytoskeleton (Hoboken) 67:431–441
69. Tischfield MA, Baris HN, Wu C, Rudolph G, Van Maldergem L, He W, Chan WM, Andrews C, Demer JL, Robertson RL, Mackey DA, Ruddle JB, Bird TD, Gottlob I, Pieh C, Traboulsi EI, Pomeroy SL, Hunter DG, Soul JS, Newlin A, Sabol LJ, Doherty EJ, de Uzcategui CE, de Uzcategui N, Collins ML, Sener EC, Wabbels B, Hellebrand H, Meitinger T, de Berardinis T, Magli A, Schiavi C, Pastore-Trossello M, Koc F, Wong AM, Levin AV, Geraghty MT, Descartes M, Flaherty M, Jamieson RV, Moller HU, Meuthen I, Callen DF, Kerwin J, Lindsay S, Meindl A, Gupta ML Jr, Pellman D, Engle EC (2010) Human TUBB3 mutations perturb microtubule dynamics, kinesin interactions, and axon guidance. Cell 140:74–87
70. Sudo H, Baas PW (2010) Acetylation of microtubules influences their sensitivity to severing by katanin in neurons and fibroblasts. J Neurosci 30:7215–7226

71. Konishi Y, Setou M (2009) Tubulin tyrosination navigates the kinesin-1 motor domain to axons. Nat Neurosci 12:559–567
72. Hammond JW, Huang CF, Kaech S, Jacobson C, Banker G, Verhey KJ (2010) Posttranslational modifications of tubulin and the polarized transport of kinesin-1 in neurons. Mol Biol Cell 21:572–583
73. Lacroix B, van Dijk J, Gold ND, Guizetti J, Aldrian-Herrada G, Rogowski K, Gerlich DW, Janke C (2010) Tubulin polyglutamylation stimulates spastin-mediated microtubule severing. J Cell Biol 189:945–954
74. Verhey KJ, Gaertig J (2007) The tubulin code. Cell Cycle 6:2152–2160
75. Janke C, Kneussel M (2010) Tubulin post-translational modifications: encoding functions on the neuronal microtubule cytoskeleton. Trends Neurosci 33:362–372
76. Peris L, Wagenbach M, Lafanechere L, Brocard J, Moore AT, Kozielski F, Job D, Wordeman L, Andrieux A (2009) Motor-dependent microtubule disassembly driven by tubulin tyrosination. J Cell Biol 185:1159–1166
77. Divinski I, Mittelman L, Gozes I (2004) A femtomolar acting octapeptide interacts with tubulin and protects astrocytes against zinc intoxication. J Biol Chem 279:28531–28538
78. Gozes I, Divinski I (2007) NAP, a neuroprotective drug candidate in clinical trials, stimulates microtubule assembly in the living cell. Curr Alzheimer Res 4:507–509
79. Divinski I, Pilzer I, Gozes I (2008) Primary cell cultures and cell lines. Neuromethods 39:21
80. Brenneman DE, Spong CY, Hauser JM, Abebe D, Pinhasov A, Golian T, Gozes I (2004) Protective peptides that are orally active and mechanistically nonchiral. J Pharmacol Exp Ther 309:1190–1197
81. Romijn HJ, Habets AM, Mud MT, Wolters PS (1981) Nerve outgrowth, synaptogenesis and bioelectric activity in fetal rat cerebral cortex tissue cultured in serum-free, chemically defined medium. Brain Res 254: 583–589
82. Lamoureux P, Steel VL, Regal C, Adgate L, Buxbaum RE, Heidemann SR (1990) Extracellular matrix allows PC12 neurite elongation in the absence of microtubules. J Cell Biol 110:71–79
83. Smith PK, Krohn RI, Hermanson GF, Mallia AK, Gartner FH, Provenzano MD, Fujimoto EK, Goeke NM, Olson BJ, Klenk DC (1985) Measurement of protein binding using bicinchoninic acid. Anal Biochem 150:76–85
84. Wiechelman KJ, Braun RD, Fitzpatrick JD (1988) Investigation of the bicinchoninic acid protein assay: identification of the groups responsible for color formation. Anal Biochem 175:231–237
85. Vassal E, Barette C, Fonrose X, Dupont R, Sans-Soleilhac E, Lafanechere L (2006) Miniaturization and validation of a sensitive multiparametric cell-based assay for the concomitant detection of microtubule-destabilizing and microtubule-stabilizing agents. J Biomol Screen 11:377
86. Mandel S, Rechavi G, Gozes I (2007) Activity-dependent neuroprotective protein (ADNP) differentially interacts with chromatin to regulate genes essential for embryogenesis. Dev Biol 303:814–824
87. Dresner E, Agam G, Gozes I (2011) Activity-dependent neuroprotective protein (ADNP) expression level is correlated with the expression of the sister protein ADNP2: deregulation in schizophrenia. Eur Neuropsychopharmacol 21:355–361
88. Smith-Swintosky VL, Gozes I, Brenneman DE, D'Andrea MR, Plata-Salaman CR (2005) Activity-dependent neurotrophic factor-9 and NAP promote neurite outgrowth in rat hippocampal and cortical cultures. J Mol Neurosci 25:225–238
89. Zemlyak I, Manley N, Sapolsky R, Gozes I (2007) NAP protects hippocampal neurons against multiple toxins. Peptides 28: 2004–2008
90. Gozes I, Divinski I (2004) The femtomolar-acting NAP interacts with microtubules: novel aspects of astrocyte protection. J Alzheimers Dis 6:S37–S41
91. Vulih-Shultzman I, Pinhasov A, Mandel S, Grigoriadis N, Touloumi O, Pittel Z, Gozes I (2007) Activity-dependent neuroprotective protein snippet NAP reduces tau hyperphosphorylation and enhances learning in a novel transgenic mouse model. J Pharmacol Exp Ther 323:438–449
92. Matsuoka Y, Gray AJ, Hirata-Fukae C, Minami SS, Waterhouse EG, Mattson MP, LaFerla FM, Gozes I, Aisen PS (2007) Intranasal NAP administration reduces accumulation of amyloid peptide and tau hyperphosphorylation in a transgenic mouse model of Alzheimer's disease at early pathological stage. J Mol Neurosci 31:165–170
93. Matsuoka Y, Jouroukhin Y, Gray AJ, Ma L, Hirata-Fukae C, Li HF, Feng L, Lecanu L, Walker BR, Planel E, Arancio O, Gozes I, Aisen PS (2008) A neuronal microtubule-interacting agent, NAPVSIPQ, reduces tau

pathology and enhances cognitive function in a mouse model of Alzheimer's disease. J Pharmacol Exp Ther 325:146–153

94. Shiryaev N, Jouroukhin Y, Giladi E, Polyzoidou E, Grigoriadis NC, Rosenmann H, Gozes I (2009) NAP protects memory, increases soluble tau and reduces tau hyperphosphorylation in a tauopathy model. Neurobiol Dis 34:381–388
95. Shiryaev N, Pickman R, Giladi E, Gozes I (2011) Protection against tauopathy by the drug candidates NAP (davunetide) and D-SAL: biochemical, cellular and behavioral aspects. Curr Pharm Des 17:2603–2612
96. Gozes I (2010) Tau pathology and future therapeutics. Curr Alzheimer Res 7:685–696
97. Li X, Kumar Y, Zempel H, Mandelkow EM, Biernat J, Mandelkow E (2011) Novel diffusion barrier for axonal retention of Tau in neurons and its failure in neurodegeneration. EMBO J 30:4825–4837
98. Shprung T, Gozes I (2009) A novel method for analyzing mitochondrial movement: inhibition by paclitaxel in a pheochromocytoma cell model. J Mol Neurosci 37:254–262
99. Zhang B, Maiti A, Shively S, Lakhani F, McDonald-Jones G, Bruce J, Lee EB, Xie SX, Joyce S, Li C, Toleikis PM, Lee VM, Trojanowski JQ (2005) Microtubule-binding drugs offset tau sequestration by stabilizing microtubules and reversing fast axonal transport deficits in a tauopathy model. Proc Natl Acad Sci USA 102:227–231
100. LaPointe NE, Morfini G, Pigino G, Gaisina IN, Kozikowski AP, Binder LI, Brady ST (2009) The amino terminus of tau inhibits kinesin-dependent axonal transport: implications for filament toxicity. J Neurosci Res 87:440–451
101. Massaad CA, Pautler RG (2011) Manganese-enhanced magnetic resonance imaging (MEMRI). Methods Mol Biol 711:145–174
102. Merenlender-Wagner A, Pikman R, Giladi E, Andrieux A, Gozes I (2010) NAP (davunetide) enhances cognitive behavior in the STOP heterozygous mouse—a microtubule-deficient model of schizophrenia. Peptides 31:1368–1373
103. Sudo H, Baas PW (2011) Strategies for diminishing katanin-based loss of microtubules in tauopathic neurodegenerative diseases. Hum Mol Genet 20:763–778
104. Gozes I (2011) Microtubules (tau) as an emerging Therapeutic Target: NAP (davunetide). Curr Pharm Des 17:3413–3417
105. Gozes I (2011) NAP (davunetide) provides functional and structural neuroprotection. Curr Pharm Des 17:1040–1044
106. Gozes I (2011) Microtubules, schizophrenia and cognitive behavior: preclinical development of davunetide (NAP) as a peptide-drug candidate. Peptides 32:428–431
107. Conde C, Caceres A (2009) Microtubule assembly, organization and dynamics in axons and dendrites. Nat Rev Neurosci 10:319–332
108. Janke C, Chloe Bulinski J (2011) Post-translational regulation of the microtubule cytoskeleton: mechanisms and functions. Nat Rev Mol Cell Biol 12:773–786

Chapter 9

Neurofilaments: Properties, Functions, and Regulation

Rodolphe Perrot and Joel Eyer

Abstract

Neuronal intermediate filaments are the most prominent cytoskeleton component of adult neurons in both central and peripheral nervous system. They include neurofilament triplet proteins, peripherin, α-internexin, nestin, and synemin. Although it was initially thought that neuronal intermediate filaments serve a primarily structural function, it has since been demonstrated that they constitute a dynamic network involved in neuronal differentiation, axon outgrowth, and regeneration. Finally, they emerged as a primary cause of some neurodegenerative diseases. Here, we focus on the properties, function, and regulation of neuronal intermediate filaments as well as their relationship to different neurodegenerative diseases.

Key words: Neuronal intermediate filaments, Composition, Molecular biology, Function, Diseases

1. Introduction

Intermediate filaments (IFs) are a ubiquitous component of a wide variety of eukaryotic cells. They represent a family of closely related, nonpolar cytoskeletal filaments intermediate in size (10-nm diameter) between microtubules (MTs) (25-nm diameter) and actin microfilaments (MFs) (7-nm diameter), the other two cytoskeletal components of the cell. In adult central (CNS) and peripheral nervous system (PNS), IFs are the most plentiful cytoskeletal components of large myelinated axons (1). They are principally composed of the NF triplet proteins and also α-internexin in CNS and peripherin in PNS (2, 3). NFs play a central role in growth and maintenance of axonal caliber. Therefore, they indirectly determine the conduction velocity of nerve impulses in large myelinated axons (4). They also contribute to the dynamic properties of the axonal cytoskeleton during neuronal differentiation, axon outgrowth, regeneration, and guidance (5).

Rolf Dermietzel (ed.), *The Cytoskeleton: Imaging, Isolation, and Interaction*, Neuromethods, vol. 79,
DOI 10.1007/978-1-62703-266-7_9, © Springer Science+Business Media, LLC 2013

An increasing body of evidences supports the view that the most common mechanism of chronic neurodegenerative disorders involves abnormal protein production, processing or misfolding, and subsequent accumulation in nervous system. Alterations in the metabolism and/or organization of neuronal IFs are frequently associated, directly or indirectly, with various neurodegenerative diseases, including amyotrophic lateral sclerosis (ALS), Charcot-Marie-Tooth disease (CMT), giant axonal neuropathy (GAN), neuronal intermediate filament inclusion disease (NIFID), Parkinson's disease (PD), diabetic neuropathy, dementia with Lewy bodies, and spinal muscular atrophy (6). While IF abnormalities in neurodegenerative disorders could simply reflect a pathological consequence of neuronal dysfunction, recent studies using transgenic mouse models suggested that IF disorganization itself can also produce deleterious effects and therefore could contribute to the neurodegeneration process.

2. Composition of Neurofilaments

Fibrous networks within neurons, initially named neurofibrils, were first described in the nineteenth century. With the development of electron microscopy, it was shown that neurofibrils were comprised of ~10-nm diameter filaments, which were called NFs and later classified in the IF family. IF proteins form a large family of proteins of various size (40–280 kDa) and primary structure, expressed differentially according to cell types (for recent reviews see refs. (7–9)). They are classified into six types and neurons express differentially several IF proteins depending on their developing stage or their localization in the nervous system: neurofilament (NF) triplet proteins (called NFL (light, 68 kDa), NFM (medium, 160 kDa), and NFH (heavy, 205 kDa); type IV), α-internexin (66 kDa; type IV), peripherin (57 kDa; type III), nestin (200 kDa; type IV), vimentin (57 kDa, type III) syncoilin isoforms (Syncl (64 kDa), Sync2 (64 kDa); type III), and synemin isoforms (Low synemin (41 kDa), Middle or beta synemin (150 kDa), High or alpha synemin (180 kDa); type IV). Originally, it was assumed that NFs were composed only of NFL, NFM, and NFH, but recent studies indicated that other proteins such as α-internexin and peripherin are also co-assembled with NF (2, 10) (Fig. 1). This chapter is focused on NF triplet proteins hereafter referred to as NF subunits or NF proteins, but it is important to keep in mind that other IF proteins also participate to the formation of neuronal IFs.

NFs represent the main cytoskeletal elements in mature neurons. They account for 13% of total proteins and 54% of the Triton-insoluble proteins (11). The molecular weights of NFL, NFM, and NFH predicted from the DNA sequences are, respectively, 61.5,

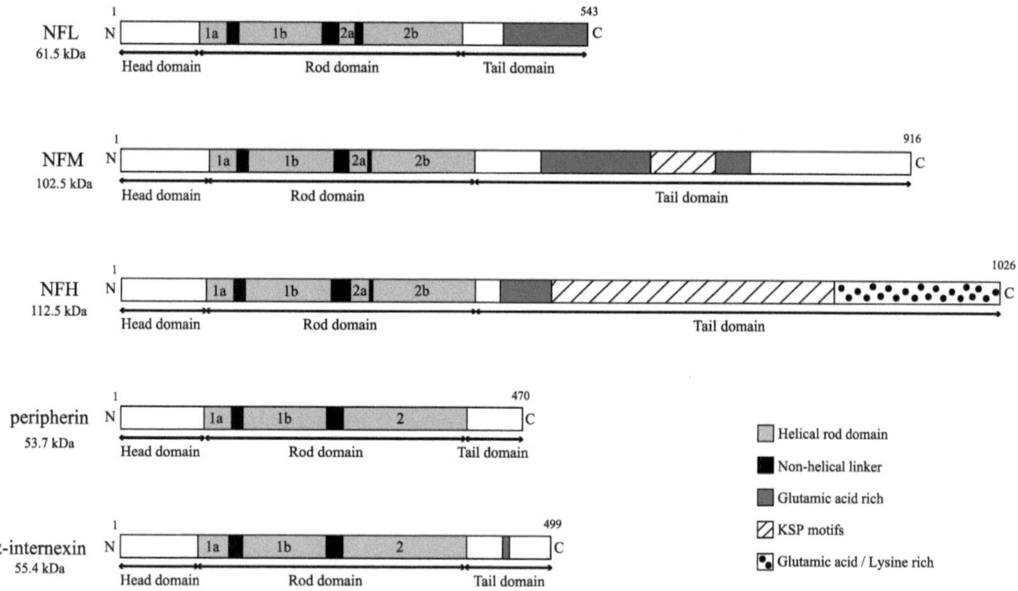

Fig. 1. Schematic representation of human NFL, NFM, NFH, peripherin, and α-internexin proteins. These proteins share a highly conserved central α-helical domain of approximately 310 amino acids that is flanked by non-α-helical amino and carboxy-terminal end domains called, respectively, heads and tails. The tail domains are of variable size and contain many glutamic acid residues for NFL and NFM and also lysine for NFH. The molecular weights predicted from the DNA sequences are shown. C, carboxy-terminal; N, amino-terminal.

102.5, and 112.5 kDa in human. However, due to their high content of negatively charged amino acids (glutamic acids) in their sequences and their extensive posttranslational modifications (phosphorylation and glycosylation), they display higher apparent molecular weights on SDS-PAGE (68, 160, and 205 kDa, for, respectively, NFL, NFM, and NFH). NFM and NFH alone or combinations of NFM and NFH fail to form filaments in the absence of NFL. NFL alone is able to form homopolymers in vitro and in cells transfected with NFL but not in rodents (12–18). As opposed to mouse or rat NFL, human NFL has been reported to form homopolymers in vivo (12). In rodent, NFs are obligate heteropolymers composed of the NFL, NFM, and NFH subunits (19) with a subunit stoechiometry of 4:2:1 (20). This ratio varies during neuronal development. Moreover, other IF proteins such as peripherin in PNS and α-internexin in CNS can co-assemble with NFL, NFM, and NFH to form NFs (2, 10). Today, it must be admitted that these proteins are integral components of NFs, as well as NFL, NFM, and NFH. Yuan et al. (2) demonstrated that α-internexin can be considered as a fourth subunit of NFs in adult CNS. They co-purified α-internexin with the NF triplet proteins from Triton-insoluble fraction in amounts comparable with those of the others subunits. Moreover, α-internexin with all three NF subunits co-assemble into single filamentous network in SW13vim (−) cells and is co-localized with NFM on the same NF in optic nerve in mice. They calculated a subunit stoichiometry of 4:2:2:1 (NFL–α-internexin–NFM–NFH) in optic nerve and spinal cord. Therefore, NF must be considered as the IFs of neurons, which can be composed of up to four or more different proteins depending on neuronal type and developmental stage.

3. Structures of Neurofilaments

Although NF protein chains differ significantly from one another in terms of molecular mass, they share a common tripartite structure with a ~46 nm long central, α-helical, coiled-coil-rich region (the rod domain) composed of approximately 310 amino acids, separating non-helical amino- and carboxy-terminal regions (called the head and tail domains) (21). The central domain contains highly conserved and repeated sequences of seven residues. Typically, the first and fourth residues of these repeats are usually hydrophobic and non-polar, while the fifth and seventh residues are charged. Such motifs are organized into coiled-coil structures, and provide templates for dimerization between two IF polypeptides. The rod domain sequence can be subdivided into four segments (1A, 1B, 2A, and 2B) that are joined together by three non-α-helical linkers (L1, L12, and L2).

Head and tail domains of the NF subunits are less conserved and show a wide range of sizes. The head domain is short and rich for Ser and Thr. The tail domain is the distinctive feature of NF proteins. For NFL, this region is short and contains many glutamic acid residues (E segment), while tail domains of NFM and NFH are longer and also contain E segments. It was suggested that the E-segments may be involved in IF assembly (22). The carboxy-terminal domains of NFM and NFH also contain numerous repeats of phosphorylation sites Lys-Ser-Pro (KSP, up to 51 repeats are present in mouse NFH). In addition, a Lys-Glu-Pro-rich region (KEP segment) exists at the C-terminal ends in NFM and NFH. The Ser residue of these repeats is highly phosphorylated in vivo (23, 24).

4. Assembly of Neurofilaments

As soon as the different NF subunits are expressed they co-assemble to form an IF. This assembly does not require nucleotide binding or hydrolysis, but is strongly dependent on ionic strength, pH, and temperature (25). Regular linear disposition of charged residues in the main coiled-coil segments provides a mechanism to aid assembly through the formation of intermolecular ionic interactions.

The first step of NF formation is the dimerization of NFL with either NFM or NFH via the association of their conserved rod domains to form parallel side-to-side coiled-coil dimers. Two coiled-coil dimers line up in a half-staggered manner, forming an anti-parallel tetramer (26). Then, a rapid lateral aggregation of about eight tetramers occurs to form a unit-length-filament (ULF) of approximately 55 nm in length (27). An elongation step occurs through the axial aggregation of ULF to form immature NF about 16 nm in diameter and many micrometers in length. Radial compaction then takes place resulting in close packing of the molecular filaments to constitute the final 10-nm filament (28, 29).

It has long been considered that NFL subunits constitute the core of the NF, whereas NFM and NFH are arranged around this core. However, immunoelectron microscopy using antibodies directed against head, rod, or tail domains of individual NF subunits showed that all three NF proteins are incorporated integrally into filaments (30, 31). The core of the filament is composed of the central α-helical rod domain of the three proteins, while the sidearm polypeptides arise from the long flexible carboxyl terminal tails of the proteins (15, 32). The sidearm structure, together with the core of the filament, forms the NF brush architecture and modulates the interaction between NFs and other cytoskeletal components or organelles for establishing axonal cytoarchitecture. By using simulation approach, Kim et al. (33) recently tested whether change of NF stoichiometry provokes structural changes of the NF brush. They found that large changes of the in vivo NF stoichiometry have only little effect on the simulated structure of NF sidearms independent of phosphorylation and ionic strength. This suggests that the axonal cytoskeleton is remarkably stable. NFM and NFH tail domains form cross-bridges that create a constant space between NFs.

Head and rod domains of NFL and NFM, and especially their posttranslational modifications, are essential for the NF assembly (34–38). Phosphorylation of Ser-51 and Ser-55 on NFL (39) and phosphorylation of Ser-23 on NFM by protein kinase A (PKA) and protein kinase C (PKC) (40) were shown to regulate in vivo NF assembly. Indeed, the phosphorylation of NFL head domain prevents their assembly, or cause disassembly when incorporated into filaments (41). It is also interesting to note the rapid turnover of the phosphate present on Ser-55 of NFL following NFL synthesis in neurons (39), suggesting a possible role in the blockade of NF assembly before their transport into neurites. The generation of a transgenic mouse with a mutant NFL transgene in which Ser-55 was mutated to Asp to mimic permanent phosphorylation resulted in pathological accumulation of NFs in brain neuronal cell bodies (42). Finally, polymerization of NFL protein in vitro was inhibited by phosphorylation of NFL head domain by protein kinase N (PKN) (43). Together, these results indicate that the transient phosphorylation of head domains in perikarya prevents the polymerization of NF subunits (36). NFM and NFH head domains are also modified by *O*-linked *N*-acetylglucosamine on sites close to phosphorylation sites (44). Similar sites are present in the NFH head and tail domains (45). This proximity could suggest that these posttranslational modifications may influence each other and play a yet unknown role in filament assembly.

Phosphatidylinositol phosphates (PtdInsPs) are additional factors capable of regulating the cytoplasmic assembly of NFs. Indeed, murine NFL directly binds PtdInsPs via its positively charged Arg^{54} (46, 47), and this binding promotes NFL self-assembly in vitro. Mutant NFL (R53A/R54A) proteins lacking binding affinity to PtdInsPs did not have the same effect, but these mutant proteins showed greater self-assembly than the wild-type in the absence of any PtdInsP. Altogether, these results suggest that Arg^{54} plays an essential role in NFL self-assembly by binding with PtdInsPs.

Finally, it was shown that NUDEL, a mammalian homologue of the *Aspergillus nidulans* nuclear distribution molecule NudE, is involved in NF assembly (48). This protein associates directly with soluble pool of NFL and indirectly with NFH subunit. By interacting with NFL, NUDEL promotes the incorporation of NF subunits in the network during NF assembly but does not assemble with NF proteins. Moreover, genetic knockdown of NUDEL disrupts NF stoichiometry, which, in turn, results in impaired NF assembly and transport (48).

In vitro, interactions occurring between NFs allow the formation of a viscoelastic network resistant to important deformation, suggesting the importance of cross-bridges for NF mechanical properties (49, 50). Kreplak et al. (51) used atomic force microscopy to test the mechanical properties of single NF. They showed that NFs can be stretched more than threefold, with an average of 2.6-fold, suggesting that NFs may indeed function as mechanical shock absorbers in vivo.

5. Neurofilament Protein Expression

In human, genes coding for NFL and NFM (*NEFL* and *NEFM* genes) are very closely linked on chromosome 8 (8p21) (52, 53), while NFH gene (*NEFH* gene) is located on chromosome 22 (22q12.2) (54). In mice, *NEFL* and *NEFM* genes are located on chromosome 14 (14 D3 for *NEFL* and 14 D1 for *NEFM*), while *NEFH* gene is located on chromosome 11 (11 A1-A5). The expression of the three NF subunits is under tight spatial and temporal control, especially during nervous system development, and coincides with neuron differentiation (5). This control is governed by a complex interplay of transcriptional and posttranscriptional gene regulatory mechanisms, including regulation of mRNA localization, stability, and translational efficiency (for recent review see ref. (55)). Peripherin is the first IF expressed in differentiated neurons during early neurite outgrowth. Then, peripherin becomes replaced by NFL when axons reach their targets (56). This exchange of NFL for peripherin is thought to reflect different structural properties of the two subunits, with NFL more conducive for a stable cytoskeleton and expansion of axon caliber in adults and peripherin for growing neuronal processes (57). NFL expression is rapidly followed by that of NFM (58–63). The appearance of NFH occurs after synaptogenesis and is accompanied by parallel increases in expressions of NFL and NFM during myelination (58). At this advanced stage, NFs of mature composition may further enhance axon stability and caliber. Interestingly, the increased NF density is accompanied by a decreased MT density (64), suggesting a fine balance between these two networks. In cultured neuroblastoma cells, the level of mRNAs coding for NFs is proportional to their protein levels, and NFL and NFM mRNAs appear several days before the expression of NFH mRNA. The sequential appearance of NF proteins follows that of their mRNAs, with an early expression of NFL and NFM and a later expression of NFH (65). Apart from Giasson and Mushynski (66), several studies observed that the expression levels of NFL and NFM are mutually regulated and independently from NFH (67, 68). This could be related to their genomic localization, but the molecular mechanisms are still unknown.

The expression of NF proteins has also been studied in the human brain, at the different stages of development. Zecevic et al. (69) demonstrated that the human primordial plexiform layer which gives rise to Layer I and the subplate was immunoreactive for phosphorylated NFs (SMI-31) at 4–5 gestation weeks (GW). The multipolar Cajal–Retzius cells in Layer I (70) and the horizontal plexus which they give rise to (71, 72) were positive for SMI-31 in different parts of the human cortex between 20 and 24 GW (73, 74). Further, Haynes et al. (73) have demonstrated the presence of NFM in the white matter of the human parietal lobe as early as 23 GW, whereas NFH can be localized to axons in this part of the cortex at 24 GW.

Axotomy induces an immediate decline of NF mRNAs and proteins in PNS (75–77) and CNS (78–81), leading to reduced levels of axonally transported NFs in injured neurons (76, 82, 83). Then, during regeneration of injured axons in PNS the expression of NF subunits is strongly upregulated (83–86). This reversion does not occur when regeneration is prevented (87, 88) as well as in mammalian CNS axons, which normally do not regenerate (80, 81). However, in transected axons of the spinal cord from lamprey, where NFM expression is initially suppressed, Jacobs et al. (89) observed an upregulation of this subunit only in axons that successfully regenerate, while NFM levels remain low in those that do not regenerate.

A similar effect was recently highlighted for the other NF subunits expressed by lamprey (90). Indeed, after spinal cord transection, expression of NF proteins first decreases, and then is recovered selectively in neurons whose axons have a high probability of regenerating. Similarly, increased amounts of NFM mRNA and protein are observed during successful optic nerve regeneration in *Xenopus laevis* (91). Altogether, these results are consistent with a possible role of NFs in the mechanism of axon regeneration. In agreement with this, sciatic nerve crush experiments have revealed that axonal degenerative and regenerative mechanisms are vastly different in the absence, or presence, of NFL (68, 92). NFL interacts with Schwann cells via Mtmr2 proteins in PNS (93) and it cannot be excluded that NFL behavior during the regenerating process in myelinated axons may to some extent reflect activity that is modulated by structural proteins found in myelin.

6. Post-translational Modifications of Neurofilaments

6.1. Phosphorylation of Neurofilaments

The phosphorylation is the best documented posttranslational modification of NF proteins. Multiple aspects of NF's biology, including their assembly and their axonal transport, are regulated by their phosphorylation status. Moreover, aberrant NF phosphorylation is a pathological hallmark of many human neurodegenerative disorders (for detailed reviews see refs. (94, 95).

NF proteins are the most extensively phosphorylated proteins in neurons with up to 51 sites of phosphorylation located on the C-terminal domain of the mouse NFH (23, 96–99). This phosphorylation is topographically regulated, with a proximo-distal gradient consisting by an intense phosphorylation in axons and little or no phosphorylation in cell bodies and dendrites (5, 100–103). Phosphorylation sites are located on the amino-terminal and carboxy-terminal domains of the three NF subunits. These sites are the targets of, respectively, second messenger-dependent kinases (104–106) and second messenger-independent kinases (24, 105, 106). Phosphorylation of head domain arises mainly in cell body soon after the synthesis of NF proteins, while the phosphorylation of tail domains coincides with their entry into the axon (107–109).

As mentioned above, phosphorylation of NFL and NFM head domains by PKA, PKC, and PKN prevents the assembly of NFs or leads to their disassembly (39, 40, 43). In vitro studies also revealed Rho-associated kinases and calcium/calmodulin-dependent protein kinase II (CaMKII) phosphorylation sites in the NFL head domain (110, 111). NFM and NFH tail domains are the most extensively phosphorylated regions. Most of these phosphorylation sites are KSP repeats, although other Ser/Thr containing motifs are also phosphorylated. Many roles have been attributed to the phosphorylation of KSP motifs, including the formation of cross-bridges between NFs or with MTs, the expansion of the axonal caliber, the slowing of the NF axonal transport and the integration of NFs in a stationary pool (16, 52, 100, 112–117). Phosphorylation of KSP sites depends on two families of Pro-directed kinases: the cyclin-kinase Cdk5 and the microtubule-associated protein (MAP) kinases. Signals transduction cascades leading to the activation of these kinases could be triggered by growth factors (118, 119), Ca^{2+} influx (120), integrins (121), and myelination (101). The link between the myelination and the phosphorylation of NFM and NFH sidearms was first suggested by the decreased phosphorylation of NFs in the dysmyelinated mouse mutant *Trembler* (122) and by their poor phosphorylation in non-myelinated axonal domains like the initial segment and nodes of Ranvier (122–125). The binding of myelin-associated glycoprotein (MAG) to axonal receptors was proposed to activate a signalization cascade leading to the phosphorylation of NFs in myelinated regions (126, 127), but the molecular cascade of this process and its regulation are still unclear.

Cdk5 preferentially phosphorylates KSPXK motifs of NFH in vitro and in vivo (128–135), preventing the binding of dephosphorylated NFH to MTs (136). But the majority of KSP repeats in rat/mouse NF tail domains are phosphorylated by MAP kinases, including extracellular signal-regulated kinases 1 and 2 (Erk1/2) which phosphorylates KSPXXK and KSPXXXK motifs of NFH (135), stress-activated protein kinase (SAPK) which is responsible for NFH tail domain phosphorylation on KSPXE motifs in cell body under stress-activated conditions (66, 137), glycogen synthetase kinase 3 (GSK3) which phosphorylates some of the KSP sites in bovine NFM (138), and few sites on NFH (129), p38 kinase (139, 140), and c-Jun N terminus kinase 1 and 3 (JNK1/3) (137, 141, 142). Ser/Thr residues in the glutamic acid region of the three subunits tail domains are phosphorylated by casein kinase I (CKI) (105, 143–147), while casein kinase II (CKII) also phosphorylates Ser-473 on the short tail domain of NFL (148, 149).

Phosphorylation of head and tail domains are intimately related. Zheng et al. (150) showed that phosphorylation of NFM head domain by PKA reduces the phosphorylation of tail domain by MAP kinases in vitro and in vivo. Moreover, mutation of Ser-1, 23 and 46 residues to Ala in the head domain of NFM prevents PKA phosphorylation in transfected NIH3T3 cells and fails to inhibit tail domain phosphorylation by MAP kinases (150). These results highlight a regulatory mechanism by which phosphorylation of NF head domains could prevent NF assembly and C-terminal phosphorylation in cell body, protecting the neuron from abnormal accumulation of phosphorylated NFs in perikarya.

The phosphorylation state of NF proteins in the different neuronal compartments depends on a dynamic balance between the activities of kinases and phosphatases. Since NF head domain phosphorylation inhibits NF assembly (39–41), their dephosphorylation is necessary to allow the polymerization of NF proteins prior to their transport into the axon. The extensive enzymatic dephosphorylation of NFs induces a progressive loss of their capacity to interconnect in vitro into a reticulated network, measured by the formation of highly viscous gels in purified preparations of NFs (151). Finally, dephosphorylation of NF tail domains facilitates their degradation at the terminals (152) and regulates their interaction with other cytoskeletal proteins. Dephosphorylation of head and tail domains of NF subunits is mainly (60%) catalyzed by phosphatase 2A (PP2A) (153–156). PP1 also contributes to the dephosphorylation of NFs but to a lower extent (10–20%) (155).

An interesting finding was that insulin synthesized by neurons is a factor that regulates NF and tau phosphorylation via MAP kinases and JNK, promoting neuronal differentiation (157, 158). The analysis of the effects of insulin deficiency within the brain in insulin knockout mice (I−/−) revealed that lack of insulin brain stimulation induces JNK hyperphosphorylation followed by hyperphosphorylation of tau and NF, and ultrastructural cellular damage, which over time may induce decrease in cognition and learning disabilities (159).

6.2. Glycosylation and Glycation of Neurofilaments

NFs are also posttranslationally modified by attachment of *O*-linked *N*-acetylglucosamine (*O*-GlcNAc) to individual Ser and Thr residues. *O*-GlcNAc is a common modification of cytosolic and nuclear proteins that regulates protein stability, subcellular localization, and protein–protein interactions (160). Like phosphorylation, *O*-glycosylation is dynamic and often reciprocal to phosphorylation at the same sites or adjacent to them. Dong et al. (44, 45) identified several *O*-GlcNAc sites on NFL head domain (Thr-21, Ser-27, Ser-34, and Ser-48), NFM head (Thr-19, Ser-34, and Thr-48), and tail domains (Thr-431). NFH is also extensively modified by *O*-GlcNAc in the head domain (Thr-53, Ser-54, and Ser-56) and at multiple sites within the KSP repeat motifs in the tail domain, although the exact sites remain to be identified. In purified NF proteins, the *O*-GlcNAc modifications occur at a stoechiometry of approximately 0.1, 0.15, and 0.3 mol of GlcNAc/mol of, respectively, NFL, NFM, and NFH (44, 45). The function of these modifications is still elusive, but several clues suggest a role in the NF assembly. For example, all *O*-glycosylation sites within head domains are located in regions essential for in vivo NF assembly, close to the phosphorylation sites involved in this process. *O*-glycosylation of NF head domains could reciprocally modulate its phosphorylation, and consequently the assembly and dynamics of NFs. Further investigations are necessary to elucidate the precise mechanism regulating NFs *O*-glycosylation, the relation between NF phosphorylation and *O*-glycosylation and the distribution of *O*-glycosylated NFs. Antibodies that specifically recognize *O*-glycosylated epitopes in NF subunits could be an important tool to elucidate these questions. To this end, Lüdemann et al. (161) generated a monoclonal antibody specifically directed against an *O*-glycosylated epitope in the tail domain of NFM. They showed that *O*-glycosylated NFM is enriched in the axons of human neurons in situ, together with hyperphosphorylated NFs, indicating a synchronous phosphorylation and *O*-glycosylation of the tail domain of NFM within the axon. However, the *O*-glycosylation of NFM and the activity of MAP kinases are reversely regulated, suggesting reciprocal regulation between phosphorylation of the KSP region and *O*-glycosylation. The first evidence of NF glycation, also called nonenzymatic glycosylation, was reported in peripheral nerves in diabetes mellitus (162). A possible role for this modification in familial and sporadic ALS was also suggested (163).

6.3. Nitration, Oxidation, and Ubiquitination of Neurofilaments

In addition to phosphorylation and *O*-glycosylation, NFs undergo nitration, oxidation, and ubiquitination. Nitration of NFL subunit was reported in the normal rat brain using a proteomic analysis (164). NF nitration was also detected in NF-rich inclusions in motoneurons of sporadic ALS cases (165). This modification is catalyzed by superoxide dismutase 1 (SOD1) in vitro on four Tyr residues of NFL, one in the head domain (Tyr-17) and three in the rod domain (Tyr-138, Tyr-177, and Tyr-265) (166). It is interesting to note that Tyr-17 is essential for the polymerization of NFs, while the other three Tyr residues are located within the coiled-coil structure of the rod domain and are likely involved in intermolecular hydrophobic interactions. The nitration change normally hydrophobic residues into negatively charged hydrophilic residues, thereby disrupting the assembly and stability of NFs. Consequently, it was proposed that SOD1-catalyzed nitration of NFs may contribute to motoneurons dysfunction in ALS (166, 167).

The oxidation of NFs arises during aging, Wallerian degeneration, and was found in neurodegenerative disorders. Oxidized NFs are more susceptible to calpain proteolysis and form dense aggregates and bundles of laterally aggregated filaments (168). The incubation of disassembled NFL with SOD1 and H_2O_2 causes the formation of dityrosine crosslinks and the aggregation of NFL protein proportionally to the concentration of hydrogen peroxide (169). Antioxidant molecules inhibit these effects. Finally, ubiquitination of NF proteins has been suggested by Gou and Leterrier (170) as a possible mechanism involved in NFs degradation.

7. Degradation of Neurofilaments

After their assembly in the perikaryon, NFs are slowly transported by the slow axonal transport toward the nerve terminal (171) where they are degraded. A degradation of NFs over the entire length of axons in mouse sciatic nerve has also been reported during Wallerian degeneration in transected fragments of nerve (172) and in a context of axonal NFs deficiency (173). Nixon and Logvinenko (117) suggested that such a degradation of NF proteins in axons could account for nonhomogeneous distribution of NFs in axons.

Calcium-activated proteases such as calpain are found in human tissues and degrade NFs from squid, rat, bovine, and worm (174–178). NF proteins that are usually absent in synaptic terminals accumulate following leupeptin treatment, a protease inhibitor. This indicates that their normal absence in synapses is due to a degradation by calcium-activated proteases (179). The initial evidence that calcium plays a role in the degradation of NFs came from observations showing disintegration of NFs in rat peripheral nerve fibers after exposure to calcium (180). Purification of calcium-activated neutral proteases in rat peripheral nerve or spinal cord resulted in the identification of calpain I (or μ-calpain) and calpain II (or m-calpain), as defined, respectively, by the μM or mM levels of calcium required for their activation (181, 182). Calpain II is present in glial cells, while calpain I is predominantly expressed in neurons (183). At endogenous sub-micromolar calcium concentrations, Nixon et al. (184) demonstrated a limited proteolysis of NFM as a posttranslational modification during the axonal transport.

Most calcium-activated neutral proteases show a high degree of substrate specificity with IFs (185). Unlike many proteases, calpain specificity appears to be determined by conformational factors and primary amino acid sequences. It usually catalyzes limited cleavage of its substrates. Participation of calpain and calcium-activated neutral proteases in NFs turnover is plausible as reflected by the occurrence in normal brain of characteristic NF protease-resistant fragments throughout the neuraxis. These fragments could be retrogradely transported, and could regulate the synthesis, assembly and delivery of NFs in accordance with their turnover level at remote sites (178, 179). Inhibition of calpain proteolytic activity in transected axon abolishes growth cone formation suggesting a central role in the reorganization of the axonal cytoskeleton during its transition from a stable differentiated state into a dynamically extending structure (186). NFs are also degraded by nonspecific proteases like lysosomial cathepsin D, trypsin, and α-chymotrypsin. Such trypsin proteolytic strategies were used to analyze the spatial architecture of NFs (187, 188). Cathepsin D plays an important role in NF metabolism. The content of cathepsin D is probably more than 1,000-fold greater than that of a calcium-activated neutral proteases (189).

Purified brain cathepsin D was shown to degrade NF proteins from rat, mouse, bovine, and human tissues, and some characteristic fragments produced by this hydrolysis were shown to be normally present in brain (190–192).

In addition to their participation in the turnover process, proteases produce NF-derived peptides that could be active. The hypothesis of a possible regulation of gene expression by NF proteins led Traub et al. (193) to show that subunit proteins of NF bind to RNA and single-stranded DNA. The DNA-binding sites are located in the amino-terminal domain (194) and are preserved during the digestion of NFs by calcium-activated proteases. However, the capacity of NFs to affect the DNA or RNA fragments to which they bind is unknown.

Posttranslational modifications can regulate proteolysis of NFs, as illustrated for phosphorylation which protect NFs from proteolysis by calpains (152, 195). Aluminum, a neurotoxin which causes NF protein phosphorylation and accumulation in neuronal perikarya (196, 197), inhibits calpain-mediated proteolysis of NFs (198). These results raise the possibility that kinase or phosphatase activity might determine the rate of turnover of NF proteins. It was also suggested that ubiquitination of NFs facilitates their proteolysis (170). Wang et al. (199) recently examined in detail the role of ubiquitin-proteasome system (UPS) in NFs degradation. They showed the association of the E3 ligase carboxyl-terminus of Hsc70 interacting protein (CHIP) with NFM, thereby enhancing NFM ubiquitination and degradation after 2,5-hexanedione (HD) treatment. The co-transfection of HEK293 cells with Myc-CHIP and Flag-NFM plasmids revealed that high expression of CHIP down-regulated levels of NFM and promoted its ubiquitination. In contrast, application of proteasome inhibitor MG132 partly reversed the decrease of NFM. These results indicated that the CHIP ubiquitin ligase complex interacted with and repressed NFs by targeting NFs for ubiquitin-mediated proteolysis (199).

Calcium-activated neutral proteases play important roles in tissue injury. Many pathological states induce an increase of free calcium within the axon leading to a massive proteolysis of NFs (200, 201). The degradation of NFs in transected rat sciatic nerve is reduced if the influx of calcium into the axoplasm is prevented or if calcium-activated neutral proteases activities are inhibited (172). Such enzymatic fragmentation in transected nerves generates protease-resistant NF fragments which may represent, after externalization into the endoneurium, a mechanism responsible for the generation of autoantibodies to NF proteins detected in Parkinson's disease and in several neurological diseases (202–205).

Abnormal aggregations of NFs are a hallmark of several human neuropathological situations. In ALS, they accumulate in cell bodies or in the proximal part of axons from motor neurons (206, 207). They accumulate in Lewy bodies of PD (208), in neurofibrillary tangles of Alzheimer's Disease (AD) (209), and following intoxication by aluminum, HD, acrylamide or β, β -Iminodipropionitrile (IDPN) (210–213). In all these pathological situations the cellular and molecular mechanisms used to eliminate the neurofilamentous aggregates are still unknown. It has been shown that trypsin-like proteases are expressed in neurons (214–216), and they accumulate within pathological neurofilamentous aggregates (217). Tsuji et al. (218) showed increased levels of calpains in the cytosolic fraction of AD brains when compared to control brains. Fasani et al. (219) showed that NFs isolated from NFH-LacZ transgenic mice (in which NFs are sequestered in cell bodies (220)) are more sensitive to exogenous trypsin and α-chymotrypsin than normal NFs. Moreover, an increased trypsin immunolabeling is detected in perikarya from such mice compared to wild-type animals. These results suggest that when NFs are sequestered in the cell body, their amount is controlled by an increased susceptibility to trypsin-like proteolysis, and an increased production of proteases (219).

As mentioned above, it has long been considered that NF degradation only occurs at the axon terminals. However, a degradation of NFs over the entire length of axons in mouse sciatic nerve was recently highlighted by Millecamps et al. (173). They generated transgenic mice with doxycycline-regulated expression of human NFL (hNFL) with or without endogenous mouse NFL proteins (respectively, tTA;hNFL;NFL$^{+/-}$ and tTA;hNFL;NFL$^{-/-}$ mice). The doxycycline administration in drinking water of tTA;hNFL;NFL$^{-/-}$ mice silences the expression of hNFL and the preexisting protein subsequently disappears in their sciatic nerve, with an estimated half-life of 3 weeks. This loss is synchronized over the entire length of the sciatic nerve, suggesting an homogeneous and not a local degradation of hNFL. In contrast, no detectable loss of hNFL protein was observed in the presence of stationary NF network in sciatic nerve from tTA;hNFL;NFL$^{+/-}$ mice, even after 4.5 months of doxycyline treatment. After 8 months of treatment, 35% of hNFL protein was still remaining (173). These results show that a preexisting NF network is a key determinant of half-lives of NF proteins by reducing their turnover rate and support the view that NF proteins can spend several months, if not years, in long NF-rich peripheral axons. This long life makes them potential targets for oxidation or other harmful modifications, which in turn may cause NF disorganization.

8. Interaction of Neurofilaments with Proteins and Organelles

As a part of a complex and dynamic network, NFs interact with several partners and these interactions are principally mediated through NF-associated proteins that can modulate function (enzyme) and structure (linker protein) of NFs. Linker proteins are responsible for the interaction between the different filaments or organelles, whereas enzymes (principally kinases and phosphatases) modulate NF architecture, assembly and spacing.

In neurons, NFs and MTs are two major components of the cytoskeleton. Dynamic interactions between these elements are crucial for the axoskeleton and are regulated mostly by phosphorylation. This was documented by several biochemical studies showing that tubulin and/or MTs are able to interact directly with NFs both in vitro and in vivo (221–227). This direct interaction was also illustrated by quick freeze deep etch electron microscopy (32, 228). Dephosphorylation of NFs by alkaline phosphatase promotes the NF–MT interaction mediated by the carboxy-terminal domain of NFH (229), suggesting spatiotemporal regulation of NF–MT interaction by kinases. Tau protein kinase II-specific phosphorylation of NFH tail domain has been described to dissociate NFs and MTs (136). If NFs interact with MTs, NF amino-terminal head domains also bind unassembled tubulin, inhibiting MT polymerization (230). This suggests a model in which axonal NFs modulate local MT assembly.

Although these data showed a direct interaction between NFs and MTs, connections between these structures appeared to primarily be mediated indirectly. For example, MAPs are proteins that link MTs to NFs. MAP2 plays an important role in the stabilization of MTs (231) and is also able to bind NFL (223) to form cross-bridges between NFs and MTs in dendrites (232). In addition to MAP2, a second MAP that may interact with both NFs and MTs is STOP protein (for Stable Tubule Only Polypeptide, named also MTAP6 or MAP6). STOP proteins constitute a family of cytoskeleton-associated proteins responsible for the MT cold stability. They were initially isolated from rat brain cold-stable MTs (233, 234), and were shown to induce MT cold stability in vitro when added to labile MTs (235, 236), or when expressed in cells normally devoid of stable MTs (237). They are abundant in neurons, which contain a large amount of stable MTs (238, 239), and associate preferentially to cold- and drug-stable polymers (240). STOP proteins were shown to be associated with NFs by both biochemical co-purification of NFs, and their co-precipitation with NFs in axonal spheroids of ALS (241). In NFH-LacZ transgenic mice STOP proteins were also found to co-accumulate with NFs in the perikaryon. While further investigations are necessary to understand the mechanism of STOP–NF interactions and their functional relevance, these data indicate that STOP proteins could be considered as a cytoskeletal integrator and a marker of spheroids in ALS (241).

Several studies indicate a direct interaction between motor protein dynein and kinesin with NFs (242–247). Kinesins are a class of protein that mediate active transport toward the plus end (growing end) of MTs, while dyneins mediate the active transport toward the MT minus end. The interaction between these molecular motors and NFs is mediated principally by the phosphorylated carboxy-terminal domains of NFM and NFH (242, 244, 247). These motor proteins contribute to the transport of NFs along axons and dendrites (for review see ref. (248)). For example, co-immunoprecipitation experiments with anti-dynein antibody induced selectively co-precipitation of phosphorylated NFs, while anti-kinesin selectively co-precipitated hypophosphorylated NFs (247). Recently, atomic force microscopy allowed a direct evaluation of the interaction between NFs and cytoplasmic dynein (244). Yeast two-hybrid and affinity chromatography assays also identified a direct binding between dynein intermediate chain and NFM (244) possibly involved in the saltatory bidirectional axonal transport of NFs in the neuron. Myosin Va is another motor protein that interacts with NFs. In contrast to kinesin and dynein/dynactin motors that are powered by MT-dependent mechanisms, myosin motor proteins move their cargoes by a "hand-over-hand" mechanism along MFs (249). Myosin Va in neurons is believed to transport synaptic vesicles, endoplasmic reticulum (ER), mitochondria, and membrane bound vesicles along axons and within synaptic terminals, and to facilitate the accumulation of mRNA/protein complexes in dendritic spines (250). The N-terminal motor domain of myosin Va binds directly NFL rod domain, which is essential for maintaining normal myosin Va levels, while loss of myosin Va leads to altered NF organization in axons (251, 252). Moreover, the loss of NFL and myosin Va from axons reduces the axonal content of ER, and redistributes ER toward the actin-rich subaxolemmal region (252). These results provided evidence that myosin Va binding to NFL modulates the distribution of vesicular organelles in axons, suggesting a novel function for NFs as a scaffold in axons for maintaining the content and proper distribution of vesicular organelles, mediated in part by myosin Va.

In addition to MT and MF cytoskeleton networks, NFs can be highly crosslinked and connected to organelles, membranes, and other parts of the cytoskeleton. These connections are mediated in part by large scaffolding proteins. Plectin, a member of the plakin family, is suspected for cross-linking NFs and MTs in neurons (253). BPAG1/dystonin (Bullous Pemphigoid Antigen 1,280 kDa) has originally been described as a major cytoskeletal integrator connecting actin filaments to NFs and peripherin-containing filaments (254, 255). In BPAG1−/− mice, electron-microscope analysis along sensory axons revealed regional swellings filled with lysosomal vesicles and disorganized arrays of NFs (254, 256). Co-transfection of BPAG1n with NFL and NFH in SW13 cells show that BPAG1n is able to link NFs and actin networks (254). Natural mutations of BPAG1 (257) cause the mouse model of neurological disorder dystonia muscularum (dt/dt) characterized by a disorganized cytoskeleton in the sensory nervous system. It was thus suggested that a loss of BPAG1 interactions with IFs is important to the pathogenesis in dt/dt mice. However, Eyer et al. (258) showed that pathogenesis of axonopathy in dt/dt mice was independent of axonal NFs as demonstrated by matting dt/dt mice and NFH-LacZ transgenic mice in which NFs aggregated in the cell body of neurons (220). Similar results were obtained by matting dt/dt mice with KO NFL mice (259), confirming that the presence or absence of NFs in the axon is inconsequential to the development of the dt/dt disorder. Moreover, the ability of the neuronal splice isoform of BPAG1 to connect actin filaments to NFs was called into question by works of Leung et al. (260). Indeed, they found that the only neuronal isoform of BPAG1 (named BPAG1a) lacks an IF binding domain. This called into question whether BPAG1 actually does serve as an IF-associated protein in neurons. However, BPAG1a has an actin binding domain and an MT binding domain, suggesting that this protein can play a role in maintaining the structural organization of the neuronal cytoskeleton (261).

Electron microscopical investigation in Dieters' neurons from lateral vestibular nucleus of rabbit brain allowed to visualize MF-mediated interaction of NFs with the plasma membrane and the nuclear pores (262). These authors suggested that NFs may play a linking role between plasma membrane and nucleus. As a matter of fact, during membrane depolarization early transduction signal mediator such as Ca^{2+} or cAMP could mediate NFs rearrangement and contribute to a modification of the DNA transcription at the site of the nuclear pore complex (262). Moreover, it has been shown that NFs were able to bind DNA, RNA, and histone H1 nuclear protein, suggesting a possible role of NFs in the regulation of transcription processes (193, 194, 263).

Interaction of NFs with vesicles is mediated by associated proteins. Biochemical preparations such as synapsin I immunoprecipitation (264), or nearest neighbor analysis for brain synapsin I (265), have shown that this protein can link nonsecretory vesicles directly to NFL subunit. By using electron microscopy on transected squid giant axons, it has been shown that NFs formed dynamic complex with smooth endoplasmic reticulum. Modification and rearrangement of such complex could be involved in some neurodegenerative diseases (266). Cross-bridges between mitochondria and NFs were also emphasized by ultrastructural studies (228, 267). Since mitochondria do not move along NFs in neuronal processes deprived of MTs and MFs (268), it was postulated that NFs serve as a docking site for these organelles and regulate their spatial distribution along axons. In agreement with this, durations and frequencies of pauses during the axonal transport of mitochondria are reduced in axons deprived of NFs in vitro (269). However, the fact that the proportion of stationary mitochondria was unchanged in axons deprived of NFs compared to control axons suggested that NFs would serve as a transient but not as a stable anchoring site for mitochondria. The overexpression of NFH in culture cells induces the selective perikaryal retention of mitochondria (270, 271), suggesting that regulated NF-mitochondria binding is required for normal translocation of mitochondria in axons. Wagner et al. (272) provided evidences that NFM and NFH sidearms mediate this interaction between NFs and mitochondria and proposed that porin molecules or other cytoskeleton-binding proteins of the mitochondrial outer membrane mediate this interaction. They also revealed that the binding of NFs to mitochondria depends on the potential of the mitochondrial membranes, suggesting that the conformation or the organization of the partner of NFs on the mitochondrial outer membrane is modified by the membrane potential of the organelle. Finally, it seems that phosphorylation of NFs may regulate their binding affinity for mitochondria in a potential-dependent manner (272).

9. Role of Neurofilaments in Development and Maintenance of Axons

NFs play an essential role in growth and maintenance of axonal caliber. This was first suggested when Friede and Samorajski (273) observed that increased NF numbers and densities are correlated with increased axonal calibers. Moreover, the axonal radial growth coincides with the entry of NFs into axons during axonal development or regeneration (274, 275). The recent use of various animal models clearly showed the importance of NFs in the control of the axonal diameter. These models include mice knockout for NF genes, mice expressing human, mouse and modified NF subunits. But the first evidence of the implication of NFs in axonal radial growth in an animal model was obtained in Japanese quails. Indeed, an important axonal atrophy was observed in *quiver* quails characterized by the absence of NFs in their axons caused by nonsense mutation in *NFL* gene (276–278). The first evidence in mice was provided by the production of transgenic mice overexpressing human NFH (279). These mice develop perikaryal accumulations of NFs, resulting in a deficiency of axonal NFs and in an axonal atrophy. This was further confirmed by the generation of NFH-LacZ transgenic mice in which expression of an NFH-β-galactosidase fusion protein provokes the perikaryal aggregation of NFs, leading to a 50% reduction of axonal calibers (220, 280). Finally, the targeted disruption of *NFL* gene in mice caused the lack of axonal NFs and strongly reduced diameter of myelinated axons (68). Multiple studies analyzed the importance of the number and stoichiometric proportion of each NF subunit in the axonal caliber determination. Transgenic mice overexpressing murine NFL have a two- to threefold increase in the number of NFs but the diameter of their axons is only slightly modified (281, 282), suggesting that the number of NFs by itself is not the main determinant of axonal diameter. Yet, triple heterozygous knockout mice (NFL+/–, NFM+/–, NFH+/–) in which integrity of NFs network and normal subunit stoichiometry are preserved exhibit a 40% decrease of NFs content and a 50% decrease of axonal diameter in L5 ventral root (283). The individual increase in each of the three NF subunits inhibits radial axonal growth, and the simultaneous increase of NFM and NFH exacerbates this axonal atrophy (282). In contrast, the co-overexpression of either NFL/NFM or NFL/NFH increases the axonal caliber (282, 284), suggesting that NFL in combination with either NFM or NFH is sufficient to promote radial growth. These results collectively indicate that both number of NFs and a precise stoichiometry of their subunits are essential in the expansion of axonal diameter. To determine the specific contributions of NFM and NFH subunits in the axonal size, Elder et al. (67) made null mutant mice deficient for NFM. The axonal caliber was strongly reduced as well as NFL mRNA and protein levels. Modification of NFM expression was also reported in NFL–/– mice (68), suggesting that the levels of NFL and NFM are mutually regulated, reinforcing the view that the stoichiometry of each NF subunit is crucial for the establishment of a proper axonal caliber. Three different NFH-null mice were produced and revealed that this subunit contributes to a lesser extent than NFM to the determination of the axonal diameter (285–287). However, subtle differences exist between these models. Indeed, Elder et al. (285) reported no modification in NFM protein level but a significant reduction of the axonal caliber, while Rao et al. (286) and Zhu et al. (287) observed an upregulation of NFM (respectively, 100% and 20%) but only minor modification of the axonal diameter, with a slight decrease in the large-caliber axons. As suggested by Hirokawa and Takeda (288), this divergence could be explained in part by the chronological differences between the data ((286) and (287): 3 months; (285): 4 months), suggesting a later effect of NFH on the axonal radial growth. It is also important to note that Rao et al. (286) and Zhu et al. (287) reported a compensatory increase in MT density and NFM phosphorylation in their NFH-null mice, complicating the conclusions about the exact role of NFH in axonal radial growth. Moreover, these analyses of NFH deleted mice have focused on proximal axonal segments from 1 to 4 months. To better understand the role of NFH in structuring axons, Shen et al. (289) examined the effects of NFH loss in proximal and distal segments of the sciatic nerve and along the entire length of the phrenic nerve. Deletion of NFH reduced axonal diameters and NF number in proximal nerve segments. However, NF spacing was greater in proximal versus distal phrenic nerve segments. These results suggest that loss of NFH reduces radial growth in proximal axonal segments by reducing the accumulation of NFs. As NFH expression is developmentally delayed (58, 59, 61), it seems that without NFH, the NF network is established in a distal to proximal gradient perhaps to allow distal axonal segments to develop prior to proximal segments (289).

It has long been suspected that the phosphorylation status of NFM and NFH carboxy-terminal domains controls axon caliber by regulating NF transport (115, 116) and/or interfilament spacing (99, 101, 275, 290). KSP repeated motifs on NFM and NFH sidearms are variably phosphorylated, principally after the entry of NFs into the axon (100, 102, 103, 109). It has been suggested that phosphorylation of KSP repeats could increase the total negative charges on sidearms and thus causes their lateral extension by repulsive interactions (52, 100), increasing NFs spacing and axonal caliber. In agreement with this assumption, Brown and Hoh (291) used atomic force microscopy to show the presence of a weak repulsive force around the core of the filament. Moreover, this repulsive force is absent in homopolymers of NFL or trypsinized native filaments which lack the sidearms present in native filaments and attenuated when the filaments are enzymatically dephosphorylated (292). As NFH has more KSP repeats than NFM (51 vs. 7 in mice), it was thought that this subunit should contribute more to the axonal radial growth. This idea was reinforced by the observation that phosphorylated NFH sidearms are less pliant and larger structures than dephosphorylated sidearms (293). Moreover, a 50% reduction in the level of axonal NFH in mice overexpressing NFM decreases axonal caliber (294) and modest increases in NFH slightly enhance radial growth in transgenic mice (295). However, production of NFH-null mice (285–287), as well as mice expressing NFH deprived of its carboxy-terminal domain (NFH$^{\mathrm{tail}\Delta}$ mice) (296), demonstrated no major modification in NFs spacing and axonal radial growth, except a delay in the acquisition of a normal axonal caliber. However, as mentioned above, NFH-null mice are characterized by a compensatory increase for MT density and NFM phosphorylation. Similarly, the phosphorylation level of NFM is higher in NFH$^{\mathrm{tail}\Delta}$ mice (296), making difficult to determine the exact implication of NFH in axonal caliber. In contrast, disruption of the *NFM* gene (67), or deletion of its carboxy-terminal domain in NFM$^{\mathrm{tail}\Delta}$ mice (286, 297), reduced the interfilament spacing and axonal caliber, showing a preponderant role of NFM in determining axonal diameter. It is interesting to note that axonal caliber in double transgenic NFM/H$^{\mathrm{tail}\Delta}$ mice from 6 months of age is similar to that of NFM$^{\mathrm{tail}\Delta}$ mice, while NFs spacing is reduced and NF network is disorganized (297). This suggests a role for NFH tail domain in NF–NF interactions but not in the control of axonal caliber.

Since NFM subunits appeared essential for the acquisition of normal axonal caliber, Garcia et al. (298) challenged the role of

NFM phosphorylation as a determinant of radial axonal growth. By employing a gene replacement technique that produces mice in which all Ser of NFM's KSP repeats have been replaced with phosphorylation-incompetent Ala, they showed that the phosphorylation of NFM KSP repeats is not required for radial axonal growth. Indeed, axonal caliber and motor neuron conduction velocity of mice expressing KSP phospho-incompetent NFM were indistinguishable from wild-type mice. Moreover, this substitution did not alter spacing between adjacents NFs. Thus, it appeared that phosphorylation of NFM KSP repeats is not an essential component for the acquisition of normal axonal caliber, contrasting with the long-standing view that correlates axonal growth with NF phosphorylation. These results are also in contrast to those of previous studies employing computer modeling to investigate the role of individual NF sidearms and their phosphorylation in lateral extension (299, 300). These studies showed an increased extension of NFM sidearms upon phosphorylation. However, these analyses were carried out on human NFs using the sequences and/or stoichiometry of human NF, whereas the experimental study by Garcia et al. (298) investigated mouse NF which differ in stoichiometry and in charge distribution. To better understand the effect of mouse NFM phosphorylation on NFs spacing, Stevenson et al. (301) applied a computational method to reveal phosphorylation-mediated conformational changes in mouse NF architecture by using the sequence and stoichiometry of mouse NF proteins. They showed that the phosphorylation of mouse NFM does not change the radial extension of NFM sidearms under a salt-free condition and in ionic solution, supporting the notion that NFM KSP phosphorylation has no effect on the axonal diameter in the mouse background. On the other hand, significant phosphorylation-mediated conformational changes were found in NFH sidearms under the salt-free condition, while the changes in ionic solution are not significant. However, NFH sidearms are found at the periphery of mouse NF architecture, implying a role in linking neighboring filaments (301).

Axonal radial growth initiates with, and is dependent on, formation of compact myelin, which begins on the first postnatal day, and is complete at 2 weeks in mice (302). In vitro, myelination induces the increase of axonal caliber (303) but demyelination in vivo causes a local axonal atrophy (304, 305) and modifications in the axoskeletal organization (274). A severe axonal atrophy is also observed in hypomyelinated tracts from *shiverer* mutant mice (306). The view that myelination regulates NF phosphorylation, which in turn influences radial growth emerged from the analysis of the dysmelinated mouse mutant *Trembler*. This mutant revealed a decreased NFs phosphorylation correlated with an increased NF density and reduced axonal calibers (122). Moreover, NFM and NFH sidearms are highly phosphorylated in myelinated axonal segments, and poorly phosphorylated in nonmyelinated domains like the initial segment and the node of Ranvier where axonal caliber is reduced (122–125). The axonal radial growth does not require the formation of a compact myelin but only the axon ensheathment by the myelin-forming cell (275), suggesting that a molecule localized in the adaxonal membrane of the glial cell regulates the axonal caliber. Because MAG is enriched in this region, and starts to be expressed when the axonal caliber expands, it has been suggested that MAG plays a crucial role for the axonal size expansion (307, 308). A defect in the radial growth of myelinated axons was observed in MAG-deficient mice, together with a reduction of NFs phosphorylation in myelinated fibers (127). It was proposed that the binding of MAG to axonal receptors activates a signalization cascade leading to the phosphorylation of NFs, and consequently induces the axonal radial growth in myelinated regions (126, 127, 297, 309). However, we recently showed that the relative extent of caliber reduction at nodes of Ranvier is similar between axons containing or not NFs (280), suggesting that NFs and consequently their phosphorylation are not responsible for the axonal caliber difference between internodes and nodes. Moreover, as seen above, the analysis of mice expressing full-length phosphor-incompetent NFM revealed no role for KSP phosphorylation in axonal radial growth (298), challenging the long-standing myelin-dependent NF phosphorylation hypothesis of radial growth, which seemed to be supported by previous models. It remains to determine why the NFM tail domain is crucial for axonal radial growth and which amino acids are involved in this process.

NFs constitute the main cytoskeletal network maintaining the structural integrity of neurons and exhibit dynamic changes during axonal and dendritic growth. However, the mechanisms underlying axonal development and maintenance remain poorly understood. It was recently identified that multisynthetase complex p43 (MSC p43) is essential for NF assembly and axon maintenance (310). The MSC p43 protein is a cofactor of the macromolecular ARS complex that is predominantly expressed in central neurons and interacts with NFL in vivo. Overexpression of MSC p43 led to a decreased level of NF protein phosphorylation, whereas MSC p43 depletion caused hyperphosphorylation of NF proteins and disorganization of the axonal NF network in primary cultured neurons and motor axons, suggesting that NF phosphorylation is normally regulated at an optimal level. Mice lacking MSC p43 exhibited axon degeneration in motor neurons, along with resultant defects of neuromuscular junctions, muscular atrophy, and motor dysfunction. Thus, it appeared that MSC p43 is indispensable in maintaining axonal integrity by regulating NF assembly. Its dysfunction may underlie the NF disorganization and axon degeneration associated with motor neuron degenerative diseases.

10. Neurofilaments Contribute to the Axonal Conduction Properties

The diameter of myelinated axons is one major property that influences the rate of impulse propagation along the axon. Ultrastructure of myelinated fibers is optimized for maximal conduction velocity through the axonal caliber (311–313), internodal length (314–317), myelin thickness (318–320), as well as geometry and molecular organization of the nodes of Ranvier (321, 322). As internodal length and myelin thickness are proportional to the axonal size (323–325), NFs are key players for modulating the axonal conduction.

Several animal models with abnormal NF expression or distribution highlighted implication of NFs in the conduction properties. *Quiver* quails expressing a mutated NFL display a reduced conduction velocity proportional to the decrease of the axonal caliber which lack NFs (326). Similarly, consistent with axonal atrophy, lower conduction velocities are observed in NFL−/− mice and NFM−/− mice (327), in mice expressing human NFH (hNFH mice) (328), in $NFM^{tail\Delta}$ mice (297), and in NFH-LacZ transgenic mice (280, 329). These modifications are not restricted to the conduction velocity but other electrophysiological parameters are also affected in these animals. These include altered auditory evoked potentials in *quiver* quails (330), prolongation of refractory period in NFM−/− mice (327), decreased resting membrane potential, prolonged duration of action potential and decreased inward and outward rectification in hNFH mice (328), reduced amplitude of the compound action potential and abnormalities of somesthesic and auditive evoked potentials in NFH-LacZ mice (280), suggesting multiple implications of NFs in conduction. This is reinforced by the fact that conduction velocity, refractory period as well as correlation between the rate of rise and decay of action potential and conduction velocity are significantly modified in NFH−/− mice despite normal axonal diameter, g-ratio, and internodal length (327) providing strong evidence that NF are involved in defining not only the structural but also the functional integrity of myelinated axons. Kriz et al. (327, 328) proposed that NFH may have a specific role in modulating ion channel function, but the exact molecular mechanism is still unclear even if it appeared that localization of Na^+ and K^+ channels in, respectively, node and juxtaparanode is unaffected by the lack of axonal NFs (280). It should also be mentioned that in contrast to NFH−/− mice, conduction velocity is not altered in $NFH^{tail\Delta}$ mice (297), indicating no implication of NFH sidearm in this parameter. Finally, it cannot be excluded that the absence of axonal NFs also affects the axoplasmic resistance.

The conduction velocity also depends on myelin thickness and internodal length, which are proportional to axonal diameter. Optimum conduction velocities are achieved when internodal lengths are ~100 times the axonal caliber and for g-ratio values comprised between 0.6 and 0.7 (314, 319, 320). To determine whether changes in myelin sheath dimensions can contribute to electrophysiological defects observed in the absence of axonal NFs, we measured g-ratio and internodal length in NFH-LacZ transgenic mice (280). It appeared that myelin thickness is differently regulated in CNS and PNS in response to reduced axon calibers. The caliber-reduced axons from NFH-LacZ mice are invested with proportionally thinner myelin in CNS without modifications of the g-ratio, while in PNS axons are overmyelinated compared to their reduced diameter (g-ratio of 0.52 vs. 0.63 in transgenic and control PNS). A similar disparity was reported in NFM−/− and NFH−/− mice (331). Surprisingly, the axonal atrophy in both CNS and PNS from NFH-LacZ mice does not affect the internodal length. Consequently, internodal lengths are ~200 times the diameter of NFs-deficient axons. According to Rushton (319), such a ratio increases internal resistance and reduces the capacity to activate sodium channels. These results indicate that myelin dimensions are not optimal for conduction in the absence of axonal NFs.

The geometry and composition of nodes of Ranvier are also crucial for the propagation of the nerve impulse. Both in PNS and CNS, nodes are characterized by an important constriction of the axon reaching approximately 30 to 15% of the internodal size (332, 333). This constriction promotes higher conduction velocities by reducing the nodal capacity through a smaller nodal area. It also reduces the contribution of the paranodal axolemmal membrane by restricting conductance along the periaxonal pathway (334). Two models have been considered to explain the nodal constriction (335). First, the contraction model suggests that myelinating cells exert a pressure via their paranodal loops sufficient to reduce the axonal diameter and/or to limit the axonal radial growth at nodes. The high content of contractile proteins (filamentous actin, myosin) and mitochondria in paranodal loops consolidates this assumption (332, 336). Secondly, the NF model suggests that reduction of axon caliber at nodes is due to a densely packed hypophosphorylated NF network while larger internodes contain spaced hyperphosphorylated NFs (122–125). However, it was showed that the extent of axonal constriction at nodes is similar with and without axonal NFs (280), indicating that they are not required for the establishment of the nodal ultrastructure and thus arguing in favor of the contraction model.

In conclusion, NFs are crucial for the correct conduction of the nerve impulse and therefore their defects could contribute to neurodegenerative processes. They increase the conduction velocity by promoting the axonal radial growth and are essential to achieve optimal g-ratio and internodal length. On the other hand they are not necessary for the formation and maintenance of ultrastructure and molecular organization of nodes of Ranvier.

11. Neurofilament Axonal Transport

Neurons are highly polarized cells and their axonal length can reach more than one meter in humans. Most proteins are synthesized in the soma and transported into axonal and dendritic processes through a mechanism called axonal transport. Weiss and Hiscoe (337) demonstrated for the first time this process using ligation of sciatic nerves. This leads to an axonal swelling proximal to the ligation and to an axonal shrinkage in distal region. When ligation is removed, material accumulated in proximal region moved down the nerve at 1–2 mm per day. These observations emphasized the existence of a flow of material from the cell body to the nerve terminals. In the 1960s, an important step in characterizing the axonal transport was achieved using classical in vivo metabolic labeling studies. Radiolabeled (^{3}H) amino acids or (^{35}S) methionine were injected into sciatic nerve or retinal ganglion cells from living animals to label newly synthetized proteins (338–340). After a number of days, the nerve is dissected into ordered segments from the cell body to the nerve terminal, and cytoskeletal fractions are prepared from each segment. By measuring the distance traveled by labeled proteins, one is able to calculate an inferred rate of transport. It has been shown that the axonal transport is divided in two major categories (341): the fast axonal transport (~250–400 mm/day in mammals), that conveys mitochondria, neurotransmitters, channel proteins, lysosomes, and endosomes (342–344), and the slow axonal transport (~0.1–4 mm/day) that conveys axonal cytoskeletal and cytosolic proteins. The slow axonal transport can also be divided into two rate components: the slow component a (SCa; 0.1–1 mm/day), containing NFs and MTs and the slow component b (SCb; 2–4 mm/day), containing actin, spectrin, and other cytoplasmic proteins (171, 345).

It was first speculated that the various rates of transport were due to the association of cargoes with different molecular motors. Members of the kinesin family and cytoplasmic dynein were identified as the major motors responsible for the fast axonal transport in, respectively, the anterograde (toward the terminal end) and retrograde (toward the cell body) directions (346, 347). The identity of motors responsible for the slow axonal transport of NFs was partially elucidated by two studies analyzing the transport of green fluorescent protein (GFP)-tagged NF subunits. Constructs were transfected into cultured sympathetic neurons which have a limited endogenous NF network, enabling transport of motile GFP-labeled NFs to be more easily imaged using microscopy focused on the gaps between stationary GFP-tagged NFs (348, 349). Surprisingly, the authors observed that both GFP-NFM and GFP-NFH move at rates of up to 1 μm/s, corresponding to the rates of molecular motors ensuring the fast axonal transport.

This suggests that motors used for slow and fast axonal transport are identical. However, contrary to fast axonal transport of cargoes that moved continuously, NFs are transported intermittently in axons because their fast movements are interrupted by prolonged pauses. Only a small fraction of NFs is moving at any given time since it was evaluated that NFs spent 97% of their time pausing (350). This study also suggested that there may be two distinct stationary states for NFs, distinguished by average pause duration of 30 s or 60 min. Thus, the overall speed of axonal transport would not depend on the motors involved but on the duration of association between cargoes and motors. Components moving with the fast axonal transport could be attached to the motor for a long period, while interaction between components moving with the slow axonal transport and their motors would be short. Another interesting discovery emerging from these studies is the description of a bidirectional transport of the NFs, but with a predominant anterograde prevalence. Transport of NFs at fast rates was also demonstrated in extruded squid axoplasm (351) and along MTs in vitro (242). These data collectively indicate that the overall slow axonal transport of NFs is the result of a combination between fast bidirectional movements and long-lasting pauses.

The implication of the MT motor proteins kinesin and dynein in axonal transport of the NFs was later confirmed by various observations. First, several evidences suggest a direct interaction between NFs and kinesin or dynein (242, 244–246, 352). The inhibition or depletion of dynein (247, 353) affects the retrograde but also the anterograde transport of NFs and ultimately induces the formation of focal accumulations of NFs in neurites, while microinjection of anti-kinesin and anti-dynein antibody affect anterograde and retrograde NFs transport in cultured dorsal root ganglia neurons (243). Finally, targeted disruption of neuronal kinesin heavy chain KIF5A in mice did not affect fast axonal transport but induced the accumulation of NFs in cell bodies of peripheral sensory neurons (354), and mutation in kinesin-1A decreased both anterograde and retrograde NF transport flux by decreasing the frequency of anterograde and retrograde movements in cultured mouse cortical neurons (355).

Millecamps et al. (173) have provided new insights into the mechanisms of axonal transport of NFs. As described above, they generated transgenic mice with doxycycline controlled expression of human NFL (hNFL), with or without endogenous mouse NFL proteins (respectively, tTA;hNFL;NFL$^{+/-}$ and tTA;hNFL;NFL$^{-/-}$ mice). They showed that the presence of the axonal NF array strongly slows down the axonal transport of NFs. Indeed, when doxycycline treatment of tTA;hNFL;NFL$^{-/-}$ mice is stopped, the reappearance of hNFL occurred in synchrony along the sciatic nerve within 1 week, indicating a fast axonal transport of NFs in axons deprived of stationary NF network. They estimated a rate of ~10 mm/day in axons with low NF content compared to ~1 mm/day in axons with a dense NF array. The authors suggested that these differences could be due to a decreased number of pauses during the travel of NFs and/or to decreased interactions with a stationary pool.

The forms in which NFs are transported (subunits or polymers) and the contribution of their phosphorylation was subjects of intense investigations for many years. Concerning the first point, one model suggested that NFs polymerize in cell bodies immediately after synthesis, and they are subsequently transported as filaments. On the opposite, the subunit model proposed that polymers are essentially stationary and that NFs could be transported as free subunits or small oligomers that integrate stationary cytoskeletal polymers. Both hypothesis were based on various observations (for review see refs. (356, 357)). For example, the metabolic labeling of cytoskeletal proteins with (^{35}S) methionine shows a synchronized movement of the three NF subunits, in favor of the polymer transport model (171, 358). Moreover, the majority of NFs in axons are polymers and it is difficult to detect free monomers. Conversely, in fluorescence recovery after photobleaching (FRAP) experiments, no movement of the bleached area was observed and the recovery of fluorescence is gradual (359), supporting the subunit transport hypothesis. In the same way, the existence of a subunit/small oligomer transport is suggested by the finding that NFM is transported along axons after injection of a recombinant adenovirus encoding tagged-NFM protein in transgenic mice deprived of axonal NFs (360), even if a diffusion of this protein cannot be completely excluded (361). An important advance was made by analyzing the transport of GFP-NFM and GFP-NFH in transfected neurons (348, 349). In these studies, intact filaments (1–15 μm of length) represent 95% of the moving structures observed, strongly supporting the polymer model. The transport of NF polymers was recently confirmed by Yan and Brown (362) who showed that NFs move in the form of assembled polymers in axons of cultured neonatal mouse sympathetic neurons and, more interestingly, that moving and stationary NFs are complex heteropolymers also containing peripherin and α-internexin along >85% of their length (3). Yuan et al. (363) also demonstrated that α-internexin is a key determinant for the axonal transport of NFs in CNS. They observed that NFM can be transported into axons of the optic nerve in the absence of NFL and NFH, while the additional deletion of α-internexin abolishes this transport, indicating that NFM monomers alone are not efficiently transported. The finding that a small number of IFs are present in optic axons from NFL−/−;NFH−/− mice but not from α-internexin−/−;NFL−/− mice, suggests that NFM is able to associate with α-internexin to form IFs. The colocalization of α-internexin and NFM on the same filament (2) reinforced this assumption. Thus, it seems that hetero-oligomer composed of at least NFM and α-internexin is the minimal form in which NF proteins can be transported in CNS axons. It is also interesting to note that the deletion of NFH or α-internexin does not affect the NF transport (296, 363), while the loss of both proteins causes the selective acceleration of the axonal transport of NFL and NFM subunits, suggesting redundant roles of α-internexin and NFH in axonal transport of NFs.

The predominant form of transporting NF subunits varies according to the neuronal differentiation state, leaving open the possibility that short NFs and punctate structures each contribute in distinct manners to the establishment and/or regional remodeling of the NF network (246, 351, 364). It was recently investigated whether kinesin or dynein preferentially translocated short NFs or punctate structures into axonal neurites. Lee et al. (365) showed that inhibition of dynein function, through dynamitin overexpression or intracellular delivery of anti-dynein antibody, selectively reduced delivery of GFP-tagged short NFs into the axonal hillock, with a corresponding increase in the delivery of puncta, suggesting that dynein selectively delivered short NFs into axonal neurites. By contrast, intracellular delivery of anti-kinesin antibody inhibited anterograde transport of short NFs and puncta to an equal extent. These findings suggest that anterograde axonal transport of linear NFs is more dependent upon association with translocating MTs (which are themselves translocated by dynein) than is transport of NF puncta or oligomers.

Another controversial subject concerns the role played by phosphorylation in the axonal transport of NFs. There is strong evidence to support a role for phosphorylation of NF sidearm as an important regulatory mechanism. Phosphorylation of NFs, in particular NFH, has long been considered to decrease their transport rate. Progressive phosphorylation of the C-terminal sidearms of NFH and NFM occurs during their translocation into and along axons, and is regionally and temporally associated with slowing of axonal NF transport (366, 367). The phosphorylation level of NFs correlates with a slow transport of radiolabeled NFs in projections of the rat L5 dorsal root ganglion (113), while hypophosphorylated NFs in mouse optic axons move faster than hyperphosphorylated ones (114). Similarly, lack of NFH accelerates NF transport (287), while increased NFH expression selectively slows NF transport (295). Finally, NFs containing NFH mutated to generate constitutively nonphosphorylated NFH move faster than those containing wild-type NFH, while NFs containing NFH mutated to generate constitutively phosphorylated NFH move slower (112). All these studies support the view that the rate of NFs transport is inversely correlated to their phosphorylation state. However, the analysis of NFs axonal transport in optic nerve from $NFM^{tail\Delta}$ mice and $NFH^{tail\Delta}$ mice shows no modification in the rate of NFs transport (296, 368, 369), suggesting that phosphorylation of

NFM and NFH sidearm is not directly involved in this process. Alternatively, a yet unknown molecule could bind differently to these mutated NFs and may modulate their transport. It should also be noted that phosphorylation of both NFM and NFH subunits are tightly regulated. Indeed, phosphorylation of NFM is significantly increased in NFH−/− mice and in $NFH^{tail\Delta}$ mice (287, 296), while an increased phosphorylation of NFH is observed in NFM−/− mice and in $NFM^{tail\Delta}$ mice (67, 368). The measure of NFs transport in $NFM^{tail\Delta}$ mice and $NFH^{tail\Delta}$ mice could thus be biased by this compensatory phosphorylation.

Two possibilities could explain how phosphorylation regulates the axonal transport of NFs. First, phosphorylation of NFs controls attachment or detachment of molecular motors. In particular phosphorylation of NFs promotes their release from kinesin (245, 352) and increase their affinity for dynein (247). Consequently, the phosphorylation of NFs would cause their slowing by reducing their anterograde transport by the kinesin and by supporting their retrograde transport by the dynein. Thus, the long pauses during their travel would correspond to the moments when NFs are dissociated from their motors. This model is consistent with the intermittent movement of NFs into axons (348, 349) and with mathematical modeling of NFs axonal transport (370). Another mechanism by which C-terminal phosphorylation of NF subunits could mediate the axonal transport of NFs is by promotion of divalent cation-mediated crosslinking of phosphorylation-NF sidearms that leads to the formation of NF bundles that are too large to undergo transport (371, 372). Recently, Lee et al. attempted to dissect the impact of phosphorylation on NF C-terminal sidearm interactions by expressing isolated sidearms with and without site-directed mutagenesis of cdk5 consensus sites (373). The absence of the rod domain afforded the opportunity to monitor the impact of these mutations on NF transport and interaction with the resident NF bundle in the absence of NF assembly. They observed that mutations of cdk5 consensus sites did not alter transport but pseudo-phosphorylated mutants displayed a greater association with axonal NF bundles. By contrast, these same mutations in full-length NFH altered NF transport as well as bundling. Since isolated sidearms cannot assemble, they can only interact with NFs via a single sidearm–sidearm interaction, while assembled NFs can form multiple such interactions. These finding suggest that individual sidearm–sidearm interactions are dynamic and do not persist long enough to slow NF transport, and that bundle formation and maintenance depends upon both the long half-life of NF polymers and the establishment of multiple phosphorylation-dependent sidearm-mediated interactions among NFs (373). Both mechanisms are not mutually exclusive since they are consistent with the observation that hypophosphorylated NFs display the most rapid transport rate (since they are more likely to be associated with kinesin and are least likely to be incorporated into NF bundles) and that extensively phosphorylated NFs display the slowest transport rate (since they are least likely to be associated with kinesin and are more likely to be associated with NF bundles).

12. Neuronal Intermediate Filaments and Neurodegenerative Diseases

12.1. Amyotrophic Lateral Sclerosis

Amyotrophic lateral sclerosis (ALS), also known as motor neuron disease (MND) or Lou Gehrig's disease, is an age-dependent fatal neurodegenerative disease that manifests as a result of progressive and selective death of upper and lower motor neurons of the CNS, leading to weakness and atrophy of the muscles followed by progressive paralysis, while the cognitive functions of the patients remain usually unaffected. ALS has a worldwide prevalence of 1–2 per 100,000. There is no cure and death usually occurs by denervation of respiratory muscles within 3–5 years from the onset of symptoms (374). Approximately 90% of ALS cases are considered sporadic while 10% are inherited in an autosomal dominant pattern. 20% of all the familial cases are due to mutations in the genes encoding the antioxidant protein Cu/Zn superoxide dismutase 1 (SOD1), the most abundant cytosolic enzyme (375), the nuclear factor TAR DNA binding protein (TDP-43) (376, 377), and fused in sarcoma/translated in liposarcoma gene (FUS/TLS) (378, 379).

The discovery of NF gene variants in ALS patients suggested the involvement of NFs in the pathogenesis of the disease. Indeed, codon deletions or insertions in the KSP repeat motifs of NFH have been identified in a small number of sporadic cases of ALS, including a large deletion of five KSP repeats (380–382). However, two others studies failed to identify variants in the NF genes linked to sporadic and familial ALS (383, 384), suggesting that mutations in the NF genes are not a systematic common cause of ALS but could be a risk factor for sporadic ALS. Peripherin mutations have also been identified in three sporadic ALS patients (385–387), including a frameshift mutation in the PRPH gene able to disrupt the NF network assembly in vitro, reinforcing the view that NF disorganization may contribute to pathogenesis. These results suggest that peripherin mutations may be responsible for a small percentage of ALS cases. It is also of interest to note the presence of high NFL and NFH levels and autoantibodies against NFL in cerebrospinal fluid (CSF) of ALS patients (388–390).

Both sporadic and familial ALS are characterized by the presence in motor neurons of axonal spheroids and perikaryal accumulations composed of NFs and/or peripherin (206). NFs in perikaryal aggregates are extensively phosphorylated, a process that occurs normally only within the axon (391). The NF protein accumulations are characterized by forming initially in the distal axon and retrogradly transported to the cell body, correlated with severity of disease. The mechanisms governing the formation of IF aggregates are still not clearly established but evidences that IF accumulations could result from abnormal stoichiometry of IF proteins or from defects of axonal transport exist.

Several observations suggested that modification in NF stoichiometry induce accumulation of NFs. The overexpression of different NF subunits in mice provokes the formation of NF aggregates reminiscent of ALS (279, 392, 393). Remarkably, the motor neuron disease caused by excess human NFH (hNFH) can be rescued by overexpression of hNFL in a dosage-dependent fashion (284). Overexpression of peripherin in mice also provokes the formation of cytoplasmic protein aggregates and the subsequent selective loss of motor neurons during aging (394, 395). This loss is preceded by axonal transport defects and formation of axonal spheroids (396). Because NFL mRNA levels are reduced in cases of ALS (397), Beaulieu et al. (394) generated double transgenic mice overexpressing peripherin and deficient for NFL (Per;NFL−/− mice), mimicking more closely the neuronal IF conditions found in ALS. Here, the onset of peripherin-mediated disease is accelerated by the deficiency of NFL. Without NFL, peripherin interacts with NFM and NFH to form disorganized IF structures. This could explain why the number of IF inclusion bodies is increased in Per;NFL−/− mice, leading to an earlier neuronal death and to defects of fast axonal transport in cultured Per;NFL−/− neurons (269). In contrast, peripherin toxicity can be attenuated by coexpression of NFL or NFH (398, 399), illustrating once again the importance of IF protein stoichiometry. Although peripherin expression was not changed in these mice, NFH overexpression shifted the intracellular localization of inclusion bodies from the axonal to the perikaryal compartment of motor neurons, suggesting that the toxicity of peripherin inclusions may be related to their axonal localization, possibly by altering the axonal transport. However, it should be noted that peripherin is not a key contributing factor to the neuronal death in disease caused by SOD1 mutations because absence or overexpression of peripherin in $SOD1^{G37R}$ mice do not affect the onset and progression of motor neuron disease (400).

ALS motor neurons also show a selective decrease in the levels of NFL, α-internexin and peripherin mRNA, while in familial ALS the levels of peripherin mRNA appear to be abnormally elevated (397, 401, 402). This suggests a change in the stoichiometry of cytoskeletal protein expression, which could be conducive to the formation of neurofilamentous aggregates in ALS. This decrease of IF mRNA could be due in part to modification in their stability. Several NFL mRNA binding proteins have been identified in human, including 14-3-3 proteins (403), TDP43 (404), both mutant and wild-type SOD1 (405) and Rho guanine nucleotide exchange factor (RGNEF) (406). These proteins are incorporated in ALS intraneuronal aggregates and affect the stability of NFL mRNA, indicating that a greater proportion of NFL may be targeted for translational silencing and/or degradation in ALS. In agreement with this, mice expressing human TDP-43 displayed reduced NF mRNAs and protein contents, inducing a decrease of caliber of their motor axons (407). The involvement of TDP-43 in ALS pathogenesis was reinforced by the discovery of several mutant forms of this protein in familial and sporadic ALS (408) and by the recent generation of TDP-43 mutant mice that develop many age-related pathological and biochemical changes reminiscent of human ALS including ubiquitinated TDP-43 positive inclusions, IF abnormalities, axonopathy, and neuroinflammation (408). Peripherin intra-isoform ratio changes could also be involved in the pathogenesis of ALS. Indeed, it was showed that such changes result in the formation of peripherin inclusions in vitro and in vivo (409–411). Aggregate-inducing Per28 is upregulated in patients with ALS, at both the mRNA and protein levels, and is associated with round inclusions in disease pathology (411). The Per61 splice variant is neurotoxic and has been observed in ALS mouse models and human patients (410). Oxidative stress was also associated with changes in peripherin isoform expression and aggregation (412). Indeed, exposure to increasing concentrations of H_2O_2 induced changes in the expression of Per-58 and Per-45 that were accompanied by a dose-dependent increase in peripherin aggregation. This may have relevance to the pathophysiology of ALS, where the cooccurrence of oxidative stress, peripherin-immunoreactive inclusions, and peripherin isoform expression abnormalities are found.

Perturbations of the axonal transport of NFs and organelles are one of the earliest pathological changes seen in several transgenic mouse models of ALS (413–416). The premature phosphorylation of NF tail-domains in motor neurons cell bodies could directly mediate their accumulation in this region. Glutamate excitotoxicity, another pathogenic process in ALS, may induce abnormal phosphorylation of NFs. Treatment of primary neurons with glutamate activates members of the mitogen-activated protein kinase family which phosphorylate NFs with ensuing slowing of their axonal transport (417). In addition, glutamate leads to caspase cleavage and activation of protein kinase N1 (PKN1), an NF head-rod domain kinase (418). This cleaved form of PKN1 disrupts NF organization and axonal transport. Excitotoxicity mediated by non-*N*-methyl-d-aspartic acid (NMDA) receptor is also associated with the aberrant colocalization of phosphorylated and dephosphorylated NF proteins (419). Inhibition of Pin1, a prolyl isomerase, was suggested as a possible therapeutic target to reduce pathological accumulation of phosphorylated NFs. Pin1 associates with phosphorylated NFH in neurons and is colocalized in ALS-affected spinal cord neuronal inclusions (420). Its inhibition rescues neurons from glutamate-induced cellular death. Finally, it was recently reported that riluzole protects against glutamate-induced slowing of NF axonal transport by decreasing perikaryal NF side-arm phosphorylation (421), probably via the inhibition of ERK and p38 activities, two NF kinases activated in ALS. Alterations of the anterograde or retrograde molecular motors may also be responsible for aggregation of IFs. Mutation of dynein or p150^{glued} (422), overexpression of dynamitin (353), and absence of kinesin heavy chain isoform 5A (KIF5A) (354) induce NF accumulations in mice. Recent studies suggest that inhibition of retrograde transport is more susceptible to cause accumulation of NFs than inhibition of anterograde transport. The inhibition of dynein by increasing the level of dynamitin induces aberrant focal accumulation of NFs within axonal neurites, whereas inhibition of kinesin inhibits anterograde transport but does not induce similar focal aggregations (423). Similarly, the neuron-specific expression of Bicaudal D2 N terminus (BICD2-N), a motor-adaptor protein, impairs dynein–dynactin function, causing the appearance of giant NF swellings in the proximal axons (424). However these mice did not develop signs of motor neuron degeneration and motor abnormalities.

Impairment of the UPS was also reported in lumbar spinal cord from SOD1^{G93A} mice (425). These mice were crossed with transgenic mice that express a fluorescently tagged reporter substrate of the UPS. In double-transgenic Ub(G76V)-GFP/SOD1G93A mice an increase in Ub(G76V)-GFP reporter was detectable in a few spinal motor neurons without modification in the levels of reporter transcript, suggesting that the accumulation of Ub(G76V)-GFP was due to deficient reporter degradation. Moreover, one-third of the cells with accumulation of the UPS reporter substrate also displayed an accumulation of phosphorylated NFs like those normally present only in axons. These data suggest that UPS impairment occurs in motor neurons of mutant SOD1-linked ALS mice and may play a role in the disease progression (425).

Neuronal IF abnormalities in ALS may also occur as a result of posttranslational protein modifications. Indeed, advanced glycation endproducts were detected in NF aggregates of motor neurons in familial and sporadic ALS (163). A strong decrease of O-linked *N*-acetylglucosamine immunoreactivity of NF proteins was also reported in a transgenic rat model for ALS, while phosphorylation was increased (161). Finally, it was showed that SOD1 can catalyze nitration of tyrosines by peroxynitrite in the rod and head domains of NFL (166). However, no significant changes were detected in the nitration of NFL isolated from cervical spinal cord tissue of sporadic ALS cases (426).

Finally, it seems that non-neuronal cells could be directly involved in the formation of cytoskeletal aggregates within proximal axon from motor neurons. Indeed, cultured mouse spinal motor neurons in contact with non-neuronal cells displayed swellings that were morphologically and neurochemically comparable to axonal spheroids that develop in vivo in ALS transgenic mouse models (427). These swellings contained NFL, NFM, NFH, α-internexin, and peripherin and induced the accumulation of mitochondria and vesicle-like structures, suggesting a disruption of the axonal transport. Moreover, the severity of this axonopathy correlated with the phenotype of the glial cells, with a significant increase being induced by a glial feeder layer expressing mutant SOD1 or that was pre-aged prior to plating the motor neurons (427).

To further determine whether NFs are directly involved in SOD1-mediated disease, mice expressing mutant SOD1 were mated with transgenic mice deficient for axonal NFs. The withdrawal of NFs from the axonal compartment and their perikaryal accumulation induced by the expression of NFH-β-galactosidase fusion protein conferred no beneficial effect to $SOD1^{G37R}$ mice (258), indicating that axonal NFs are not necessary for SOD1-mediated disease. This was also observed in $SOD1^{G85R}$ mice deprived of NFL, but the absence of axonal NFs in these animals prolongs their life span by approximately 15% (428). Surprisingly, overexpression of mouse NFL or mouse NFH in $SOD1^{G93A}$ mice (429), and overexpression of hNFH in $SOD1^{G37R}$ mice (430) also increase their life span by, respectively, 15% and 65%. This suggests a protective effect of NF perikaryal accumulation in motor neuron disease caused by mutant SOD1. While the mechanism of protection is unclear, it seems that perikaryal accumulation of NFs rather than their axonal deficiency is responsible for slowing disease in these models. Indeed, the formation of large perikaryal aggregates and a massive depletion of axonal NFs due to the expression of the human NFH43 allele cause more positive effects than human NFH44 allele, which induces smaller aggregates and more axonal NFs (430). Moreover, the disruption of one allele for each NF gene induces a 40% decrease of axonal NF proteins content and an important axonal atrophy without perikaryal accumulation of NFs in $SOD1^{G37R}$ mice, but it does not extend their life span nor does it alleviate the loss of motor axons (283). Several hypotheses were proposed to explain this protective effect of perikaryal aggregates in SOD1-mediated disease. Through their multiple calcium-binding sites NFs may act as calcium chelators. Supporting this hypothesis, a significant neuroprotection was obtained by overexpressing the calcium-binding protein calbindin-D28k in cultured motor neurons (431). It was also proposed that perikaryal accumulations of NFs in motor neurons may alleviate ALS pathogenesis by acting as a phosphorylation sink for cyclin-dependent kinase 5 dysregulation induced by mutant SOD1, thereby reducing the excessive phosphorylation of tau and other neuronal substrates (283). This was supported by the fact that NF accumulations contain hyperphosphorylated NFM and NFH subunits in ALS patients (391) and in SOD1 mutant mice (432). However, genetic elimination of NFM and NFH sidearms and their phosphorylation sites in SOD1 mutant mice led to a delay of disease rather than the acceleration predicted by a kinase dysregulation model (433), indicating that perikaryal phosphorylation of NFs is not an essential contributor to reduced toxicity of SOD1 mutants. Alternatively, axonal removal of NFs could enhance axonal transport, which is impaired in SOD1 mice, by providing a more flexible axoplasm.

Finally, it was shown that NFs are involved in the localization of NMDA receptors in the neuronal plasma membrane by interacting with the NMDA NR1 subunit (434). Thus, accumulation of NFs could interfere with glutamate receptor function and prevent glutamate excitotoxicity. However, neurons bearing NF aggregates demonstrate increased intracellular calcium levels and enhanced cell death in response to NMDA receptor activation without increased NMDA receptor expression. These results suggest that the presence of NF aggregates renders motor neurons more susceptible to NMDA-mediated excitotoxicity (435).

The clinical diagnostic of ALS is considered difficult and, to date there is no reliable test for diagnosis. However, the in vivo detection of neurodegeneration in its early phase is possible using body fluid biomarkers. Degeneration of neurons and their axons results in disintegration of the cellular membrane and release of components of the cytosol into the extracellular fluid and from there into the adjacent body fluid compartment from where they can be sampled and quantified. Of the many components released, NF proteins have been established and validated as a sensitive and specific protein biomarker for neuroaxonal degeneration (436). A robust diagnostic biomarker(s) for ALS would accelerate and simplify the investigative process and result in the earlier implementation of appropriate therapies. Cytoskeletal proteins including NF proteins and tau have been shown to be elevated in the CSF of neurodegenerative diseases and proposed as biomarkers for ALS (388, 437, 438). Levels of phosphorylated NFH were shown to be significantly increased in the CSF of ALS patients when compared to disease mimics (439), but the optimal biomarker was found to be a ratio of phosphorylated NFH to complement C3 levels (440), indicating that the CSF levels of phosphorylated NFH and C3 may aid in the diagnosis of ALS and supports prospective studies to further evaluate the clinical utility of these candidate biomarkers.

12.2. Charcot-Marie-Tooth Disease

CMT represents a heterogeneous group of inherited peripheral neuropathies affecting both motor and sensory neurons as well as the muscles. With an estimated prevalence of 1 in 2,500 individuals, CMT is the most common inherited neurological disorder. CMT is characterized by an insidious onset and slowly progressive weakness and atrophy of the distal limb muscles. First signs typically appear in the first or second decade of life, although it may be detected in infancy. On the basis of electrophysiological properties and histopathology, CMT was originally subclassified into CMT1 and CMT2. CMT1 is a demyelinating disease with reduced nerve conduction velocity, whereas CMT2 is an axonal neuropathy with relatively normal nerve conduction velocity. CMT patients show a high degree of heterogeneity, due to mutations in multiple genes. This led to the distinction of other subtypes of CMT, including CMT3 (or Dejerine-Sottas disease, a particularly severe demyelinating form of CMT), CMT4 (autosomal recessive form of demyelinating CMT), and CMTX (X-linked form of CMT with both demyelinating and axonal features).

Vogel et al. (441) reported the presence of NF accumulations in CMT. Evidence for the involvement of IFs in the pathogenesis of CMT was provided by the identification of more than 20 mutations in gene coding for NFL in patients with CMT1F and CMT2E. Mutations in *NEFL* gene on chromosome 8 are responsible for approximately 2% of CMT cases and a high percentage of CMT2 cases. These mutations are located throughout the three functional domains of this protein (head, rod, and tail) and consist of substitutions, deletions, and frame-shift mutations. The first two CMT-associated *NEFL* mutations, NFL^{P8R} and NFL^{Q333P}, were identified in, respectively, a Belgian and a Russian family. They disrupt NF assembly and axonal transport in vitro (442, 443). This effect was dominant, since wild-type NFL could not rescue the assembly defect. These mutations also cause the sequestration of mitochondria in cell bodies and proximal axons, the fragmentation of the Golgi apparatus and the degeneration of neuritic processes in cultured neurons. Filament formation was also abolished in SW13 cells by the rod domain A148V mutation (444). These data provide possible mechanisms by which these mutants could be involved in axonal degeneration and CMT pathogenesis.

The Pro-22 residue of NFL is also the target of several mutations: P22R, P22S, and P22T. The P22R mutation, identified in a Korean family, is associated with demyelinating neuropathy features of CMT1F (445). The P22S substitution was first described in a Slovenian CMT2 family (446), then in an Italian family developing a primary axonopathy characterized by giant axons with swellings composed essentially of aggregated NFs (447). Interestingly, clinical and electrophysiological studies from patients with P22S mutation revealed a mixed axonal and demyelinating neuropathy (448), emphasizing the complexity of genotype–phenotype correlations in CMT. Finally, the P22T mutation was detected in unrelated Japanese patients with CMT disease (449). The formation of NF aggregates in patients expressing NFL^{P22S} and NFL^{P22T} mutant proteins could be explain by the ability of these mutations to abolish the phosphorylation of the adjacent Thr21 by cdk5 and ERK1/2, which normally suppressed filament assembly (140). But the phosphorylation of NFL head domain by PKA inhibited aggregate formation in vitro and alleviated aggregates in cortical neurons, providing a potential therapeutic approach to dissociate NF aggregates in CMT disease (140).

The screening of 323 patients with CMT or related peripheral neuropathies allowed the identification of six disease-associated missense mutations and one 3-bp in-frame deletion in the *NEFL* gene (450). Other mutations were also detected in Korean CMT patients (451), in a German family (452), and four mutations in the head and rod domains of NFL, including a L268P substitution and a del322Cys_326 Asn deletion, were identified by the screening of 177 patients (453). Most of these mutated proteins (except E7K and D469N) form aggregates, and thus could alter the axonal transport following their abnormal aggregation in cell bodies and axons. A duplication–insertion mutation of NFL in a patient with CMT was also reported (454), which probably provoked neuronal degeneration through both aggregation and destabilization of the IF network. Finally, new mutations in the *NEFL* gene were identified following the screening of 223 Japanese CMT patients (455). Four heterozygous missense mutations (P8L, E90K, N98S, and E396K) were detected in five unrelated patients as well as a homozygous nonsense mutation (E140Stop) in one patient. All these patients displayed moderate delayed nerve conduction velocities, possibly caused by a loss of large diameter fibers. This study suggested that nonsense *NEFL* mutations probably cause a recessive phenotype, while missense mutations cause a dominant phenotype (455). The majority of NFL mutations are linked to axonal forms of CMT but their implication in demyelinating CMT cannot be excluded since nerves from patients expressing NFL^{L268P} or NFL^{E90K} showed evidence of Schwann cell abnormalities (450, 453).

The generation of the first mouse model of the CMT2E disease was recently reported (456). These mice express the $hNFL^{P22S}$ mutant protein specifically in the nervous system and develop key feature of CMT2E disease at 9 months of age, including aberrant hindlimb posture, motor deficits, hypertrophy of muscle fibers, and loss of muscle innervation without neuronal loss. Axonal NF density appeared normal but MT density was decreased. Thus, the effects of $hNFL^{P22S}$ could arise from cytoskeletal alterations that provoke axonal transport defects due to MT abnormalities. To address whether CMT2E disease is potentially reversible, this mouse model was based on the tetracycline-responsive gene system that allows the suppression of mutant $hNFL^{P22S}$ expression in mature neurons through administration of doxycycline. Remarkably, a 3-month treatment of these mice with doxycycline after disease onset efficiently downregulated expression of $hNFL^{P22S}$ and reversed the neurological phenotype (456), providing hope that future therapeutic strategies might not only stop progress of CMT2E disease but also reverse it. A novel line of CMT2E mice was recently developed by Shen et al. (457). These mice constitutively express human NFL with a glutamic acid to lysine mutation at position 397 (E397K). They recapitulate many of the overt phenotypes observed in CMT2E patients as early as 4 months of age, including aberrant hind limb posture, digit deformities, reduced voluntary locomotor activity, reduced motor nerve conduction velocities (MNCVs), and muscle atrophy. However, no significant denervation was reported and animals showed relatively mild signs of nerve pathology. Examination of the neuronal cytoskeleton revealed NF disorganization in motor and sensory roots, and ectopic accumulations of phosphorylated NFs in motor neuron cell bodies as early as 1 month (457).

Mutations of myotubularin-related protein 2 (MTMR2) (CMT4B), heat-shock protein B1 (HSPB1) (CMT2F), or HSPB8 (CMT2L) can also cause NFL aggregation (458–461), indicating that mutation of NFs is not the only mechanism inducing their accumulation in CMT. Co-expression of Wt HSPB1 with P8R or Q333P CMT mutant NFL reduced their aggregation, induced reversal of mutant NFL aggregates and decreased mutant NFL-induced loss of motor neuron viability (462). On the opposite, mutant HSPB1 has dominant effect on disruption of NF assembly and aggregation of NFL protein (463). In the same way, mutant MTMR2 induces abnormal NFL assembly in transfected cells (460). Zhai et al. (462) showed that deletion of NFL markedly reduces degeneration and loss of motor neurons induced by mutant HSPB1. Another study showed that expression of NFL^{P8R} or NFL^{Q333P} in cultured motor neurons causes the rounding of mitochondria and reduction of axonal diameter before disruption of the NF network (464). Cotransfection of HSPB1 helped to maintain normal NF network, axonal caliber, and mitochondrial morphology. On the other hand, the cotransfection of HSPA1 was effective in neurons expressing NFL^{Q333P}, but not NFL^{P8R}, suggesting that chaperone-based therapies have potential for the treatment of CMT2E but their efficacy would depend on the profile of HSPs induced and the type of *NEFL* mutation.

A model system using a mouse muscle cell line expressing chimeric NF proteins has been developed to explore the mechanism of protein aggregation (465). Formation of protein aggregates in these cells leads to reduced cell viability and activated caspases, which mimic the pathology of CMT and may be a useful model for future studies to investigate the mechanism responsible for NFL disruption.

12.3. Giant Axonal Neuropathy

GAN is a severe neurodegenerative disorder affecting both PNS and CNS with early onset and a fatal outcome in young adults. Phenotypic variability has been reported but typical clinical features include distal limb weakness, areflexia, and a marked gait disturbance. As the disorder progresses, CNS involvement includes electroencephalographic abnormalities, mental retardation, speech defect, seizures, and defective upper motor neuron function.

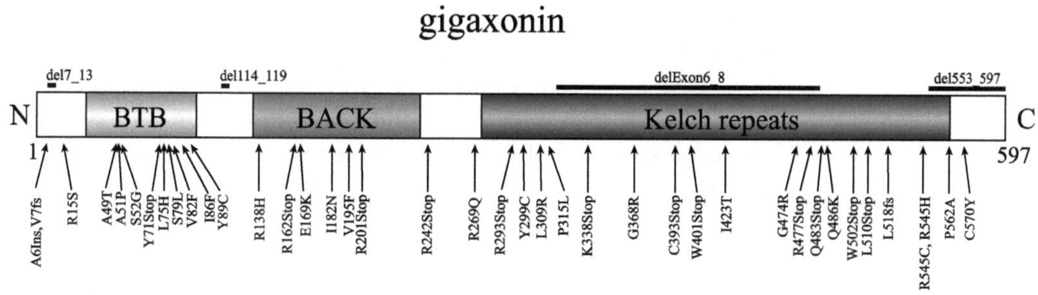

Fig. 2. Mapping of gigaxonin mutations identified in GAN patients. Schematic representation of human gigaxonin protein. BTB, BACK, and C-terminal Kelch domains are represented. Mutations associated with GAN are found in each functional domain. C, carboxy-terminal; N, amino-terminal.

GAN is caused by mutations in the *GAN* gene encoding the ubiquitously expressed protein gigaxonin (Fig. 2). Gigaxonin belongs to a protein family that is characterized by an N-terminal BTB (broad-complex, Tramtrack, and Bric a brac) domain and six kelch repeats (466). BTB/kelch proteins are organizers of the cytoskeletal network and closely linked to the ubiquitin degradation pathway. More than 45 mutations of the gigaxonin have been identified to date. They are localized throughout the *GAN* gene and are thought to lead a loss of function of the encoded protein.

Defined as an axonal neuropathy by electrophysiological examination, GAN is characterized by the presence of abnormally packed IFs producing focal enlargements in the distal regions of axons associated with a reduced number of MTs (467). In contrast, axonal segments proximal to the swellings exhibit a reduced number of NFs (468). Disorganization and accumulation of other types of IFs are also found in skin fibroblasts, Schwann cells, and muscle fibers (469–471), suggesting a critical role of gigaxonin in maintaining IF organization. A decreased inter-NF distance was observed in sural nerve axons of a GAN patient and, more surprisingly, the mean diameter of NFs was increased (12.4 nm in GAN compared with 10.1 nm in controls) (472). Although the mechanism leading to the distal axonal accumulation of NFs is still unclear, an acceleration of their axonal transport was observed in optic nerve from experimentally induced GAN rat model, concomitant with a proximal decreased content of NFs and their distal accumulation (473). The authors proposed that acceleration of NF transport in the presence of a normal rate of NF protein synthesis and insertion into transport system would lead to the formation of distal axonal swellings with packed NFs.

Gigaxonin was proposed to promote the ubiquitin-mediated degradation of microtubule-associated protein 1B (MAP1B) (474), tubulin folding cofactor B (TBCB) (475), and MAP8 (476). Disease-associated gigaxonin mutations perturb its association with these partners while gigaxonin ablation results in their accumulation

(475–477). This raised the possibility that IF accumulation in GAN results from an MT reorganization/destabilization. However, it is intriguing to note that these proteins have opposite effects on MT network: MAP1B is a MT-stabilizing phosphoprotein, whereas overexpression of TBCB depolymerizes MTs. Using primary fibroblasts derived from skin biopsies of GAN patients with aberrant aggregates of vimentin, Cleveland et al. (478) demonstrated that gigaxonin mutations do not affect MT density or TBCB levels. Moreover, the prolonged depletion of the MT network did not induce GAN-like aggregates of vimentin in normal fibroblasts. These results indicated that the generalized disorganization of IFs in GAN patients may not involve TBCB-mediated MT disassembly and must be regulated by a yet unidentified mechanism (478).

Mice deleted in exons 3–5 of the *GAN* gene were produced in order to determine how loss of gigaxonin's function leads to GAN and to evaluate the efficacy of GAN treatments that are currently under development (476). Despite the development of a progressive deterioration for motor function, these animals displayed normal life span and fertility, and giant axons were never seen. Nevertheless these mice exhibited enlarged axons with densely packed NF, leading to the segregation of axonal organelles, a feature characteristic of human GAN pathology. This was accompanied by an axonal loss at the age of 9–12 months. However, it should be noted that some null mice showed no overt neurological phenotypes, suggesting that some genetic modifiers may exist (479). Another mouse model deleted in exons 3–5 of the *GAN* gene was generated by gene targeting in parallel to the first model (480), and showed that gigaxonin depletion induces different effects in mice depending on their genetic background. 129/SvJ mice developed mild but persistent motor deficits starting at 60 weeks of age, while sensory deficits were demonstrated in C57BL/6 animals. No apparent neurodegeneration was observed in knockout mice, but dysregulation of NFs in proximal and distal axons was massive. Indeed, NFs were not only more abundant but they also showed the abnormal increase in diameter and misorientation that are characteristics of the human pathology. Ganay et al. (480) also compared their mutant mice with the first GAN model generated by Ding et al. (476). It is interesting to note that in their hands these mice did not display the early and severe motor deficits reported previously. Finally, another mouse model with deletion of exon 1 of the *GAN* gene was generated (481), which exhibited no overt phenotype and no giant axons but developed accumulations of IF proteins. These aggregates were composed of non-phosphorylated NFH and α-internexin and formed preferentially in cerebral cortex and thalamus. Small aggregates of NFL and peripherin also formed in cell bodies of dorsal root ganglion neurons. Moreover, increased levels of neuronal IF proteins were detected in various regions of the nervous system, confirming the importance of gigaxonin in modulating the levels and organization of IF proteins. Altogether, these studies showed that the absence of gigaxonin in mice results in a milder version of the GAN disease at the behavioral level, associated with a severe disorganization of the NF network that recapitulates what is observed in patients (480).

12.4. Neuronal Intermediate Filament Inclusion Disease

NIFID is a recently described uncommon neurological disorder of early onset with a heterogeneous clinical phenotype, including sporadic frontotemporal dementia associated with a pyramidal and/or extrapyramidal movement disorder. The symptoms comprise behavioral and personality changes, which can be associated to memory loss, cognitive impairment, language deficits, and motor weakness. The cytopathological characteristics consist of neuronal loss, gliosis, swollen neurons, and presence of large inclusions in the cell body of many neurons that are immunoreactive for all of the class IV neuronal IFs and especially for α-internexin (482, 483). These inclusions containing α-internexin but not tau or synuclein distinguish NIFID from most cases of frontotemporal lobar degeneration. This raises the question whether α-internexin-positive neuronal inclusions in NIFID reflect any selective neuronal dysfunction, and as such if they are associated with some specific clinical symptoms. Genetic screening revealed no pathogenic variants for all type IV neuronal IFs, SOD1, NUDEL, and gigaxonin (484, 485). Interestingly, the number of IFs aggregates is high in areas with reduced neuronal loss, and low in sites of intense neuronal degeneration. Cairns et al. (482) proposed that the formation of these inclusions is an early event in the pathogenesis of NIFID, and these aggregates are then released and degraded into the extracellular space following degeneration of the neuron. The mechanism of IF aggregation and the role they play in neuronal dysfunction and cell death are still unclear. Although immunoreactivity for IFs was initially described as the defining pathological feature of NIFID, not all the inclusions in NIFID are IF-positive. It now appears that aggregates of FUS (fused in sarcoma) protein are a more consistent feature of NIFID. Indeed, intracellular accumulations of FUS are more often encountered than IF inclusions and all neurons that contained abnormal IF aggregates also contained FUS inclusions (486). The authors interpreted this finding as suggesting that FUS plays a more central role in the pathogenesis of NIFID and that the abnormal accumulation of IFs is likely a secondary phenomenon.

12.5. Diabetic Neuropathy

Diabetes is the leading cause of peripheral neuropathy worldwide. Diabetic neuropathies are complex, heterogeneous disorders that affect dorsal root ganglia and sensory axons more than motor fibers. Although its pathogenesis has not been fully elucidated, diabetic neuropathy is characterized by slower conduction velocity, impairment of axonal transport, axonal atrophy, and reduced capacity for nerve regeneration. All these features of nerve function depend on the integrity of the axonal cytoskeleton and particularly on NFs. In agreement with this, multiple abnormalities of NF biology were identified in models of diabetes. An impairment of the axonal transport of NFs, actin and tubulin concomitant with a proximal increase and a distal decrease of axonal calibers were observed in rats with streptozotocin-induced diabetes and in BioBreeding rats (a model of spontaneous type I diabetes) (487, 488). The distal axonal atrophy is accompanied by a concomitant NF loss in this region (489), and accumulations of highly phosphorylated NF epitopes are present in proximal axonal segments of dorsal root ganglia sensory neurons from diabetic patients (490). An increase of NF phosphorylation, correlated with activation of JNK, was also detected in lumbar dorsal root ganglia from rat models (491). Finally, there were a substantial decline in the mRNA levels of all three NF subunits as well as reduced NF numbers and densities within large myelinated sensory fibers of long-term diabetic models (492). All these results suggest that NF abnormalities may contribute to the development of diabetic neuropathy or may be affected by this disease. However, slowing of conduction velocity in diabetic models occurs much earlier than loss of NF investment or axonal atrophy (492). To further elucidate the contribution of NFs to diabetic neuropathy pathogenesis, the effect of streptozotocin-induced diabetes was analyzed in NFH-LacZ transgenic mice characterized by axons lacking NFs (329). Interestingly, diabetic mice lacking NFs developed progressive slowing of conduction velocity in their motor and sensory fibers and displayed decreased nerve action potential amplitudes earlier than diabetic mice with normal IF cytoskeleton. Moreover, superimposing diabetes on axons without NFs also accentuated axonal atrophy. Administration of insulin that restored normal glycemia reversed conduction slowing and restored sensory axon caliber. These findings indicate that changes in NF expression, transport or post-translational modifications cannot account alone for neurological features of diabetic neuropathy, but these IFs may help axons to better resist the negative effects of diabetes (329).

12.6. Parkinson Disease

Parkinson disease is a progressive disorder of the CNS characterized by degeneration of dopaminergic neurons of the substantia nigra, resulting in decreased dopamine availability. A neuropathological hallmark of PD is the formation of filamentous neuronal inclusions called Lewy bodies, containing a number of proteins, including α-synuclein protein, NF proteins, ubiquitin and proteasome subunits, torsinA, and parkin (493, 494). Various features distinguish NFs in PD, including inappropriate phosphorylation and proteolysis in Lewy bodies (495, 496), and decreased NFL and NFH mRNA levels (208). A point mutation in the *NEFM* gene was reported in a French-Canadian patient who developed the disease at the age of 16 (497). This mutation consisted in a substitution of Ser for Gly at residue 336, a highly conserved region in the rod domain 2B of NFM, and was argued to disrupt NF assembly. Although three other unaffected family members also carried this mutation, the authors had then proposed that aberrations in neuronal IFs could lead to the development of the pathology seen in PD. However, the G336S mutation does not disrupt the assembly and the distribution of NFs in vitro (498) and the screenings of PD patients of similar or different ethnic background failed to identify this mutations (499, 500). Nevertheless, they identified in NFM gene from two patients a Pro to Gln substitution at residue 725 and a deletion of Val in position 829, two highly conserved sites, suggesting that NF gene mutations could act as susceptibility factors for PD.

12.7. Toxic Agents That Disorganize the Neurofilament Network

IDPN (β, β′-Iminodipropionitrile) is a toxin that segregates MTs from NFs and thereby causes their abnormal accumulation. Its administration affects the slow axonal transport of NFs (501), and induces their accumulation in the proximal region of axons (210) and segregation of NFs and MTs distal to this region (213, 502). A decrease in axonal diameter has been reported distal to the swollen proximal segments (503). This proximal neurofilamentous axonopathy induced by IDPN may have a potential impact on the function or stability of the neuromuscular junction. The aggregation of NFs occurs through abnormal cross-linking of hyperphosphorylated NF (504). The susceptibility of various neurons to these effects depends on their NF content; NF-rich large caliber axons being the most affected (505). Unlike normal mice, NFH null mutant mice do not develop swellings of motor axons when treated with IDPN (287), demonstrating that NFH protein is a key mediator of IDPN-induced axonopathy.

NFs could also be the targets of *n*-hexane-induced neuropathy. Occupational or experimental exposure to the hexacarbons *n*-hexane or methyl *n*-butyl ketone produces nerve damage classified as a central-peripheral distal axonopathy (506). Giant axon swelling containing massive accumulation of NFs was first described as the morphological hallmark of HD axonopathy. However, recent researches have suggested that the swellings are a nonspecific effect related to subchronic exposure to HD, and that atrophied axons in peripheral and central nerves are instead the defining morphologic feature of HD neuropathy (507). HD-induced neuropathy is characterized by a significant reduction in the levels of NF proteins (508, 509), which could partly be responsible for HD-induced axon atrophy. Several studies proposed that alterations of the UPS could be involved in HD-induced neuropathy. Components of UPS system (e.g., ubiquitin-activating enzyme E1, CHIP, and proteasome) were significantly increased in sciatic nerve of HD-intoxicated rats (199). Exposure of astrocyte cultures to HD led to the formation of IFs that contained ubiquitin–protein conjugates. In addition, ubiquitin mRNA levels were found to be elevated approximately threefold by HD treatment (510). In the same way, NFM ubiquitination and degradation are increased after HD treatment, and the application of proteasome inhibitor MG132 partly reversed HD-induced decrease of NFM (199). These results suggested that regulation of NFs levels via ubiquitin-dependent proteolysis provided a potential mechanism for HD-induced neuropathy. However, other studies support the notion that NF accumulation could be epiphenomenal and thus not be directly involved in the pathogenesis of HD-induced neuropathy. For example, the temporal onset of neurotoxicity by HD does not correlate with the onset or the magnitude of NF accumulation (511) and the progressive NF accumulation does not correlate with transient and repeated block of fast anterograde axonal transport by HD (512, 513). Finally, the comparison of changes in fast axonal transport, the development of symptoms, and pathology in transgenic mice lacking axonal NFs with those of their normal littermates revealed no significant difference between mice with or without axonal NFs after HD treatment (514–516). The authors concluded that modification of NFs and their accumulation contribute little or nothing to the pathogenesis of symptoms and pathology in HD neurotoxicity. The injection of aluminum chloride in spinal cord of rabbits causes the formation of NF tangles in neuronal perikarya and proximal parts of dendrites (212). NF transport is maintained in the distal region of the axon, resulting in lack of NFs in axonal segments immediately distal to the block (196). NFs sequestered in cell bodies are highly phosphorylated (517), which may account for their abnormal axonal transport. Moreover, aluminum inhibits NF degradation and dephosphorylation (198, 518, 519) and reduces the assembly of newly synthesized NF subunits into NFs (520).

The effects of acrylamide are very similar to those of aluminum. During the development of experimental acrylamide neuropathy, NF transport is inhibited and NFs accumulate in proximal axons with formation of axonal swellings (211). The administration of acrylamide also increases the expression of mRNA for NF proteins and the phosphorylation of NFs in rats (521–523), but decreases their degradation (524). However, acrylamide-induced neurotoxicity is not initiated exclusively through its action on axonal NFs because NF-deficient mutant *quiver* quails, crayfish (a species lacking NFs), and NFH-LacZ mice are sensitive to neurotoxic effects of acrylamide (515, 525, 526).

Arsenic also disrupts the organization of NF network. Protein analysis of sciatic nerves from rats treated with arsenite (the inorganic form of arsenic) showed disappearance of NFs (527). DeFuria and Shea (528) demonstrate that arsenite decreases NF transport and induces the perikaryal accumulation of phosphorylated NFs in NB2/d1 cells and in cultured dorsal root ganglion neurons. These effects were prevented by inhibiting JNK and GSK-3β.

Sarin and *soman* are two highly toxic organophosphorus nerve agents that inhibit acetylcholinesterase function at neuronal synapses and cause many toxic effects including death within minutes. A strong increase of NFH phosphorylation was reported in cerebral cortex and cerebellum of the rat brain after sarin and soman exposure (529). Soman exposure also affects content and activity of CaMKIIα and PKC in different brain regions (530, 531). Hyperphosphorylation of NFs by these abnormally activated kinases could induce the aggregation of these proteins thereby disrupting the neuronal structure as well as function.

Finally, lead exposure during the development of a chicken model of auditory temporal processing results in decreased amount and phosphorylation of NFM within the axons connecting auditory nuclei in the avian brainstem (532). During the mouse development, lead exposure increases phosphorylation of both NFM and NFH within auditory brainstem nuclei. Moreover, neuritic beadings immunolabeled for NFs are observed following lead exposure both in vivo and in vitro, suggesting an impairment of the axonal transport (533).

References

1. Julien JP, Mushynski WE (1998) Neurofilaments in health and disease. Prog Nucleic Acid Res Mol Biol 61:1–23
2. Yuan A, Rao MV, Sasaki T et al (2006) Alpha-internexin is structurally and functionally associated with the neurofilament triplet proteins in the mature CNS. J Neurosci 26: 10006–10019
3. Yan Y, Jensen K, Brown A (2007) The polypeptide composition of moving and stationary neurofilaments in cultured sympathetic neurons. Cell Motil Cytoskeleton 64:299–309
4. Perrot R, Berges R, Bocquet A et al (2008) Review of the multiple aspects of neurofilament functions, and their possible contribution to neurodegeneration. Mol Neurobiol 38: 27–65
5. Nixon RA, Shea TB (1992) Dynamics of neuronal intermediate filaments: a developmental perspective. Cell Motil Cytoskeleton 22:81–91
6. Perrot R, Eyer J (2009) Neuronal intermediate filaments and neurodegenerative disorders. Brain Res Bull 80:282–295
7. Herrmann H, Bar H, Kreplak L et al (2007) Intermediate filaments: from cell architecture to nanomechanics. Nat Rev Mol Cell Biol 8:562–573
8. Kim S, Coulombe PA (2007) Intermediate filament scaffolds fulfill mechanical, organizational, and signaling functions in the cytoplasm. Genes Dev 21:1581–1597
9. Oshima RG (2007) Intermediate filaments: a historical perspective. Exp Cell Res 313: 1981–1994
10. Beaulieu JM, Robertson J, Julien JP (1999) Interactions between peripherin and neurofilaments in cultured cells: disruption of peripherin assembly by the NF-M and NF-H subunits. Biochem Cell Biol 77:41–45
11. Morris JR, Lasek RJ (1982) Stable polymers of the axonal cytoskeleton: the axoplasmic ghost. J Cell Biol 92:192–198
12. Carter J, Gragerov A, Konvicka K et al (1998) Neurofilament (NF) assembly; divergent characteristics of human and rodent NF-L subunits. J Biol Chem 273:5101–5108
13. Gardner EE, Dahl D, Bignami A (1984) Formation of 10-nanometer filaments from the 150 K-dalton neurofilament protein in vitro. J Neurosci Res 11:145–155
14. Geisler N, Weber K (1981) Self-assembly in Vitro of the 68,000 molecular weight component of the mammalian neurofilament triplet proteins into intermediate-sized filaments. J Mol Biol 151:565–571
15. Hisanaga S, Hirokawa N (1988) Structure of the peripheral domains of neurofilaments revealed by low angle rotary shadowing. J Mol Biol 202:297–305
16. Hisanaga S, Hirokawa N (1990) Molecular architecture of the neurofilament. II. Reassembly process of neurofilament L protein in vitro. J Mol Biol 211:871–882

17. Jacomy H, Zhu Q, Couillard-Despres S et al (1999) Disruption of type IV intermediate filament network in mice lacking the neurofilament medium and heavy subunits. J Neurochem 73:972–984
18. Liem RK, Hutchison SB (1982) Purification of individual components of the neurofilament triplet: filament assembly from the 70 000-dalton subunit. Biochemistry 21:3221–3226
19. Lee MK, Xu Z, Wong PC et al (1993) Neurofilaments are obligate heteropolymers in vivo. J Cell Biol 122:1337–1350
20. Scott D, Smith KE, O'Brien BJ et al (1985) Characterization of mammalian neurofilament triplet proteins. Subunit stoichiometry and morphology of native and reconstituted filaments. J Biol Chem 260:10736–10747
21. Geisler N, Kaufmann E, Fischer S et al (1983) Neurofilament architecture combines structural principles of intermediate filaments with carboxy-terminal extensions increasing in size between triplet proteins. EMBO J 2: 1295–1302
22. Birkenberger L, Ip W (1990) Properties of the desmin tail domain: studies using synthetic peptides and antipeptide antibodies. J Cell Biol 111:2063–2075
23. Julien JP, Mushynski WE (1982) Multiple phosphorylation sites in mammalian neurofilament polypeptides. J Biol Chem 257:10467–10470
24. Julien JP, Mushynski WE (1983) The distribution of phosphorylation sites among identified proteolytic fragments of mammalian neurofilaments. J Biol Chem 258: 4019–4025
25. Angelides KJ, Smith KE, Takeda M (1989) Assembly and exchange of intermediate filament proteins of neurons: neurofilaments are dynamic structures. J Cell Biol 108: 1495–1506
26. Heins S, Wong PC, Muller S et al (1993) The rod domain of NF-L determines neurofilament architecture, whereas the end domains specify filament assembly and network formation. J Cell Biol 123:1517–1533
27. Herrmann H, Haner M, Brettel M et al (1999) Characterization of distinct early assembly units of different intermediate filament proteins. J Mol Biol 286:1403–1420
28. Parry DA, Strelkov SV, Burkhard P et al (2007) Towards a molecular description of intermediate filament structure and assembly. Exp Cell Res 313:2204–2216
29. Herrmann H, Aebi U (2004) Intermediate filaments: molecular structure, assembly mechanism, and integration into functionally distinct intracellular scaffolds. Annu Rev Biochem 73:749–789
30. Balin BJ, Clark EA, Trojanowski JQ et al (1991) Neurofilament reassembly in vitro: biochemical, morphological and immunoelectron microscopic studies employing monoclonal antibodies to defined epitopes. Brain Res 556:181–195
31. Balin BJ, Lee VM (1991) Individual neurofilament subunits reassembled in vitro exhibit unique biochemical, morphological and immunological properties. Brain Res 556: 196–208
32. Hirokawa N, Glicksman MA, Willard MB (1984) Organization of mammalian neurofilament polypeptides within the neuronal cytoskeleton. J Cell Biol 98:1523–1536
33. Kim S, Chang R, Teunissen C et al (2011) Neurofilament stoichiometry simulations during neurodegeneration suggest a remarkable self-sufficient and stable in vivo protein structure. J Neurol Sci 307:132–138
34. Chin SS, Macioce P, Liem RK (1991) Effects of truncated neurofilament proteins on the endogenous intermediate filaments in transfected fibroblasts. J Cell Sci 99(Pt 2):335–350
35. Ching GY, Liem RK (1993) Assembly of type IV neuronal intermediate filaments in nonneuronal cells in the absence of preexisting cytoplasmic intermediate filaments. J Cell Biol 122:1323–1335
36. Ching GY, Liem RK (1999) Analysis of the roles of the head domains of type IV rat neuronal intermediate filament proteins in filament assembly using domain-swapped chimeric proteins. J Cell Sci 112(Pt 13):2233–2240
37. Gill SR, Wong PC, Monteiro MJ et al (1990) Assembly properties of dominant and recessive mutations in the small mouse neurofilament (NF-L) subunit. J Cell Biol 111:2005–2019
38. Wong PC, Cleveland DW (1990) Characterization of dominant and recessive assembly-defective mutations in mouse neurofilament NF-M. J Cell Biol 111: 1987–2003
39. Sihag RK, Nixon RA (1991) Identification of Ser-55 as a major protein kinase A phosphorylation site on the 70-kDa subunit of neurofilaments. Early turnover during axonal transport. J Biol Chem 266:18861–18867
40. Sihag RK, Jaffe H, Nixon RA et al (1999) Serine-23 is a major protein kinase A phosphorylation site on the amino-terminal head domain of the middle molecular mass subunit of neurofilament proteins. J Neurochem 72:491–499

41. Hisanaga S, Gonda Y, Inagaki M et al (1990) Effects of phosphorylation of the neurofilament L protein on filamentous structures. Cell Regul 1:237–248
42. Gibb BJ, Brion JP, Brownlees J et al (1998) Neuropathological abnormalities in transgenic mice harbouring a phosphorylation mutant neurofilament transgene. J Neurochem 70:492–500
43. Mukai H, Toshimori M, Shibata H et al (1996) PKN associates and phosphorylates the head-rod domain of neurofilament protein. J Biol Chem 271:9816–9822
44. Dong DL, Xu ZS, Chevrier MR et al (1993) Glycosylation of mammalian neurofilaments. Localization of multiple O-linked N-acetylglucosamine moieties on neurofilament polypeptides L and M. J Biol Chem 268:16679–16687
45. Dong DL, Xu ZS, Hart GW et al (1996) Cytoplasmic O-GlcNAc modification of the head domain and the KSP repeat motif of the neurofilament protein neurofilament-H. J Biol Chem 271:20845–20852
46. Kim SK, Choi JH, Suh PG et al (2006) Pleckstrin homology domain of phospholipase C-gamma1 directly binds to 68-kDa neurofilament light chain. Exp Mol Med 38: 265–272
47. Kim SK, Kim H, Yang YR et al (2011) Phosphatidylinositol phosphates directly bind to neurofilament light chain (NF-L) for the regulation of NF-L self assembly. Exp Mol Med 43:153–160
48. Nguyen MD, Shu T, Sanada K et al (2004) A NUDEL-dependent mechanism of neurofilament assembly regulates the integrity of CNS neurons. Nat Cell Biol 6:595–608
49. Leterrier JF, Kas J, Hartwig J et al (1996) Mechanical effects of neurofilament cross-bridges. Modulation by phosphorylation, lipids, and interactions with F-actin. J Biol Chem 271:15687–15694
50. Rammensee S, Janmey PA, Bausch AR (2007) Mechanical and structural properties of in vitro neurofilament hydrogels. Eur Biophys J 36:661–668
51. Kreplak L, Bar H, Leterrier JF et al (2005) Exploring the mechanical behavior of single intermediate filaments. J Mol Biol 354: 569–577
52. Myers MW, Lazzarini RA, Lee VM et al (1987) The human mid-size neurofilament subunit: a repeated protein sequence and the relationship of its gene to the intermediate filament gene family. EMBO J 6:1617–1626
53. Strausberg RL, Feingold EA, Grouse LH et al (2002) Generation and initial analysis of more than 15,000 full-length human and mouse cDNA sequences. Proc Natl Acad Sci U S A 99:16899–16903
54. Lees JF, Shneidman PS, Skuntz SF et al (1988) The structure and organization of the human heavy neurofilament subunit (NF-H) and the gene encoding it. EMBO J 7: 1947–1955
55. Thyagarajan A, Strong MJ, Szaro BG (2007) Post-transcriptional control of neurofilaments in development and disease. Exp Cell Res 313:2088–2097
56. Escurat M, Djabali K, Gumpel M et al (1990) Differential expression of two neuronal intermediate-filament proteins, peripherin and the low-molecular-mass neurofilament protein (NF-L), during the development of the rat. J Neurosci 10:764–784
57. Oblinger MM, Wong J, Parysek LM (1989) Axotomy-induced changes in the expression of a type III neuronal intermediate filament gene. J Neurosci 9:3766–3775
58. Carden MJ, Trojanowski JQ, Schlaepfer WW et al (1987) Two-stage expression of neurofilament polypeptides during rat neurogenesis with early establishment of adult phosphorylation patterns. J Neurosci 7: 3489–3504
59. Pachter JS, Liem RK (1984) The differential appearance of neurofilament triplet polypeptides in the developing rat optic nerve. Dev Biol 103:200–210
60. Shaw G, Osborn M, Weber K (1981) An immunofluorescence microscopical study of the neurofilament triplet proteins, vimentin and glial fibrillary acidic protein within the adult rat brain. Eur J Cell Biol 26:68–82
61. Shaw G, Weber K (1982) Differential expression of neurofilament triplet proteins in brain development. Nature 298:277–279
62. Walker KL, Yoo HK, Undamatla J et al (2001) Loss of neurofilaments alters axonal growth dynamics. J Neurosci 21:9655–9666
63. Zhao Y, Szaro BG (1995) The optic tract and tectal ablation influence the composition of neurofilaments in regenerating optic axons of Xenopus laevis. J Neurosci 15:4629–4640
64. Cuenca N, Fernandez E, de Juan J et al (1987) Postnatal development of microtubules and neurofilaments in the rat optic nerve: a quantitative study. J Comp Neurol 263:613–617
65. Breen KC, Anderton BH (1991) Temporal expression of neurofilament polypeptides in differentiating neuroblastoma cells. Neuroreport 2:21–24
66. Giasson BI, Mushynski WE (1997) Study of proline-directed protein kinases involved in

phosphorylation of the heavy neurofilament subunit. J Neurosci 17:9466–9472
67. Elder GA, Friedrich VL Jr, Bosco P et al (1998) Absence of the mid-sized neurofilament subunit decreases axonal calibers, levels of light neurofilament (NF-L), and neurofilament content. J Cell Biol 141:727–739
68. Zhu Q, Couillard-Despres S, Julien JP (1997) Delayed maturation of regenerating myelinated axons in mice lacking neurofilaments. Exp Neurol 148:299–316
69. Zecevic N, Milosevic A, Rakic S et al (1999) Early development and composition of the human primordial plexiform layer: an immunohistochemical study. J Comp Neurol 412: 241–254
70. Sarnat HB, Flores-Sarnat L (2002) Cajal-Retzius and subplate neurons: their role in cortical development. Eur J Paediatr Neurol 6:91–97
71. Marin-Padilla M (1990) Three-dimensional structural organization of layer I of the human cerebral cortex: a Golgi study. J Comp Neurol 299:89–105
72. Radnikow G, Feldmeyer D, Lubke J (2002) Axonal projection, input and output synapses, and synaptic physiology of Cajal-Retzius cells in the developing rat neocortex. J Neurosci 22:6908–6919
73. Haynes RL, Borenstein NS, Desilva TM et al (2005) Axonal development in the cerebral white matter of the human fetus and infant. J Comp Neurol 484:156–167
74. Verney C, Derer P (1995) Cajal-Retzius neurons in human cerebral cortex at midgestation show immunoreactivity for neurofilament and calcium-binding proteins. J Comp Neurol 359:144–153
75. Goldstein ME, Weiss SR, Lazzarini RA et al (1988) mRNA levels of all three neurofilament proteins decline following nerve transection. Brain Res 427:287–291
76. Oblinger MM, Lasek RJ (1988) Axotomy-induced alterations in the synthesis and transport of neurofilaments and microtubules in dorsal root ganglion cells. J Neurosci 8:1747–1758
77. Wong J, Oblinger MM (1987) Changes in neurofilament gene expression occur after axotomy of dorsal root ganglion neurons: an in situ hybridization study. Metab Brain Dis 2:291–303
78. Hoffman PN, Pollock SC, Striph GG (1993) Altered gene expression after optic nerve transection: reduced neurofilament expression as a general response to axonal injury. Exp Neurol 119:32–36
79. McKerracher L, Essagian C, Aguayo AJ (1993) Temporal changes in beta-tubulin and neurofilament mRNA levels after transection of adult rat retinal ganglion cell axons in the optic nerve. J Neurosci 13:2617–2626
80. Mikucki SA, Oblinger MM (1991) Corticospinal neurons exhibit a novel pattern of cytoskeletal gene expression after injury. J Neurosci Res 30:213–225
81. Tetzlaff W, Alexander SW, Miller FD et al (1991) Response of facial and rubrospinal neurons to axotomy: changes in mRNA expression for cytoskeletal proteins and GAP-43. J Neurosci 11:2528–2544
82. Hoffman PN, Lasek RJ (1980) Axonal transport of the cytoskeleton in regenerating motor neurons: constancy and change. Brain Res 202:317–333
83. Hoffman PN, Thompson GW, Griffin JW et al (1985) Changes in neurofilament transport coincide temporally with alterations in the caliber of axons in regenerating motor fibers. J Cell Biol 101:1332–1340
84. Hoffman PN, Cleveland DW (1988) Neurofilament and tubulin expression recapitulates the developmental program during axonal regeneration: induction of a specific beta-tubulin isotype. Proc Natl Acad Sci U S A 85:4530–4533
85. Muma NA, Hoffman PN, Slunt HH et al (1990) Alterations in levels of mRNAs coding for neurofilament protein subunits during regeneration. Exp Neurol 107: 230–235
86. Wong J, Oblinger MM (1990) A comparison of peripheral and central axotomy effects on neurofilament and tubulin gene expression in rat dorsal root ganglion neurons. J Neurosci 10:2215–2222
87. Jiang YQ, Pickett J, Oblinger MM (1994) Comparison of changes in beta-tubulin and NF gene expression in rat DRG neurons under regeneration-permissive and regeneration-prohibitive conditions. Brain Res 637: 233–241
88. Tetzlaff W, Bisby MA, Kreutzberg GW (1988) Changes in cytoskeletal proteins in the rat facial nucleus following axotomy. J Neurosci 8:3181–3189
89. Jacobs AJ, Swain GP, Snedeker JA et al (1997) Recovery of neurofilament expression selectively in regenerating reticulospinal neurons. J Neurosci 17:5206–5220
90. Zhang G, Jin L, Selzer ME (2011) Assembly properties of lamprey neurofilament subunits and their expression after spinal cord transection. J Comp Neurol 519:3657–3671

91. Gervasi C, Thyagarajan A, Szaro BG (2003) Increased expression of multiple neurofilament mRNAs during regeneration of vertebrate central nervous system axons. J Comp Neurol 461:262–275
92. Toth C, Shim SY, Wang J et al (2008) Ndel1 promotes axon regeneration via intermediate filaments. PLoS One 3:e2014
93. Previtali SC, Zerega B, Sherman DL et al (2003) Myotubularin-related 2 protein phosphatase and neurofilament light chain protein, both mutated in CMT neuropathies, interact in peripheral nerve. Hum Mol Genet 12: 1713–1723
94. Grant P, Pant HC (2000) Neurofilament protein synthesis and phosphorylation. J Neurocytol 29:843–872
95. Sihag RK, Inagaki M, Yamaguchi T et al (2007) Role of phosphorylation on the structural dynamics and function of types III and IV intermediate filaments. Exp Cell Res 313:2098–2109
96. Goldstein ME, Sternberger LA, Sternberger NH (1987) Varying degrees of phosphorylation determine microheterogeneity of the heavy neurofilament polypeptide (Nf-H). J Neuroimmunol 14:135–148
97. Jones SM, Williams RC Jr (1982) Phosphate content of mammalian neurofilaments. J Biol Chem 257:9902–9905
98. Lee VM, Otvos L Jr, Carden MJ et al (1988) Identification of the major multiphosphorylation site in mammalian neurofilaments. Proc Natl Acad Sci U S A 85:1998–2002
99. Pant HC, Veeranna (1995) Neurofilament phosphorylation. Biochem Cell Biol 73: 575–592
100. Glicksman MA, Soppet D, Willard MB (1987) Posttranslational modification of neurofilament polypeptides in rabbit retina. J Neurobiol 18:167–196
101. Nixon RA, Paskevich PA, Sihag RK et al (1994) Phosphorylation on carboxyl terminus domains of neurofilament proteins in retinal ganglion cell neurons in vivo: influences on regional neurofilament accumulation, interneurofilament spacing, and axon caliber. J Cell Biol 126:1031–1046
102. Oblinger MM, Brady ST, McQuarrie IG et al (1987) Cytotypic differences in the protein composition of the axonally transported cytoskeleton in mammalian neurons. J Neurosci 7:453–462
103. Sternberger LA, Sternberger NH (1983) Monoclonal antibodies distinguish phosphorylated and nonphosphorylated forms of neurofilaments in situ. Proc Natl Acad Sci U S A 80:6126–6130
104. Sihag RK, Jeng AY, Nixon RA (1988) Phosphorylation of neurofilament proteins by protein kinase C. FEBS Lett 233:181–185
105. Sihag RK, Nixon RA (1989) In vivo phosphorylation of distinct domains of the 70-kilodalton neurofilament subunit involves different protein kinases. J Biol Chem 264: 457–464
106. Sihag RK, Nixon RA (1990) Phosphorylation of the amino-terminal head domain of the middle molecular mass 145-kDa subunit of neurofilaments. Evidence for regulation by second messenger-dependent protein kinases. J Biol Chem 265:4166–4171
107. Nixon RA, Lewis SE (1986) Differential turnover of phosphate groups on neurofilament subunits in mammalian neurons in vivo. J Biol Chem 261:16298–16301
108. Nixon RA, Lewis SE, Dahl D et al (1989) Early posttranslational modifications of the three neurofilament subunits in mouse retinal ganglion cells: neuronal sites and time course in relation to subunit polymerization and axonal transport. Brain Res Mol Brain Res 5:93–108
109. Nixon RA, Lewis SE, Marotta CA (1987) Posttranslational modification of neurofilament proteins by phosphate during axoplasmic transport in retinal ganglion cell neurons. J Neurosci 7:1145–1158
110. Hashimoto R, Nakamura Y, Goto H et al (1998) Domain- and site-specific phosphorylation of bovine NF-L by Rho-associated kinase. Biochem Biophys Res Commun 245:407–411
111. Hashimoto R, Nakamura Y, Komai S et al (2000) Site-specific phosphorylation of neurofilament-L is mediated by calcium/calmodulin-dependent protein kinase II in the apical dendrites during long-term potentiation. J Neurochem 75:373–382
112. Ackerley S, Thornhill P, Grierson AJ et al (2003) Neurofilament heavy chain side arm phosphorylation regulates axonal transport of neurofilaments. J Cell Biol 161:489–495
113. Archer DR, Watson DF, Griffin JW (1994) Phosphorylation-dependent immunoreactivity of neurofilaments and the rate of slow axonal transport in the central and peripheral axons of the rat dorsal root ganglion. J Neurochem 62:1119–1125
114. Jung C, Yabe JT, Lee S et al (2000) Hypophosphorylated neurofilament subunits undergo axonal transport more rapidly than more extensively phosphorylated subunits in situ. Cell Motil Cytoskeleton 47:120–129
115. Lewis SE, Nixon RA (1988) Multiple phosphorylated variants of the high molecular

mass subunit of neurofilaments in axons of retinal cell neurons: characterization and evidence for their differential association with stationary and moving neurofilaments. J Cell Biol 107:2689–2701

116. Nixon RA, Brown BA, Marotta CA (1982) Posttranslational modification of a neurofilament protein during axoplasmic transport: implications for regional specialization of CNS axons. J Cell Biol 94:150–158

117. Nixon RA, Logvinenko KB (1986) Multiple fates of newly synthesized neurofilament proteins: evidence for a stationary neurofilament network distributed nonuniformly along axons of retinal ganglion cell neurons. J Cell Biol 102:647–659

118. Li BS, Veeranna, Gu J et al (1999) Activation of mitogen-activated protein kinases (Erkl and Erk2) cascade results in phosphorylation of NF-M tail domains in transfected NIH 3T3 cells. Eur J Biochem 262:211–7

119. Pearson G, Robinson F, Beers Gibson T et al (2001) Mitogen-activated protein (MAP) kinase pathways: regulation and physiological functions. Endocr Rev 22:153–183

120. Li BS, Veeranna, Grant P et al (1999) Calcium influx and membrane depolarization induce phosphorylation of neurofilament (NF-M) KSP repeats in PC12 cells. Brain Res Mol Brain Res 70:84–91

121. Li BS, Zhang L, Gu J et al (2000) Integrin alpha(1) beta(1)-mediated activation of cyclin-dependent kinase 5 activity is involved in neurite outgrowth and human neurofilament protein H Lys-Ser-Pro tail domain phosphorylation. J Neurosci 20: 6055–6062

122. de Waegh SM, Lee VM, Brady ST (1992) Local modulation of neurofilament phosphorylation, axonal caliber, and slow axonal transport by myelinating Schwann cells. Cell 68:451–463

123. Hsieh ST, Crawford TO, Griffin JW (1994) Neurofilament distribution and organization in the myelinated axons of the peripheral nervous system. Brain Res 642:316–326

124. Mata M, Kupina N, Fink DJ (1992) Phosphorylation-dependent neurofilament epitopes are reduced at the node of Ranvier. J Neurocytol 21:199–210

125. Reles A, Friede RL (1991) Axonal cytoskeleton at the nodes of Ranvier. J Neurocytol 20:450–458

126. Dashiell SM, Tanner SL, Pant HC et al (2002) Myelin-associated glycoprotein modulates expression and phosphorylation of neuronal cytoskeletal elements and their associated kinases. J Neurochem 81:1263–1272

127. Yin X, Crawford TO, Griffin JW et al (1998) Myelin-associated glycoprotein is a myelin signal that modulates the caliber of myelinated axons. J Neurosci 18:1953–1962

128. Bajaj NP, Miller CC (1997) Phosphorylation of neurofilament heavy-chain side-arm fragments by cyclin-dependent kinase-5 and glycogen synthase kinase-3alpha in transfected cells. J Neurochem 69:737–743

129. Guidato S, Tsai LH, Woodgett J et al (1996) Differential cellular phosphorylation of neurofilament heavy side-arms by glycogen synthase kinase-3 and cyclin-dependent kinase-5. J Neurochem 66:1698–1706

130. Hisanaga S, Uchiyama M, Hosoi T et al (1995) Porcine brain neurofilament-H tail domain kinase: its identification as cdk5/p26 complex and comparison with cdc2/cyclin B kinase. Cell Motil Cytoskeleton 31:283–297

131. Lew J, Winkfein RJ, Paudel HK et al (1992) Brain proline-directed protein kinase is a neurofilament kinase which displays high sequence homology to p34cdc2. J Biol Chem 267:25922–25926

132. Sharma M, Sharma P, Pant HC (1999) CDK-5-mediated neurofilament phosphorylation in SHSY5Y human neuroblastoma cells. J Neurochem 73:79–86

133. Shetty KT, Link WT, Pant HC (1993) cdc2-like kinase from rat spinal cord specifically phosphorylates KSPXK motifs in neurofilament proteins: isolation and characterization. Proc Natl Acad Sci U S A 90:6844–6848

134. Sun D, Leung CL, Liem RK (1996) Phosphorylation of the high molecular weight neurofilament protein (NF-H) by Cdk5 and p35. J Biol Chem 271:14245–14251

135. Veeranna, Amin ND, Ahn NG et al (1998) Mitogen-activated protein kinases (Erkl,2) phosphorylate Lys-Ser-Pro (KSP) repeats in neurofilament proteins NF-H and NF-M. J Neurosci 18:4008–21

136. Miyasaka H, Okabe S, Ishiguro K et al (1993) Interaction of the tail domain of high molecular weight subunits of neurofilaments with the COOH-terminal region of tubulin and its regulation by tau protein kinase II. J Biol Chem 268:22695–22702

137. Brownlees J, Yates A, Bajaj NP et al (2000) Phosphorylation of neurofilament heavy chain side-arms by stress activated protein kinase-1b/Jun N-terminal kinase-3. J Cell Sci 113(Pt 3):401–407

138. Guan RJ, Khatra BS, Cohlberg JA (1991) Phosphorylation of bovine neurofilament proteins by protein kinase FA (glycogen synthase kinase 3). J Biol Chem 266: 8262–8267

139. Ackerley S, Grierson AJ, Banner S et al (2004) p38alpha stress-activated protein kinase phosphorylates neurofilaments and is associated with neurofilament pathology in amyotrophic lateral sclerosis. Mol Cell Neurosci 26: 354–364
140. Sasaki T, Gotow T, Shiozaki M et al (2006) Aggregate formation and phosphorylation of neurofilament-L Pro22 Charcot-Marie-Tooth disease mutants. Hum Mol Genet 15: 943–952
141. Giasson BI, Mushynski WE (1996) Aberrant stress-induced phosphorylation of perikaryal neurofilaments. J Biol Chem 271: 30404–30409
142. O'Ferrall EK, Robertson J, Mushynski WE (2000) Inhibition of aberrant and constitutive phosphorylation of the high-molecular-mass neurofilament subunit by CEP-1347 (KT7515), an inhibitor of the stress-activated protein kinase signaling pathway. J Neurochem 75:2358–2367
143. Bennett GS, Quintana R (1997) Identification of Ser-Pro and Thr-Pro phosphorylation sites in chicken neurofilament-M tail domain. J Neurochem 68:534–543
144. Dosemeci A, Floyd CC, Pant HC (1990) Characterization of neurofilament-associated protein kinase activities from bovine spinal cord. Cell Mol Neurobiol 10:369–382
145. Floyd CC, Grant P, Gallant PE et al (1991) Principal neurofilament-associated protein kinase in squid axoplasm is related to casein kinase I. J Biol Chem 266:4987–4994
146. Hollander BA, Bennett GS, Shaw G (1996) Localization of sites in the tail domain of the middle molecular mass neurofilament subunit phosphorylated by a neurofilament-associated kinase and by casein kinase I. J Neurochem 66:412–420
147. Link WT, Dosemeci A, Floyd CC et al (1993) Bovine neurofilament-enriched preparations contain kinase activity similar to casein kinase I–neurofilament phosphorylation by casein kinase I (CKI). Neurosci Lett 151:89–93
148. Nakamura Y, Hashimoto R, Kashiwagi Y et al (1999) Casein kinase II is responsible for phosphorylation of NF-L at Ser-473. FEBS Lett 455:83–86
149. Xu ZS, Liu WS, Willard M (1990) Identification of serine 473 as a major phosphorylation site in the neurofilament polypeptide NF-L. J Neurosci 10:1838–1846
150. Zheng YL, Li BS, Veeranna et al (2003) Phosphorylation of the head domain of neurofilament protein (NF-M): a factor regulating topographic phosphorylation of NF-M tail domain KSP sites in neurons. J Biol Chem 278:24026–24032
151. Leterrier JF, Eyer J (1987) Properties of highly viscous gels formed by neurofilaments in vitro. A possible consequence of a specific inter-filament cross-bridging. Biochem J 245:93–101
152. Pant HC (1988) Dephosphorylation of neurofilament proteins enhances their susceptibility to degradation by calpain. Biochem J 256:665–668
153. Sacher MG, Athlan ES, Mushynski WE (1994) Increased phosphorylation of the amino-terminal domain of the low molecular weight neurofilament subunit in okadaic acid-treated neurons. J Biol Chem 269:18480–18484
154. Saito T, Shima H, Osawa Y et al (1995) Neurofilament-associated protein phosphatase 2A: its possible role in preserving neurofilaments in filamentous states. Biochemistry 34:7376–7384
155. Strack S, Westphal RS, Colbran RJ et al (1997) Protein serine/threonine phosphatase 1 and 2A associate with and dephosphorylate neurofilaments. Brain Res Mol Brain Res 49:15–28
156. Veeranna, Shetty KT, Link WT et al (1995) Neuronal cyclin-dependent kinase-5 phosphorylation sites in neurofilament protein (NF-H) are dephosphorylated by protein phosphatase 2A. J Neurochem 64:2681–90
157. Schechter R, Abboud M, Johnson G (1999) Brain endogenous insulin effects on neurite growth within fetal rat neuron cell cultures. Brain Res Dev Brain Res 116:159–167
158. Schechter R, Yanovitch T, Abboud M et al (1998) Effects of brain endogenous insulin on neurofilament and MAPK in fetal rat neuron cell cultures. Brain Res 808:270–278
159. Schechter R, Beju D, Miller KE (2005) The effect of insulin deficiency on tau and neurofilament in the insulin knockout mouse. Biochem Biophys Res Commun 334:979–986
160. Slawson C, Hart GW (2003) Dynamic interplay between O-GlcNAc and O-phosphate: the sweet side of protein regulation. Curr Opin Struct Biol 13:631–636
161. Ludemann N, Clement A, Hans VH et al (2005) O-glycosylation of the tail domain of neurofilament protein M in human neurons and in spinal cord tissue of a rat model of amyotrophic lateral sclerosis (ALS). J Biol Chem 280:31648–31658
162. Ryle C, Leow CK, Donaghy M (1997) Nonenzymatic glycation of peripheral and central nervous system proteins in experimental diabetes mellitus. Muscle Nerve 20:577–584

163. Chou SM, Wang HS, Taniguchi A et al (1998) Advanced glycation endproducts in neurofilament conglomeration of motoneurons in familial and sporadic amyotrophic lateral sclerosis. Mol Med 4:324–332
164. Suzuki Y, Tanaka M, Sohmiya M et al (2005) Identification of nitrated proteins in the normal rat brain using a proteomics approach. Neurol Res 27:630–633
165. Chou SM, Wang HS, Taniguchi A (1996) Role of SOD-1 and nitric oxide/cyclic GMP cascade on neurofilament aggregation in ALS/MND. J Neurol Sci 139(Suppl):16–26
166. Crow JP, Ye YZ, Strong M et al (1997) Superoxide dismutase catalyzes nitration of tyrosines by peroxynitrite in the rod and head domains of neurofilament-L. J Neurochem 69:1945–1953
167. Reynolds MR, Berry RW, Binder LI (2007) Nitration in neurodegeneration: deciphering the "Hows" "nYs". Biochemistry 46: 7325–7336
168. Troncoso JC, Costello AC, Kim JH et al (1995) Metal-catalyzed oxidation of bovine neurofilaments in vitro. Free Radic Biol Med 18:891–899
169. Kim NH, Jeong MS, Choi SY et al (2004) Oxidative modification of neurofilament-L by the Cu, Zn-superoxide dismutase and hydrogen peroxide system. Biochimie 86:553–559
170. Gou JP, Leterrier JF (1995) Possible involvement of ubiquitination in neurofilament degradation. Biochem Biophys Res Commun 217:529–538
171. Hoffman PN, Lasek RJ (1975) The slow component of axonal transport. Identification of major structural polypeptides of the axon and their generality among mammalian neurons. J Cell Biol 66:351–366
172. Schlaepfer WW (1974) Calcium-induced degeneration of axoplasm in isolated segments of rat peripheral nerve. Brain Res 69:203–215
173. Millecamps S, Gowing G, Corti O et al (2007) Conditional NF-L transgene expression in mice for in vivo analysis of turnover and transport rate of neurofilaments. J Neurosci 27:4947–4956
174. Eagles PA, Gilbert DS, Maggs A (1981) The location of phosphorylation sites and Ca2+-dependent proteolytic cleavage sites on the major neurofilament polypeptides from Myxicola infundibulum. Biochem J 199: 101–111
175. Gallant PE, Pant HC, Pruss RM et al (1986) Calcium-activated proteolysis of neurofilament proteins in the squid giant neuron. J Neurochem 46:1573–1581
176. Pant HC, Gainer H (1980) Properties of a calcium-activated protease in squid axoplasm which selectively degrades neurofilament proteins. J Neurobiol 11:1–12
177. Perlmutter LS, Gall C, Baudry M et al (1990) Distribution of calcium-activated protease calpain in the rat brain. J Comp Neurol 296:269–276
178. Schlaepfer WW, Lee C, Lee VM et al (1985) An immunoblot study of neurofilament degradation in situ and during calcium-activated proteolysis. J Neurochem 44:502–509
179. Roots BI (1983) Neurofilament accumulation induced in synapses by leupeptin. Science 221:971–972
180. Schlaepfer WW (1971) Experimental alterations of neurofilaments and neurotubules by calcium and other ions. Exp Cell Res 67: 73–80
181. Murachi T (1990) Calpain and calpastatin. Rinsho Byori 38:337–346
182. Murachi T, Tanaka K, Hatanaka M et al (1980) Intracellular Ca2+-dependent protease (calpain) and its high-molecular-weight endogenous inhibitor (calpastatin). Adv Enzyme Regul 19:407–424
183. Hamakubo T, Kannagi R, Murachi T et al (1986) Distribution of calpains I and II in rat brain. J Neurosci 6:3103–3111
184. Nixon RA, Brown BA, Marotta CA (1983) Limited proteolytic modification of a neurofilament protein involves a proteinase activated by endogenous levels of calcium. Brain Res 275:384–388
185. Nelson WJ, Traub P (1982) Intermediate (10 nm) filament proteins and the Ca2+-activated proteinase specific for vimentin and desmin in the cells from fish to man: an example of evolutionary conservation. J Cell Sci 57:25–49
186. Gitler D, Spira ME (1998) Real time imaging of calcium-induced localized proteolytic activity after axotomy and its relation to growth cone formation. Neuron 20:1123–1135
187. Chin TK, Eagles PA, Maggs A (1983) The proteolytic digestion of ox neurofilaments with trypsin and alpha-chymotrypsin. Biochem J 215:239–252
188. Chin TK, Harding SE, Eagles PA (1989) Characterization of two proteolytically derived soluble polypeptides from the neurofilament triplet components NFM and NFH. Biochem J 264:53–60
189. Malik MN, Fenko MD, Iqbal K et al (1983) Purification and characterization of two forms of Ca2+-activated neutral protease from calf brain. J Biol Chem 258:8955–8962

190. Banay-Schwartz M, Dahl D, Hui KS et al (1987) The breakdown of the individual neurofilament proteins by cathepsin D. Neurochem Res 12:361–367
191. Nixon RA, Marotta CA (1984) Degradation of neurofilament proteins by purified human brain cathepsin D. J Neurochem 43:507–516
192. Suzuki H, Takeda M, Nakamura Y et al (1988) Neurofilament degradation by bovine brain cathepsin D. Neurosci Lett 89:240–245
193. Traub P, Vorgias CE, Nelson WJ (1985) Interaction in vitro of the neurofilament triplet proteins from porcine spinal cord with natural RNA and DNA. Mol Biol Rep 10: 129–136
194. Wang Q, Tolstonog GV, Shoeman R et al (2001) Sites of nucleic acid binding in type I-IV intermediate filament subunit proteins. Biochemistry 40:10342–10349
195. Goldstein ME, Sternberger NH, Sternberger LA (1987) Phosphorylation protects neurofilaments against proteolysis. J Neuroimmunol 14:149–160
196. Bizzi A, Crane RC, Autilio-Gambetti L et al (1984) Aluminum effect on slow axonal transport: a novel impairment of neurofilament transport. J Neurosci 4:722–731
197. Troncoso JC, Sternberger NH, Sternberger LA et al (1986) Immunocytochemical studies of neurofilament antigens in the neurofibrillary pathology induced by aluminum. Brain Res 364:295–300
198. Nixon RA, Clarke JF, Logvinenko KB et al (1990) Aluminum inhibits calpain-mediated proteolysis and induces human neurofilament proteins to form protease-resistant high molecular weight complexes. J Neurochem 55:1950–1959
199. Wang Q, Song F, Zhang C et al (2011) Carboxyl-terminus of Hsc70 interacting protein mediates 2,5-hexanedione-induced neurofilament medium chain degradation. Biochem Pharmacol 81:793–799
200. Banik NL, Matzelle DC, Gantt-Wilford G et al (1997) Increased calpain content and progressive degradation of neurofilament protein in spinal cord injury. Brain Res 752:301–306
201. Posmantur R, Hayes RL, Dixon CE et al (1994) Neurofilament 68 and neurofilament 200 protein levels decrease after traumatic brain injury. J Neurotrauma 11:533–545
202. Bahmanyar S, Moreau-Dubois MC, Brown P et al (1983) Serum antibodies to neurofilament antigens in patients with neurological and other diseases and in healthy controls. J Neuroimmunol 5:191–196
203. Elizan TS, Casals J, Yahr MD (1983) Antineurofilament antibodies in postencephalitic and idiopathic Parkinson's disease. J Neurol Sci 59:341–347
204. Stefansson K, Marton LS, Dieperink ME et al (1985) Circulating autoantibodies to the 200,000-dalton protein of neurofilaments in the serum of healthy individuals. Science 228:1117–1119
205. Toh BH, Gibbs CJ Jr, Gajdusek DC et al (1985) The 200- and 150-kDa neurofilament proteins react with IgG autoantibodies from chimpanzees with kuru or Creutzfeldt-Jakob disease; a 62-kDa neurofilament-associated protein reacts with sera from sheep with natural scrapie. Proc Natl Acad Sci U S A 82:3894–3896
206. Corbo M, Hays AP (1992) Peripherin and neurofilament protein coexist in spinal spheroids of motor neuron disease. J Neuropathol Exp Neurol 51:531–537
207. Hirano A (1991) Cytopathology of amyotrophic lateral sclerosis. Adv Neurol 56:91–101
208. Hill WD, Arai M, Cohen JA et al (1993) Neurofilament mRNA is reduced in Parkinson's disease substantia nigra pars compacta neurons. J Comp Neurol 329:328–336
209. Leigh PN, Dodson A, Swash M et al (1989) Cytoskeletal abnormalities in motor neuron disease. An immunocytochemical study. Brain 112(Pt 2):521–535
210. Chou SM, Hartmann HA (1965) Electron microscopy of focal neuroaxonal lesions produced by beta-beta-iminodipropionitrile (IDPN) in rats. I. The advanced lesions. Acta Neuropathol 4:590–603
211. Gold BG, Griffin JW, Price DL (1985) Slow axonal transport in acrylamide neuropathy: different abnormalities produced by single-dose and continuous administration. J Neurosci 5:1755–1768
212. Kadota T, Kadota K (1978) Neurofilament hypertrophy induced in the rabbit spinal cord after intracisternal injection of aluminum chloride (author's transl). J Toxicol Sci 3:57–67
213. Papasozomenos SC, Autilio-Gambetti L, Gambetti P (1981) Reorganization of axoplasmic organelles following beta, beta'-iminodipropionitrile administration. J Cell Biol 91:866–871
214. Gschwend TP, Krueger SR, Kozlov SV et al (1997) Neurotrypsin, a novel multidomain serine protease expressed in the nervous system. Mol Cell Neurosci 9:207–219
215. Scarisbrick IA, Isackson PJ, Ciric B et al (2001) MSP, a trypsin-like serine protease, is

abundantly expressed in the human nervous system. J Comp Neurol 431:347–361
216. Yamashiro K, Tsuruoka N, Kodama S et al (1997) Molecular cloning of a novel trypsin-like serine protease (neurosin) preferentially expressed in brain. Biochim Biophys Acta 1350:11–14
217. Chou SM, Taniguchi A, Wang HS et al (1998) Serpin=serine protease-like complexes within neurofilament conglomerates of motoneurons in amyotrophic lateral sclerosis. J Neurol Sci 160(Suppl 1):S73–S79
218. Tsuji T, Shimohama S, Kimura J et al (1998) m-Calpain (calcium-activated neutral proteinase) in Alzheimer's disease brains. Neurosci Lett 248:109–112
219. Fasani F, Bocquet A, Robert P et al (2004) The amount of neurofilaments aggregated in the cell body is controlled by their increased sensitivity to trypsin-like proteases. J Cell Sci 117:861–869
220. Eyer J, Peterson A (1994) Neurofilament-deficient axons and perikaryal aggregates in viable transgenic mice expressing a neurofilament-beta-galactosidase fusion protein. Neuron 12:389–405
221. Aamodt EJ, Williams RC Jr (1984) Association of microtubules and neurofilaments in vitro is not mediated by ATP. Biochemistry 23: 6031–6035
222. Flynn G, Purich DL (1987) GTP regeneration influences interactions of microtubules, neurofilaments, and microtubule-associated proteins in vitro. J Biol Chem 262: 15443–15447
223. Heimann R, Shelanski ML, Liem RK (1985) Microtubule-associated proteins bind specifically to the 70-kDa neurofilament protein. J Biol Chem 260:12160–12166
224. Leterrier JF, Wong J, Liem RK et al (1984) Promotion of microtubule assembly by neurofilament-associated microtubule-associated proteins. J Neurochem 43:1385–1391
225. Minami Y, Endo S, Sakai H (1984) Participation of 200 K or 150 K subunit of neurofilament in construction of the filament core with 70 K subunit and promotion of tubulin polymerization by incorporated 200 K subunit. J Biochem 96:1481–1490
226. Minami Y, Sakai H (1983) Network formation by neurofilament-induced polymerization of tubulin: 200 K subunit of neurofilament triplet promotes nucleation of tubulin polymerization and enhances microtubule assembly. J Biochem 94:2023–2033
227. Runge MS, Laue TM, Yphantis DA et al (1981) ATP-induced formation of an associated complex between microtubules and neurofilaments. Proc Natl Acad Sci U S A 78: 1431–1435
228. Hirokawa N (1982) Cross-linker system between neurofilaments, microtubules, and membranous organelles in frog axons revealed by the quick-freeze, deep-etching method. J Cell Biol 94:129–142
229. Hisanaga S, Hirokawa N (1989) The effects of dephosphorylation on the structure of the projections of neurofilament. J Neurosci 9:959–966
230. Bocquet A, Berges R, Frank R et al (2009) Neurofilaments bind tubulin and modulate its polymerization. J Neurosci 29: 11043–11054
231. Kaech S, Ludin B, Matus A (1996) Cytoskeletal plasticity in cells expressing neuronal microtubule-associated proteins. Neuron 17: 1189–1199
232. Hirokawa N, Hisanaga S, Shiomura Y (1988) MAP2 is a component of crossbridges between microtubules and neurofilaments in the neuronal cytoskeleton: quick-freeze, deep-etch immunoelectron microscopy and reconstitution studies. J Neurosci 8:2769–2779
233. Denarier E, Fourest-Lieuvin A, Bosc C et al (1998) Nonneuronal isoforms of STOP protein are responsible for microtubule cold stability in mammalian fibroblasts. Proc Natl Acad Sci U S A 95:6055–6060
234. Job D, Rauch CT, Fischer EH et al (1982) Recycling of cold-stable microtubules: evidence that cold stability is due to substoichiometric polymer blocks. Biochemistry 21: 509–515
235. Job D, Fischer EH, Margolis RL (1981) Rapid disassembly of cold-stable microtubules by calmodulin. Proc Natl Acad Sci U S A 78:4679–4682
236. Margolis RL, Rauch CT, Job D (1986) Purification and assay of a 145-kDa protein (STOP145) with microtubule-stabilizing and motility behavior. Proc Natl Acad Sci U S A 83:639–643
237. Bosc C, Cronk JD, Pirollet F et al (1996) Cloning, expression, and properties of the microtubule-stabilizing protein STOP. Proc Natl Acad Sci U S A 93:2125–2130
238. Pirollet F, Derancourt J, Haiech J et al (1992) Ca(2+)-calmodulin regulated effectors of microtubule stability in bovine brain. Biochemistry 31:8849–8855
239. Pirollet F, Rauch CT, Job D et al (1989) Monoclonal antibody to microtubule-associated STOP protein: affinity purification of neuronal STOP activity and comparison of antigen with activity in neuronal and nonneuronal cell extracts. Biochemistry 28:835–842

240. Guillaud L, Bosc C, Fourest-Lieuvin A et al (1998) STOP proteins are responsible for the high degree of microtubule stabilization observed in neuronal cells. J Cell Biol 142: 167–179
241. Letournel F, Bocquet A, Dubas F et al (2003) Stable tubule only polypeptides (STOP) proteins co-aggregate with spheroid neurofilaments in amyotrophic lateral sclerosis. J Neuropathol Exp Neurol 62:1211–1219
242. Shah JV, Flanagan LA, Janmey PA et al (2000) Bidirectional translocation of neurofilaments along microtubules mediated in part by dynein/dynactin. Mol Biol Cell 11: 3495–3508
243. Theiss C, Napirei M, Meller K (2005) Impairment of anterograde and retrograde neurofilament transport after anti-kinesin and anti-dynein antibody microinjection in chicken dorsal root ganglia. Eur J Cell Biol 84:29–43
244. Wagner OI, Ascano J, Tokito M et al (2004) The interaction of neurofilaments with the microtubule motor cytoplasmic dynein. Mol Biol Cell 15:5092–5100
245. Yabe JT, Jung C, Chan WK et al (2000) Phospho-dependent association of neurofilament proteins with kinesin in situ. Cell Motil Cytoskeleton 45:249–262
246. Yabe JT, Pimenta A, Shea TB (1999) Kinesin-mediated transport of neurofilament protein oligomers in growing axons. J Cell Sci 112(Pt 21):3799–3814
247. Motil J, Chan WK, Dubey M et al (2006) Dynein mediates retrograde neurofilament transport within axons and anterograde delivery of NFs from perikarya into axons: regulation by multiple phosphorylation events. Cell Motil Cytoskeleton 63: 266–286
248. Hirokawa N (1998) Kinesin and dynein superfamily proteins and the mechanism of organelle transport. Science 279:519–526
249. Mehta AD, Rock RS, Rief M et al (1999) Myosin-V is a processive actin-based motor. Nature 400:590–593
250. Yoshimura A, Fujii R, Watanabe Y et al (2006) Myosin-Va facilitates the accumulation of mRNA/protein complex in dendritic spines. Curr Biol 16:2345–2351
251. Rao MV, Engle LJ, Mohan PS et al (2002) Myosin Va binding to neurofilaments is essential for correct myosin Va distribution and transport and neurofilament density. J Cell Biol 159:279–290
252. Rao MV, Mohan PS, Kumar A et al (2011) The myosin Va head domain binds to the neurofilament-L rod and modulates endoplasmic reticulum (ER) content and distribution within axons. PLoS One 6:e17087
253. Errante LD, Wiche G, Shaw G (1994) Distribution of plectin, an intermediate filament-associated protein, in the adult rat central nervous system. J Neurosci Res 37:515–528
254. Yang Y, Dowling J, Yu QC et al (1996) An essential cytoskeletal linker protein connecting actin microfilaments to intermediate filaments. Cell 86:655–665
255. Leung CL, Sun D, Liem RK (1999) The intermediate filament protein peripherin is the specific interaction partner of mouse BPAG1-n (dystonin) in neurons. J Cell Biol 144:435–446
256. Guo L, Degenstein L, Dowling J et al (1995) Gene targeting of BPAG1: abnormalities in mechanical strength and cell migration in stratified epithelia and neurologic degeneration. Cell 81:233–243
257. Brown A, Bernier G, Mathieu M et al (1995) The mouse dystonia musculorum gene is a neural isoform of bullous pemphigoid antigen 1. Nat Genet 10:301–306
258. Eyer J, Cleveland DW, Wong PC et al (1998) Pathogenesis of two axonopathies does not require axonal neurofilaments. Nature 391: 584–587
259. Yang Y, Bauer C, Strasser G et al (1999) Integrators of the cytoskeleton that stabilize microtubules. Cell 98:229–238
260. Leung CL, Zheng M, Prater SM et al (2001) The BPAG1 locus: alternative splicing produces multiple isoforms with distinct cytoskeletal linker domains, including predominant isoforms in neurons and muscles. J Cell Biol 154:691–697
261. Young KG, Kothary R (2007) Dystonin/Bpag1–a link to what? Cell Motil Cytoskeleton 64:897–905
262. Metuzals J, Mushynski WE (1974) Electron microscope and experimental investigations of the neurofilamentous network in Deiters' neurons. Relationship with the cell surface and nuclear pores. J Cell Biol 61:701–722
263. Traub P, Perides G, Kuhn S et al (1987) Interaction in vitro of non-epithelial intermediate filament proteins with histones. Z Naturforsch C 42:47–63
264. Goldenring JR, Lasher RS, Vallano ML et al (1986) Association of synapsin I with neuronal cytoskeleton. Identification in cytoskeletal preparations in vitro and immunocytochemical localization in brain of synapsin I. J Biol Chem 261:8495–8504

265. Steiner JP, Ling E, Bennett V (1987) Nearest neighbor analysis for brain synapsin I. Evidence from in vitro reassociation assays for association with membrane protein(s) and the Mr = 68,000 neurofilament subunit. J Biol Chem 262:905–914
266. Metuzals J, Fishman HM, Robb IA (1995) The neurofilamentous network-smooth endoplasmic reticulum complex in transected squid giant axon. Biol Bull 189:216–218
267. Leterrier JF, Rusakov DA, Nelson BD et al (1994) Interactions between brain mitochondria and cytoskeleton: evidence for specialized outer membrane domains involved in the association of cytoskeleton-associated proteins to mitochondria in situ and in vitro. Microsc Res Tech 27:233–261
268. Morris RL, Hollenbeck PJ (1995) Axonal transport of mitochondria along microtubules and F-actin in living vertebrate neurons. J Cell Biol 131:1315–1326
269. Perrot R, Julien JP (2009) Real-time imaging reveals defects of fast axonal transport induced by disorganization of intermediate filaments. FASEB J 23:3213–3225
270. Straube-West K, Loomis PA, Opal P et al (1996) Alterations in neural intermediate filament organization: functional implications and the induction of pathological changes related to motor neuron disease. J Cell Sci 109(Pt 9):2319–2329
271. Szebenyi G, Smith GM, Li P et al (2002) Overexpression of neurofilament H disrupts normal cell structure and function. J Neurosci Res 68:185–198
272. Wagner OI, Lifshitz J, Janmey PA et al (2003) Mechanisms of mitochondria-neurofilament interactions. J Neurosci 23:9046–9058
273. Friede RL, Samorajski T (1970) Axon caliber related to neurofilaments and microtubules in sciatic nerve fibers of rats and mice. Anat Rec 167:379–387
274. Hoffman PN, Griffin JW, Price DL (1984) Control of axonal caliber by neurofilament transport. J Cell Biol 99:705–714
275. Sanchez I, Hassinger L, Paskevich PA et al (1996) Oligodendroglia regulate the regional expansion of axon caliber and local accumulation of neurofilaments during development independently of myelin formation. J Neurosci 16:5095–5105
276. Ohara O, Gahara Y, Miyake T et al (1993) Neurofilament deficiency in quail caused by nonsense mutation in neurofilament-L gene. J Cell Biol 121:387–395
277. Yamasaki H, Bennett GS, Itakura C et al (1992) Defective expression of neurofilament protein subunits in hereditary hypotrophic axonopathy of quail. Lab Invest 66:734–743
278. Yamasaki H, Itakura C, Mizutani M (1991) Hereditary hypotrophic axonopathy with neurofilament deficiency in a mutant strain of the Japanese quail. Acta Neuropathol 82: 427–434
279. Cote F, Collard JF, Julien JP (1993) Progressive neuronopathy in transgenic mice expressing the human neurofilament heavy gene: a mouse model of amyotrophic lateral sclerosis. Cell 73:35–46
280. Perrot R, Lonchampt P, Peterson AC et al (2007) Axonal neurofilaments control multiple fiber properties but do not influence structure or spacing of nodes of Ranvier. J Neurosci 27:9573–9584
281. Monteiro MJ, Hoffman PN, Gearhart JD et al (1990) Expression of NF-L in both neuronal and nonneuronal cells of transgenic mice: increased neurofilament density in axons without affecting caliber. J Cell Biol 111:1543–1557
282. Xu Z, Marszalek JR, Lee MK et al (1996) Subunit composition of neurofilaments specifies axonal diameter. J Cell Biol 133:1061–1069
283. Nguyen MD, Lariviere RC, Julien JP (2000) Reduction of axonal caliber does not alleviate motor neuron disease caused by mutant superoxide dismutase 1. Proc Natl Acad Sci U S A 97:12306–12311
284. Meier J, Couillard-Despres S, Jacomy H et al (1999) Extra neurofilament NF-L subunits rescue motor neuron disease caused by overexpression of the human NF-H gene in mice. J Neuropathol Exp Neurol 58:1099–1110
285. Elder GA, Friedrich VL Jr, Kang C et al (1998) Requirement of heavy neurofilament subunit in the development of axons with large calibers. J Cell Biol 143:195–205
286. Rao MV, Houseweart MK, Williamson TL et al (1998) Neurofilament-dependent radial growth of motor axons and axonal organization of neurofilaments does not require the neurofilament heavy subunit (NF-H) or its phosphorylation. J Cell Biol 143:171–181
287. Zhu Q, Lindenbaum M, Levavasseur F et al (1998) Disruption of the NF-H gene increases axonal microtubule content and velocity of neurofilament transport: relief of axonopathy resulting from the toxin beta, beta-iminodipropionitrile. J Cell Biol 143:183–193
288. Hirokawa N, Takeda S (1998) Gene targeting studies begin to reveal the function of neurofilament proteins. J Cell Biol 143:1–4
289. Shen H, Barry DM, Garcia ML (2010) Distal to proximal development of peripheral nerves

requires the expression of neurofilament heavy. Neuroscience 170:16–21

290. Gotow T, Takeda M, Tanaka T et al (1992) Macromolecular structure of reassembled neurofilaments as revealed by the quick-freeze deep-etch mica method: difference between NF-M and NF-H subunits in their ability to form cross-bridges. Eur J Cell Biol 58: 331–345
291. Brown HG, Hoh JH (1997) Entropic exclusion by neurofilament sidearms: a mechanism for maintaining interfilament spacing. Biochemistry 36:15035–15040
292. Kumar S, Hoh JH (2004) Modulation of repulsive forces between neurofilaments by sidearm phosphorylation. Biochem Biophys Res Commun 324:489–496
293. Aranda-Espinoza H, Carl P, Leterrier JF et al (2002) Domain unfolding in neurofilament sidearms: effects of phosphorylation and ATP. FEBS Lett 531:397–401
294. Wong PC, Marszalek J, Crawford TO et al (1995) Increasing neurofilament subunit NF-M expression reduces axonal NF-H, inhibits radial growth, and results in neurofilamentous accumulation in motor neurons. J Cell Biol 130:1413–1422
295. Marszalek JR, Williamson TL, Lee MK et al (1996) Neurofilament subunit NF-H modulates axonal diameter by selectively slowing neurofilament transport. J Cell Biol 135: 711–724
296. Rao MV, Garcia ML, Miyazaki Y et al (2002) Gene replacement in mice reveals that the heavily phosphorylated tail of neurofilament heavy subunit does not affect axonal caliber or the transit of cargoes in slow axonal transport. J Cell Biol 158:681–693
297. Garcia ML, Lobsiger CS, Shah SB et al (2003) NF-M is an essential target for the myelin-directed "outside-in" signaling cascade that mediates radial axonal growth. J Cell Biol 163:1011–1020
298. Garcia ML, Rao MV, Fujimoto J et al (2009) Phosphorylation of highly conserved neurofilament medium KSP repeats is not required for myelin-dependent radial axonal growth. J Neurosci 29:1277–1284
299. Chang R, Kwak Y, Gebremichael Y (2009) Structural properties of neurofilament sidearms: sequence-based modeling of neurofilament architecture. J Mol Biol 391:648–660
300. Zhulina EB, Leermakers FA (2007) A self-consistent field analysis of the neurofilament brush with amino-acid resolution. Biophys J 93:1421–1430
301. Stevenson W, Chang R, Gebremichael Y (2011) Phosphorylation-mediated conformational changes in the mouse neurofilament architecture: insight from a neurofilament brush model. J Mol Biol 405:1101–1118
302. Mirsky R, Jessen KR, Brennan A et al (2002) Schwann cells as regulators of nerve development. J Physiol Paris 96:17–24
303. Windebank AJ, Wood P, Bunge RP et al (1985) Myelination determines the caliber of dorsal root ganglion neurons in culture. J Neurosci 5:1563–1569
304. Aguayo AJ, Attiwell M, Trecarten J et al (1977) Abnormal myelination in transplanted Trembler mouse Schwann cells. Nature 265: 73–75
305. Pollard JD, McLeod JG (1980) Nerve grafts in the Trembler mouse. An electrophysiological and histological study. J Neurol Sci 46: 373–383
306. Kirkpatrick LL, Witt AS, Payne HR et al (2001) Changes in microtubule stability and density in myelin-deficient shiverer mouse CNS axons. J Neurosci 21:2288–2297
307. Sternberger NH, Quarles RH, Itoyama Y et al (1979) Myelin-associated glycoprotein demonstrated immunocytochemically in myelin and myelin-forming cells of developing rat. Proc Natl Acad Sci U S A 76:1510–1514
308. Trapp BD, Andrews SB, Cootauco C et al (1989) The myelin-associated glycoprotein is enriched in multivesicular bodies and periaxonal membranes of actively myelinating oligodendrocytes. J Cell Biol 109:2417–2426
309. Lunn MP, Crawford TO, Hughes RA et al (2002) Anti-myelin-associated glycoprotein antibodies alter neurofilament spacing. Brain 125:904–911
310. Zhu X, Liu Y, Yin Y et al (2009) MSC p43 required for axonal development in motor neurons. Proc Natl Acad Sci U S A 106: 15944–15949
311. Gasser HS, Grundfest H (1939) Axon diameters in relation to the spike dimensions and the conduction velocity in mammalian A fibers. Am J Physiol 127:393–414
312. Hursh JB (1939) Conduction velocity and diameter of nerve fibers. Am J Physiol 127:131–139
313. Hutchinson NA, Koles ZJ, Smith RS (1970) Conduction velocity in myelinated nerve fibres of Xenopus laevis. J Physiol 208:279–289
314. Brill MH, Waxman SG, Moore JW et al (1977) Conduction velocity and spike configuration in myelinated fibres: computed dependence on internode distance. J Neurol Neurosurg Psychiatry 40:769–774

315. Court FA, Sherman DL, Pratt T et al (2004) Restricted growth of Schwann cells lacking Cajal bands slows conduction in myelinated nerves. Nature 431:191–195
316. Goldman L, Albus JS (1968) Computation of impulse conduction in myelinated fibers; theoretical basis of the velocity-diameter relation. Biophys J 8:596–607
317. Huxley AF, Stampfli R (1949) Evidence for saltatory conduction in peripheral myelinated nerve fibres. J Physiol 108:315–339
318. Hodgkin AL (1964) The ionic basis of nervous conduction. Science 145:1148–1154
319. Rushton WA (1951) A theory of the effects of fibre size in medullated nerve. J Physiol 115:101–122
320. Smith RS, Koles ZJ (1970) Myelinated nerve fibers: computed effect of myelin thickness on conduction velocity. Am J Physiol 219: 1256–1258
321. Bhat MA, Rios JC, Lu Y et al (2001) Axon-glia interactions and the domain organization of myelinated axons requires neurexin IV/ Caspr/Paranodin. Neuron 30:369–383
322. Boyle ME, Berglund EO, Murai KK et al (2001) Contactin orchestrates assembly of the septate-like junctions at the paranode in myelinated peripheral nerve. Neuron 30:385–397
323. Fried K, Hildebrand C, Erdelyi G (1982) Myelin sheath thickness and internodal length of nerve fibres in the developing feline inferior alveolar nerve. J Neurol Sci 54:47–57
324. Friede RL, Meier T, Diem M (1981) How is the exact length of an internode determined. J Neurol Sci 50:217–228
325. Murray JA, Blakemore WF (1980) The relationship between internodal length and fibre diameter in the spinal cord of the cat. J Neurol Sci 45:29–41
326. Sakaguchi T, Okada M, Kitamura T et al (1993) Reduced diameter and conduction velocity of myelinated fibers in the sciatic nerve of a neurofilament-deficient mutant quail. Neurosci Lett 153:65–68
327. Kriz J, Zhu Q, Julien JP et al (2000) Electrophysiological properties of axons in mice lacking neurofilament subunit genes: disparity between conduction velocity and axon diameter in absence of NF-H. Brain Res 885:32–44
328. Kriz J, Meier J, Julien JP et al (2000) Altered ionic conductances in axons of transgenic mouse expressing the human neurofilament heavy gene: a mouse model of amyotrophic lateral sclerosis. Exp Neurol 163:414–421
329. Zochodne DW, Sun HS, Cheng C et al (2004) Accelerated diabetic neuropathy in axons without neurofilaments. Brain 127: 2193–2200
330. Sheykholeslami K, Kaga K, Mizutani M (2001) Auditory nerve fiber differences in the normal and neurofilament deficient Japanese quail. Hear Res 159:117–124
331. Elder GA, Friedrich VL Jr, Lazzarini RA (2001) Schwann cells and oligodendrocytes read distinct signals in establishing myelin sheath thickness. J Neurosci Res 65:493–499
332. Berthold CH, Rydmark M (1983) Electron microscopic serial section analysis of nodes of Ranvier in lumbosacral spinal roots of the cat: ultrastructural organization of nodal compartments in fibres of different sizes. J Neurocytol 12:475–505
333. Hildebrand C, Remahl S, Persson H et al (1993) Myelinated nerve fibres in the CNS. Prog Neurobiol 40:319–384
334. Halter JA, Clark JW Jr (1993) The influence of nodal constriction on conduction velocity in myelinated nerve fibers. Neuroreport 4:89–92
335. Zimmermann H (1996) Accumulation of synaptic vesicle proteins and cytoskeletal specializations at the peripheral node of Ranvier. Microsc Res Tech 34:462–473
336. Zimmermann H, Vogt M (1989) Membrane proteins of synaptic vesicles and cytoskeletal specializations at the node of Ranvier in electric ray and rat. Cell Tissue Res 258:617–629
337. Weiss P, Hiscoe H (1948) Experiments in the mechanism of nerve growth. J Exp Zool 107:315–395
338. Droz B, Leblond CP (1962) Migration of proteins along the axons of the sciatic nerve. Science 137:1047–1048
339. Lasek RJ (1967) Bidirectional transport of radioactively labelled axoplasmic components. Nature 216:1212–1214
340. Lasek RJ (1968) Axoplasmic transport of labeled proteins in rat ventral motoneurons. Exp Neurol 21:41–51
341. Grafstein B, Forman DS (1980) Intracellular transport in neurons. Physiol Rev 60: 1167–1283
342. Dahlstrom A, Haggendal J, Heiwall PO et al (1974) Intra-axonal transport of neurotransmitters in mammalian neurons. Symp Soc Exp Biol 28:229–247
343. Hollenbeck PJ (1996) The pattern and mechanism of mitochondrial transport in axons. Front Biosci 1:d91–d102
344. Lombet A, Laduron P, Mourre C et al (1985) Axonal transport of the voltage-dependent Na+ channel protein identified by its tetrodotoxin binding site in rat sciatic nerves. Brain Res 345:153–158

345. Black MM, Lasek RJ (1980) Slow components of axonal transport: two cytoskeletal networks. J Cell Biol 86:616–623
346. Brady ST (1985) A novel brain ATPase with properties expected for the fast axonal transport motor. Nature 317:73–75
347. Vale RD, Reese TS, Sheetz MP (1985) Identification of a novel force-generating protein, kinesin, involved in microtubule-based motility. Cell 42:39–50
348. Roy S, Coffee P, Smith G et al (2000) Neurofilaments are transported rapidly but intermittently in axons: implications for slow axonal transport. J Neurosci 20:6849–6861
349. Wang L, Ho CL, Sun D et al (2000) Rapid movement of axonal neurofilaments interrupted by prolonged pauses. Nat Cell Biol 2:137–141
350. Trivedi N, Jung P, Brown A (2007) Neurofilaments switch between distinct mobile and stationary states during their transport along axons. J Neurosci 27:507–516
351. Prahlad V, Helfand BT, Langford GM et al (2000) Fast transport of neurofilament protein along microtubules in squid axoplasm. J Cell Sci 113(Pt 22):3939–3946
352. Jung C, Lee S, Ortiz D et al (2005) The high and middle molecular weight neurofilament subunits regulate the association of neurofilaments with kinesin: inhibition by phosphorylation of the high molecular weight subunit. Brain Res Mol Brain Res 141:151–155
353. LaMonte BH, Wallace KE, Holloway BA et al (2002) Disruption of dynein/dynactin inhibits axonal transport in motor neurons causing late-onset progressive degeneration. Neuron 34:715–727
354. Xia CH, Roberts EA, Her LS et al (2003) Abnormal neurofilament transport caused by targeted disruption of neuronal kinesin heavy chain KIF5A. J Cell Biol 161:55–66
355. Wang L, Brown A (2010) A hereditary spastic paraplegia mutation in kinesin-1A/KIF5A disrupts neurofilament transport. Mol Neurodegener 5:52
356. Baas PW, Brown A (1997) Slow axonal transport: the polymer transport model. Trends Cell Biol 7:380–384
357. Hirokawa N, Funakoshi ST, Takeda S (1997) Slow axonal transport: the subunit transport model. Trends Cell Biol 7:384–388
358. Lasek RJ, Garner JA, Brady ST (1984) Axonal transport of the cytoplasmic matrix. J Cell Biol 99:212s–221s
359. Okabe S, Miyasaka H, Hirokawa N (1993) Dynamics of the neuronal intermediate filaments. J Cell Biol 121:375–386
360. Terada S, Nakata T, Peterson AC et al (1996) Visualization of slow axonal transport in vivo. Science 273:784–788
361. Popov S, Poo MM (1992) Diffusional transport of macromolecules in developing nerve processes. J Neurosci 12:77–85
362. Yan Y, Brown A (2005) Neurofilament polymer transport in axons. J Neurosci 25: 7014–7021
363. Yuan A, Rao MV, Kumar A et al (2003) Neurofilament transport in vivo minimally requires hetero-oligomer formation. J Neurosci 23:9452–9458
364. Yabe JT, Chan WK, Chylinski TM et al (2001) The predominant form in which neurofilament subunits undergo axonal transport varies during axonal initiation, elongation, and maturation. Cell Motil Cytoskeleton 48:61–83
365. Lee S, Sunil N, Tejada JM et al (2011) Differential roles of kinesin and dynein in translocation of neurofilaments into axonal neurites. J Cell Sci 124:1022–1031
366. Hoffman PN, Lasek RJ, Griffin JW et al (1983) Slowing of the axonal transport of neurofilament proteins during development. J Neurosci 3:1694–1700
367. Willard M, Simon C (1983) Modulations of neurofilament axonal transport during the development of rabbit retinal ganglion cells. Cell 35:551–559
368. Rao MV, Campbell J, Yuan A et al (2003) The neurofilament middle molecular mass subunit carboxyl-terminal tail domains is essential for the radial growth and cytoskeletal architecture of axons but not for regulating neurofilament transport rate. J Cell Biol 163:1021–1031
369. Yuan A, Nixon RA, Rao MV (2006) Deleting the phosphorylated tail domain of the neurofilament heavy subunit does not alter neurofilament transport rate in vivo. Neurosci Lett 393:264–268
370. Brown A, Wang L, Jung P (2005) Stochastic simulation of neurofilament transport in axons: the "stop-and-go" hypothesis. Mol Biol Cell 16:4243–4255
371. Yabe JT, Chylinski T, Wang FS et al (2001) Neurofilaments consist of distinct populations that can be distinguished by C-terminal phosphorylation, bundling, and axonal transport rate in growing axonal neurites. J Neurosci 21:2195–2205
372. Kushkuley J, Chan WK, Lee S et al (2009) Neurofilament cross-bridging competes with kinesin-dependent association of neurofilaments with microtubules. J Cell Sci 122:3579–3586

373. Lee S, Sunil N, Shea TB (2011) C-terminal neurofilament phosphorylation fosters neurofilament-neurofilament associations that compete with axonal transport. Cytoskeleton (Hoboken) 68:8–17
374. Goodall EF, Morrison KE (2006) Amyotrophic lateral sclerosis (motor neuron disease): proposed mechanisms and pathways to treatment. Expert Rev Mol Med 8:1–22
375. Rosen DR, Siddique T, Patterson D et al (1993) Mutations in Cu/Zn superoxide dismutase gene are associated with familial amyotrophic lateral sclerosis. Nature 362:59–62
376. Kabashi E, Valdmanis PN, Dion P et al (2008) TARDBP mutations in individuals with sporadic and familial amyotrophic lateral sclerosis. Nat Genet 40:572–574
377. Sreedharan J, Blair IP, Tripathi VB et al (2008) TDP-43 mutations in familial and sporadic amyotrophic lateral sclerosis. Science 319:1668–1672
378. Kwiatkowski TJ Jr, Bosco DA, Leclerc AL et al (2009) Mutations in the FUS/TLS gene on chromosome 16 cause familial amyotrophic lateral sclerosis. Science 323: 1205–1208
379. Vance C, Rogelj B, Hortobagyi T et al (2009) Mutations in FUS, an RNA processing protein, cause familial amyotrophic lateral sclerosis type 6. Science 323:1208–1211
380. Al-Chalabi A, Andersen PM, Nilsson P et al (1999) Deletions of the heavy neurofilament subunit tail in amyotrophic lateral sclerosis. Hum Mol Genet 8:157–164
381. Figlewicz DA, Krizus A, Martinoli MG et al (1994) Variants of the heavy neurofilament subunit are associated with the development of amyotrophic lateral sclerosis. Hum Mol Genet 3:1757–1761
382. Tomkins J, Usher P, Slade JY et al (1998) Novel insertion in the KSP region of the neurofilament heavy gene in amyotrophic lateral sclerosis (ALS). Neuroreport 9: 3967–3970
383. Rooke K, Figlewicz DA, Han FY et al (1996) Analysis of the KSP repeat of the neurofilament heavy subunit in familiar amyotrophic lateral sclerosis. Neurology 46:789–790
384. Vechio JD, Bruijn LI, Xu Z et al (1996) Sequence variants in human neurofilament proteins: absence of linkage to familial amyotrophic lateral sclerosis. Ann Neurol 40: 603–610
385. Gros-Louis F, Lariviere R, Gowing G et al (2004) A frameshift deletion in peripherin gene associated with amyotrophic lateral sclerosis. J Biol Chem 279:45951–45956
386. Leung CL, He CZ, Kaufmann P et al (2004) A pathogenic peripherin gene mutation in a patient with amyotrophic lateral sclerosis. Brain Pathol 14:290–296
387. Corrado L, Carlomagno Y, Falasco L et al (2011) A novel peripherin gene (PRPH) mutation identified in one sporadic amyotrophic lateral sclerosis patient. Neurobiol Aging 32(552):e1–e6
388. Brettschneider J, Petzold A, Sussmuth SD et al (2006) Axonal damage markers in cerebrospinal fluid are increased in ALS. Neurology 66:852–856
389. Niebroj-Dobosz I, Dziewulska D, Janik P (2006) Auto-antibodies against proteins of spinal cord cells in cerebrospinal fluid of patients with amyotrophic lateral sclerosis (ALS). Folia Neuropathol 44:191–196
390. Zetterberg H, Jacobsson J, Rosengren L et al (2007) Cerebrospinal fluid neurofilament light levels in amyotrophic lateral sclerosis: impact of SOD1 genotype. Eur J Neurol 14:1329–1333
391. Manetto V, Sternberger NH, Perry G et al (1988) Phosphorylation of neurofilaments is altered in amyotrophic lateral sclerosis. J Neuropathol Exp Neurol 47:642–653
392. Gama Sosa MA, Friedrich VL Jr, DeGasperi R et al (2003) Human midsized neurofilament subunit induces motor neuron disease in transgenic mice. Exp Neurol 184:408–419
393. Xu Z, Cork LC, Griffin JW et al (1993) Increased expression of neurofilament subunit NF-L produces morphological alterations that resemble the pathology of human motor neuron disease. Cell 73:23–33
394. Beaulieu JM, Jacomy H, Julien JP (2000) Formation of intermediate filament protein aggregates with disparate effects in two transgenic mouse models lacking the neurofilament light subunit. J Neurosci 20:5321–5328
395. Beaulieu JM, Nguyen MD, Julien JP (1999) Late onset of motor neurons in mice overexpressing wild-type peripherin. J Cell Biol 147:531–544
396. Millecamps S, Robertson J, Lariviere R et al (2006) Defective axonal transport of neurofilament proteins in neurons overexpressing peripherin. J Neurochem 98: 926–938
397. Bergeron C, Beric-Maskarel K, Muntasser S et al (1994) Neurofilament light and polyadenylated mRNA levels are decreased in amyotrophic lateral sclerosis motor neurons. J Neuropathol Exp Neurol 53:221–230
398. Beaulieu JM, Julien JP (2003) Peripherin-mediated death of motor neurons rescued

by overexpression of neurofilament NF-H proteins. J Neurochem 85:248–256
399. Robertson J, Beaulieu JM, Doroudchi MM et al (2001) Apoptotic death of neurons exhibiting peripherin aggregates is mediated by the proinflammatory cytokine tumor necrosis factor-alpha. J Cell Biol 155:217–226
400. Lariviere RC, Beaulieu JM, Nguyen MD et al (2003) Peripherin is not a contributing factor to motor neuron disease in a mouse model of amyotrophic lateral sclerosis caused by mutant superoxide dismutase. Neurobiol Dis 13: 158–166
401. Strong MJ, Leystra-Lantz C, Ge WW (2004) Intermediate filament steady-state mRNA levels in amyotrophic lateral sclerosis. Biochem Biophys Res Commun 316:317–322
402. Wong NK, He BP, Strong MJ (2000) Characterization of neuronal intermediate filament protein expression in cervical spinal motor neurons in sporadic amyotrophic lateral sclerosis (ALS). J Neuropathol Exp Neurol 59:972–982
403. Ge WW, Volkening K, Leystra-Lantz C et al (2007) 14-3-3 protein binds to the low molecular weight neurofilament (NFL) mRNA 3′ UTR. Mol Cell Neurosci 34:80–87
404. Strong MJ, Volkening K, Hammond R et al (2007) TDP43 is a human low molecular weight neurofilament (hNFL) mRNA-binding protein. Mol Cell Neurosci 35:320–327
405. Ge WW, Wen W, Strong W et al (2005) Mutant copper-zinc superoxide dismutase binds to and destabilizes human low molecular weight neurofilament mRNA. J Biol Chem 280:118–124
406. Volkening K, Leystra-Lantz C, Strong MJ (2010) Human low molecular weight neurofilament (NFL) mRNA interacts with a predicted p190RhoGEF homologue (RGNEF) in humans. Amyotroph Lateral Scler 11:97–103
407. Shan X, Chiang PM, Price DL et al (2010) Altered distributions of Gemini of coiled bodies and mitochondria in motor neurons of TDP-43 transgenic mice. Proc Natl Acad Sci U S A 107:16325–16330
408. Swarup V, Julien JP (2011) ALS pathogenesis: recent insights from genetics and mouse models. Prog Neuropsychopharmacol Biol Psychiatry 35:363–369
409. McLean J, Xiao S, Miyazaki K et al (2008) A novel peripherin isoform generated by alternative translation is required for normal filament network formation. J Neurochem 104:1663–1673
410. Robertson J, Doroudchi MM, Nguyen MD et al (2003) A neurotoxic peripherin splice variant in a mouse model of ALS. J Cell Biol 160:939–949
411. Xiao S, Tjostheim S, Sanelli T et al (2008) An aggregate-inducing peripherin isoform generated through intron retention is upregulated in amyotrophic lateral sclerosis and associated with disease pathology. J Neurosci 28:1833–1840
412. McLean JR, Robertson J (2011) Isoform-specific expression and ratio changes accompany oxidant-induced peripherin aggregation in a neuroblastoma cell line. Brain Res 1422:57–65
413. Collard JF, Cote F, Julien JP (1995) Defective axonal transport in a transgenic mouse model of amyotrophic lateral sclerosis. Nature 375:61–64
414. Williamson TL, Cleveland DW (1999) Slowing of axonal transport is a very early event in the toxicity of ALS-linked SOD1 mutants to motor neurons. Nat Neurosci 2:50–56
415. Zhang B, Tu P, Abtahian F et al (1997) Neurofilaments and orthograde transport are reduced in ventral root axons of transgenic mice that express human SOD1 with a G93A mutation. J Cell Biol 139:1307–1315
416. Bilsland LG, Sahai E, Kelly G et al (2010) Deficits in axonal transport precede ALS symptoms in vivo. Proc Natl Acad Sci U S A 107:20523–20528
417. Ackerley S, Grierson AJ, Brownlees J et al (2000) Glutamate slows axonal transport of neurofilaments in transfected neurons. J Cell Biol 150:165–176
418. Manser C, Stevenson A, Banner S et al (2008) Deregulation of PKN1 activity disrupts neurofilament organisation and axonal transport. FEBS Lett 582:2303–2308
419. King AE, Dickson TC, Blizzard CA et al (2007) Excitotoxicity mediated by non-NMDA receptors causes distal axonopathy in long-term cultured spinal motor neurons. Eur J Neurosci 26:2151–2159
420. Kesavapany S, Patel V, Zheng YL et al (2007) Inhibition of Pin1 reduces glutamate-induced perikaryal accumulation of phosphorylated neurofilament-H in neurons. Mol Biol Cell 18:3645–3655
421. Stevenson A, Yates DM, Manser C et al (2009) Riluzole protects against glutamate-induced slowing of neurofilament axonal transport. Neurosci Lett 454:161–164
422. Hafezparast M, Klocke R, Ruhrberg C et al (2003) Mutations in dynein link motor neuron degeneration to defects in retrograde transport. Science 300:808–812

423. Motil J, Dubey M, Chan WK et al (2007) Inhibition of dynein but not kinesin induces aberrant focal accumulation of neurofilaments within axonal neurites. Brain Res 1164: 125–131
424. Teuling E, van Dis V, Wulf PS et al (2008) A novel mouse model with impaired dynein/dynactin function develops amyotrophic lateral sclerosis (ALS)-like features in motor neurons and improves lifespan in SOD1-ALS mice. Hum Mol Genet 17:2849–2862
425. Cheroni C, Marino M, Tortarolo M et al (2009) Functional alterations of the ubiquitin-proteasome system in motor neurons of a mouse model of familial amyotrophic lateral sclerosis. Hum Mol Genet 18:82–96
426. Strong MJ, Sopper MM, Crow JP et al (1998) Nitration of the low molecular weight neurofilament is equivalent in sporadic amyotrophic lateral sclerosis and control cervical spinal cord. Biochem Biophys Res Commun 248:157–164
427. King AE, Dickson TC, Blizzard CA et al (2011) Neuron-glia interactions underlie ALS-like axonal cytoskeletal pathology. Neurobiol Aging 32:459–469
428. Williamson TL, Bruijn LI, Zhu Q et al (1998) Absence of neurofilaments reduces the selective vulnerability of motor neurons and slows disease caused by a familial amyotrophic lateral sclerosis-linked superoxide dismutase 1 mutant. Proc Natl Acad Sci U S A 95:9631–9636
429. Kong J, Xu Z (2000) Overexpression of neurofilament subunit NF-L and NF-H extends survival of a mouse model for amyotrophic lateral sclerosis. Neurosci Lett 281:72–74
430. Couillard-Despres S, Zhu Q, Wong PC et al (1998) Protective effect of neurofilament heavy gene overexpression in motor neuron disease induced by mutant superoxide dismutase. Proc Natl Acad Sci U S A 95:9626–9630
431. Roy J, Minotti S, Dong L et al (1998) Glutamate potentiates the toxicity of mutant Cu/Zn-superoxide dismutase in motor neurons by postsynaptic calcium-dependent mechanisms. J Neurosci 18:9673–9684
432. Tu PH, Raju P, Robinson KA et al (1996) Transgenic mice carrying a human mutant superoxide dismutase transgene develop neuronal cytoskeletal pathology resembling human amyotrophic lateral sclerosis lesions. Proc Natl Acad Sci U S A 93:3155–3160
433. Lobsiger CS, Garcia ML, Ward CM et al (2005) Altered axonal architecture by removal of the heavily phosphorylated neurofilament tail domains strongly slows superoxide dismutase 1 mutant-mediated ALS. Proc Natl Acad Sci U S A 102:10351–10356
434. Ehlers MD, Fung ET, O'Brien RJ et al (1998) Splice variant-specific interaction of the NMDA receptor subunit NR1 with neuronal intermediate filaments. J Neurosci 18:720–730
435. Sanelli T, Ge W, Leystra-Lantz C et al (2007) Calcium mediated excitotoxicity in neurofilament aggregate-bearing neurons in vitro is NMDA receptor dependant. J Neurol Sci 256:39–51
436. Teunissen CE, Dijkstra C, Polman C (2005) Biological markers in CSF and blood for axonal degeneration in multiple sclerosis. Lancet Neurol 4:32–41
437. Sussmuth SD, Reiber H, Tumani H (2001) Tau protein in cerebrospinal fluid (CSF): a blood-CSF barrier related evaluation in patients with various neurological diseases. Neurosci Lett 300:95–98
438. Sussmuth SD, Tumani H, Ecker D et al (2003) Amyotrophic lateral sclerosis: disease stage related changes of tau protein and S100 beta in cerebrospinal fluid and creatine kinase in serum. Neurosci Lett 353:57–60
439. Reijn TS, Abdo WF, Schelhaas HJ et al (2009) CSF neurofilament protein analysis in the differential diagnosis of ALS. J Neurol 256: 615–619
440. Ganesalingam J, An J, Shaw CE et al (2011) Combination of neurofilament heavy chain and complement C3 as CSF biomarkers for ALS. J Neurochem 117:528–537
441. Vogel P, Gabriel M, Goebel HH et al (1985) Hereditary motor sensory neuropathy type II with neurofilament accumulation: new finding or new disorder? Ann Neurol 17:455–461
442. Brownlees J, Ackerley S, Grierson AJ et al (2002) Charcot-Marie-Tooth disease neurofilament mutations disrupt neurofilament assembly and axonal transport. Hum Mol Genet 11:2837–2844
443. Perez-Olle R, Lopez-Toledano MA, Goryunov D et al (2005) Mutations in the neurofilament light gene linked to Charcot-Marie-Tooth disease cause defects in transport. J Neurochem 93:861–874
444. Lee IB, Kim SK, Chung SH et al (2008) The effect of rod domain A148V mutation of neurofilament light chain on filament formation. BMB Rep 41:868–874
445. Shin JS, Chung KW, Cho SY et al (2008) NEFL Pro22Arg mutation in Charcot-Marie-Tooth disease type 1. J Hum Genet 53:936–940
446. Georgiou DM, Zidar J, Korosec M et al (2002) A novel NF-L mutation Pro22Ser is associated with CMT2 in a large Slovenian family. Neurogenetics 4:93–96

447. Fabrizi GM, Cavallaro T, Angiari C et al (2004) Giant axon and neurofilament accumulation in Charcot-Marie-Tooth disease type 2E. Neurology 62:1429–1431
448. Bhagavati S, Maccabee PJ, Xu W (2009) The neurofilament light chain gene (NEFL) mutation Pro22Ser can be associated with mixed axonal and demyelinating neuropathy. J Clin Neurosci 16:830–831
449. Yoshihara T, Yamamoto M, Hattori N et al (2002) Identification of novel sequence variants in the neurofilament-light gene in a Japanese population: analysis of Charcot-Marie-Tooth disease patients and normal individuals. J Peripher Nerv Syst 7:221–224
450. Jordanova A, De Jonghe P, Boerkoel CF et al (2003) Mutations in the neurofilament light chain gene (NEFL) cause early onset severe Charcot-Marie-Tooth disease. Brain 126: 590–597
451. Choi BO, Lee MS, Shin SH et al (2004) Mutational analysis of PMP22, MPZ, GJB1, EGR2 and NEFL in Korean Charcot-Marie-Tooth neuropathy patients. Hum Mutat 24:185–186
452. Zuchner S, Vorgerd M, Sindern E et al (2004) The novel neurofilament light (NEFL) mutation Glu397Lys is associated with a clinically and morphologically heterogeneous type of Charcot-Marie-Tooth neuropathy. Neuromuscul Disord 14:147–157
453. Fabrizi GM, Cavallaro T, Angiari C et al (2007) Charcot-Marie-Tooth disease type 2E, a disorder of the cytoskeleton. Brain 130: 394–403
454. Leung CL, Nagan N, Graham TH et al (2006) A novel duplication/insertion mutation of NEFL in a patient with Charcot-Marie-Tooth disease. Am J Med Genet A 140:1021–1025
455. Abe A, Numakura C, Saito K et al (2009) Neurofilament light chain polypeptide gene mutations in Charcot-Marie-Tooth disease: nonsense mutation probably causes a recessive phenotype. J Hum Genet 54:94–97
456. Dequen F, Filali M, Lariviere RC et al (2010) Reversal of neuropathy phenotypes in conditional mouse model of Charcot-Marie-Tooth disease type 2E. Hum Mol Genet 19: 2616–2629
457. Shen H, Barry DM, Dale JM et al (2011) Muscle pathology without severe nerve pathology in a new mouse model of Charcot-Marie-Tooth disease type 2E. Hum Mol Genet 20:2535–2548
458. Ackerley S, James PA, Kalli A et al (2006) A mutation in the small heat-shock protein HSPB1 leading to distal hereditary motor neuronopathy disrupts neurofilament assembly and the axonal transport of specific cellular cargoes. Hum Mol Genet 15:347–354
459. Evgrafov OV, Mersiyanova I, Irobi J et al (2004) Mutant small heat-shock protein 27 causes axonal Charcot-Marie-Tooth disease and distal hereditary motor neuropathy. Nat Genet 36:602–606
460. Goryunov D, Nightingale A, Bornfleth L et al (2008) Multiple disease-linked myotubularin mutations cause NFL assembly defects in cultured cells and disrupt myotubularin dimerization. J Neurochem 104:1536–1552
461. Irobi J, Van Impe K, Seeman P et al (2004) Hot-spot residue in small heat-shock protein 22 causes distal motor neuropathy. Nat Genet 36:597–601
462. Zhai J, Lin H, Julien JP et al (2007) Disruption of neurofilament network with aggregation of light neurofilament protein: a common pathway leading to motor neuron degeneration due to Charcot-Marie-Tooth disease-linked mutations in NFL and HSPB1. Hum Mol Genet 16:3103–3116
463. Zhang R, Yang X, Zi X et al (2011) Cellular expression of (R127W) HSPB1 and its co-localization with neurofilament light chain. Zhonghua Yi Xue Yi Chuan Xue Za Zhi 28:496–500
464. Tradewell ML, Durham HD, Mushynski WE et al (2009) Mitochondrial and axonal abnormalities precede disruption of the neurofilament network in a model of charcot-marie-tooth disease type 2E and are prevented by heat shock proteins in a mutant-specific fashion. J Neuropathol Exp Neurol 68:642–652
465. Hull E, Spoja C, Cordova M et al (2008) Neurofilament protein aggregation in a cell line model system. Biochem Biophys Res Commun 366:73–79
466. Bomont P, Cavalier L, Blondeau F et al (2000) The gene encoding gigaxonin, a new member of the cytoskeletal BTB/kelch repeat family, is mutated in giant axonal neuropathy. Nat Genet 26:370–374
467. Pfeiffer J, Schlote W, Bishoff A et al (1977) Generalized giant axonal neuropathy. A filament-forming disease of neuronal, endothelial, glial and Schwann cells in a partient without kinky hair. Acta neuropathologica (Berl) 40:213–218
468. Asbury AK, Gale MK, Cox SC et al (1972) Giant axonal neuropathy–a unique case with segmental neurofilamentous masses. Acta Neuropathol 20:237–247

469. Fois A, Balestri P, Farnetani MA et al (1985) Giant axonal neuropathy. Endocrinological and histological studies. Eur J Pediatr 144: 274–280
470. Mohri I, Taniike M, Yoshikawa H et al (1998) A case of giant axonal neuropathy showing focal aggregation and hypophosphorylation of intermediate filaments. Brain Dev 20:594–597
471. Treiber-Held S, Budjarjo-Welim H, Reimann D et al (1994) Giant axonal neuropathy: a generalized disorder of intermediate filaments with longitudinal grooves in the hair. Neuropediatrics 25:89–93
472. Donaghy M, King RH, Thomas PK et al (1988) Abnormalities of the axonal cytoskeleton in giant axonal neuropathy. J Neurocytol 17:197–208
473. Monaco S, Autilio-Gambetti L, Zabel D et al (1985) Giant axonal neuropathy: acceleration of neurofilament transport in optic axons. Proc Natl Acad Sci U S A 82:920–924
474. Ding J, Liu JJ, Kowal AS et al (2002) Microtubule-associated protein 1B: a neuronal binding partner for gigaxonin. J Cell Biol 158:427–433
475. Wang W, Ding J, Allen E et al (2005) Gigaxonin interacts with tubulin folding cofactor B and controls its degradation through the ubiquitin-proteasome pathway. Curr Biol 15:2050–2055
476. Ding J, Allen E, Wang W et al (2006) Gene targeting of GAN in mouse causes a toxic accumulation of microtubule-associated protein 8 and impaired retrograde axonal transport. Hum Mol Genet 15:1451–1463
477. Allen E, Ding J, Wang W et al (2005) Gigaxonin-controlled degradation of MAP1B light chain is critical to neuronal survival. Nature 438:224–228
478. Cleveland DW, Yamanaka K, Bomont P (2009) Gigaxonin controls vimentin organization through a tubulin chaperone-independent pathway. Hum Mol Genet 18:1384–1394
479. Yang Y, Allen E, Ding J et al (2007) Giant axonal neuropathy. Cell Mol Life Sci 64:601–609
480. Ganay T, Boizot A, Burrer R et al (2011) Sensory-motor deficits and neurofilament disorganization in gigaxonin-null mice. Mol Neurodegener 6:25
481. Dequen F, Bomont P, Gowing G et al (2008) Modest loss of peripheral axons, muscle atrophy and formation of brain inclusions in mice with targeted deletion of gigaxonin exon 1. J Neurochem 107:253–264
482. Cairns NJ, Zhukareva V, Uryu K et al (2004) Alpha-internexin is present in the pathological inclusions of neuronal intermediate filament inclusion disease. Am J Pathol 164:2153–2161
483. Uchikado H, Shaw G, Wang DS et al (2005) Screening for neurofilament inclusion disease using alpha-internexin immunohistochemistry. Neurology 64:1658–1659
484. Momeni P, Cairns NJ, Perry RH et al (2006) Mutation analysis of patients with neuronal intermediate filament inclusion disease (NIFID). Neurobiol Aging 27:778, e1-778.e6
485. Dequen F, Cairns NJ, Bigio EH et al (2011) Gigaxonin mutation analysis in patients with NIFID. Neurobiol Aging 32:1528–1529
486. Neumann M, Roeber S, Kretzschmar HA et al (2009) Abundant FUS-immunoreactive pathology in neuronal intermediate filament inclusion disease. Acta Neuropathol 118:605–616
487. Medori R, Autilio-Gambetti L, Monaco S et al (1985) Experimental diabetic neuropathy: impairment of slow transport with changes in axon cross-sectional area. Proc Natl Acad Sci U S A 82:7716–7720
488. Medori R, Jenich H, Autilio-Gambetti L et al (1988) Experimental diabetic neuropathy: similar changes of slow axonal transport and axonal size in different animal models. J Neurosci 8:1814–1821
489. Yagihashi S, Kamijo M, Watanabe K (1990) Reduced myelinated fiber size correlates with loss of axonal neurofilaments in peripheral nerve of chronically streptozotocin diabetic rats. Am J Pathol 136:1365–1373
490. Schmidt RE, Beaudet LN, Plurad SB et al (1997) Axonal cytoskeletal pathology in aged and diabetic human sympathetic autonomic ganglia. Brain Res 769:375–383
491. Fernyhough P, Gallagher A, Averill SA et al (1999) Aberrant neurofilament phosphorylation in sensory neurons of rats with diabetic neuropathy. Diabetes 48:881–889
492. Scott JN, Clark AW, Zochodne DW (1999) Neurofilament and tubulin gene expression in progressive experimental diabetes: failure of synthesis and export by sensory neurons. Brain 122(Pt 11):2109–2118
493. Galloway PG, Mulvihill P, Perry G (1992) Filaments of Lewy bodies contain insoluble cytoskeletal elements. Am J Pathol 140: 809–822
494. Trimmer PA, Borland MK, Keeney PM et al (2004) Parkinson's disease transgenic mitochondrial cybrids generate Lewy inclusion bodies. J Neurochem 88:800–812
495. Forno LS, Sternberger LA, Sternberger NH et al (1986) Reaction of Lewy bodies with antibodies to phosphorylated and non-phosphorylated neurofilaments. Neurosci Lett 64:253–258

496. Pappolla MA (1986) Lewy bodies of Parkinson's disease. Immune electron microscopic demonstration of neurofilament antigens in constituent filaments. Arch Pathol Lab Med 110:1160–1163
497. Lavedan C, Buchholtz S, Nussbaum RL et al (2002) A mutation in the human neurofilament M gene in Parkinson's disease that suggests a role for the cytoskeleton in neuronal degeneration. Neurosci Lett 322:57–61
498. Perez-Olle R, Lopez-Toledano MA, Liem RK (2004) The G336S variant in the human neurofilament-M gene does not affect its assembly or distribution: importance of the functional analysis of neurofilament variants. J Neuropathol Exp Neurol 63:759–774
499. Han F, Bulman DE, Panisset M et al (2005) Neurofilament M gene in a French-Canadian population with Parkinson's disease. Can J Neurol Sci 32:68–70
500. Kruger R, Fischer C, Schulte T et al (2003) Mutation analysis of the neurofilament M gene in Parkinson's disease. Neurosci Lett 351:125–129
501. Griffin JW, Hoffman PN, Clark AW et al (1978) Slow axonal transport of neurofilament proteins: impairment of beta, beta′-iminodipropionitrile administration. Science 202:633–635
502. Soler-Martin C, Vilardosa U, Saldana-Ruiz S et al (2012) Loss of neurofilaments in the neuromuscular junction in a rat model of proximal axonopathy. Neuropathol Appl Neurobiol 38(1):61–71
503. Clark AW, Griffin JW, Price DL (1980) The axonal pathology in chronic IDPN intoxication. J Neuropathol Exp Neurol 39:42–55
504. Eyer J, McLean WG, Leterrier JF (1989) Effect of a single dose of beta, beta′-iminodipropionitrile in vivo on the properties of neurofilaments in vitro: comparison with the effect of iminodipropionitrile added directly to neurofilaments in vitro. J Neurochem 52:1759–1765
505. Griffin JW, Parhad I, Gold B et al (1985) Axonal transport of neurofilament proteins in IDPN neurotoxicity. Neurotoxicology 6: 43–53
506. Couri D, Milks M (1982) Toxicity and metabolism of the neurotoxic hexacarbons n-hexane, 2-hexanone, and 2,5-hexanedione. Annu Rev Pharmacol Toxicol 22:145–166
507. Lehning EJ, Dyer KS, Jortner BS et al (1995) Axonal atrophy is a specific component of 2,5-hexanedione peripheral neuropathy. Toxicol Appl Pharmacol 135:58–66
508. Wang QS, Hou LY, Zhang CL et al (2008) Changes of cytoskeletal proteins in nerve tissues and serum of rats treated with 2,5-hexanedione. Toxicology 244:166–178
509. Karlsson JE, Rosengren LE, Haglid KG (1991) Quantitative and qualitative alterations of neuronal and glial intermediate filaments in rat nervous system after exposure to 2,5-hexanedione. J Neurochem 57:1437–1444
510. Tuckwell DS, Laszlo L, Mayer RJ (1992) 2,5-Hexanedione-induced intermediate filament aggregates contain ubiquitin-protein conjugate immunoreactivity and resemble Rosenthal fibres. Neuropathol Appl Neurobiol 18:593–609
511. Anthony DC, Giangaspero F, Graham DG (1983) The spatio-temporal pattern of the axonopathy associated with the neurotoxicity of 3,4-dimethyl-2,5-hexanedione in the rat. J Neuropathol Exp Neurol 42:548–560
512. Sickles DW (1991) Toxic neurofilamentous axonopathies and fast anterograde axonal transport. III. Recovery from single injections and multiple dosing effects of acrylamide and 2,5-hexanedione. Toxicol Appl Pharmacol 108:390–396
513. Sickles DW (1992) Toxic neurofilamentous axonopathies and fast anterograde axonal transport. IV. In vitro analysis of transport following acrylamide and 2,5-hexanedione. Toxicol Lett 61:199–204
514. Stone JD, Peterson AP, Eyer J et al (1999) Axonal neurofilaments are nonessential elements of toxicant-induced reductions in fast axonal transport: video-enhanced differential interference microscopy in peripheral nervous system axons. Toxicol Appl Pharmacol 161: 50–58
515. Stone JD, Peterson AP, Eyer J et al (2001) Neurofilaments are nonessential to the pathogenesis of toxicant-induced axonal degeneration. J Neurosci 21:2278–2287
516. Stone JD, Peterson AP, Eyer J et al (2000) Neurofilaments are non-essential elements of toxicant-induced reductions in fast axonal transport: pulse labeling in CNS neurons. Neurotoxicology 21:447–457
517. Bizzi A, Gambetti P (1986) Phosphorylation of neurofilaments is altered in aluminium intoxication. Acta Neuropathol 71:154–158
518. Shea TB, Balikian P, Beermann ML (1992) Aluminum inhibits neurofilament protein degradation by multiple cytoskeleton-associated proteases. FEBS Lett 307:195–198
519. Shea TB, Beermann ML (1994) Multiple interactions of aluminum with neurofilament subunits: regulation by phosphate-dependent interactions between C-terminal extensions of the high and middle molecular weight subunits. J Neurosci Res 38:160–166

520. Shea TB, Wheeler E, Jung C (1997) Aluminum inhibits neurofilament assembly, cytoskeletal incorporation, and axonal transport. Dynamic nature of aluminum-induced perikaryal neurofilament accumulations as revealed by subunit turnover. Mol Chem Neuropathol 32:17–39
521. Endo H, Kittur S, Sabri MI (1994) Acrylamide alters neurofilament protein gene expression in rat brain. Neurochem Res 19:815–820
522. Gold BG, Price DL, Griffin JW et al (1988) Neurofilament antigens in acrylamide neuropathy. J Neuropathol Exp Neurol 47: 145–157
523. Howland RD, Alli P (1986) Altered phosphorylation of rat neuronal cytoskeletal proteins in acrylamide induced neuropathy. Brain Res 363:333–339
524. Tanii H, Hayashi M, Hashimoto K (1988) Neurofilament degradation in the nervous system of rats intoxicated with acrylamide, related compounds or 2,5-hexanedione. Arch Toxicol 62:70–75
525. Sickles DW, Pearson JK, Beall A et al (1994) Toxic axonal degeneration occurs independent of neurofilament accumulation. J Neurosci Res 39:347–354
526. Takahashi A, Mizutani M, Agr B et al (1994) Acrylamide-induced neurotoxicity in the central nervous system of Japanese quails. Comparative studies of normal and neurofilament-deficient quails. J Neuropathol Exp Neurol 53:276–283
527. Vahidnia A, Romijn F, Tiller M et al (2006) Arsenic-induced toxicity: effect on protein composition in sciatic nerve. Hum Exp Toxicol 25:667–674
528. DeFuria J, Shea TB (2007) Arsenic inhibits neurofilament transport and induces perikaryal accumulation of phosphorylated neurofilaments: roles of JNK and GSK-3beta. Brain Res 1181:74–82
529. RamaRao G, Acharya JN, Bhattacharya BK (2011) Changes of protein oxidation, calpain and cytoskeletal proteins (alpha tubulin and pNF-H) levels in rat brain after nerve agent poisoning. Toxicol Lett 203:227–236
530. RamaRao G, Waghmare C, Kumar Gupta A et al (2011) Soman-induced alterations of protein kinase C isozymes expression in five discrete areas of the rat brain. Drug Chem Toxicol 34:221–232
531. Ramarao G, Waghmare C, Srivastava N et al (2011) Regional alterations of JNK3 and CaMKIIalpha subunit expression in the rat brain after soman poisoning. Hum Exp Toxicol 30:448–459
532. Lurie DI, Brooks DM, Gray LC (2006) The effect of lead on the avian auditory brainstem. Neurotoxicology 27:108–117
533. Jones LG, Prins J, Park S et al (2008) Lead exposure during development results in increased neurofilament phosphorylation, neuritic beading, and temporal processing deficits within the murine auditory brainstem. J Comp Neurol 506:1003–1017

Chapter 10

From an Axon into a Growth Cone After Axotomy: A Model for Cytoskeletal Dynamics

Micha E. Spira and Hadas Erez

Abstract

Using live confocal microscope imaging of molecular probes, retrospective immunolabeling, classical electron microscopy, and cultured *Aplysia* neurons this chapter describes the experimental approaches to document the cascades of cytoskeleton remodeling that underlie the transformation of a cut axonal-end into a growth cone (GC). Rapture of the axon's plasma membrane is followed by massive influx of calcium through the cut end, leading to depolymerization of the microtubules (MTs) and the actin filaments. The elevated free intracellular calcium concentration ($[Ca^{2+}]_i$) activates calpain which cleaves the submembrane spectrin skeleton. Repair of the ruptured membrane barrier is followed by the recovery of the $(Ca^{2+})_i$ and the restructuring of the cytoskeleton in a way that totally differs from that of a differentiated axon. The typical unipolar orientation of axonal MTs is changed to form two distinct MT-based vesicle traps. One traps capture and concentrates anterogradely transported Golgi-derived vesicles while the other concentrates retrogradely transported vesicles. The trapped Golgi-derived vesicles fuse with the plasma membrane from which the submembrane spectrin skeleton was removed by calpain. Actin filaments repolymerize to form radially oriented bundles that generate the mechanical force underlying the extension of the GC's lamellipodium. Feedback interactions between the cytoskeletal elements and the transported cargo as well as interactions with membrane targets participate in the definition of cytoskeleton restructuring and the formation of competent GCs.

Key words: Axotomy, Axon growth cone, Cytoskeleton, Microtubules, Actin spectrin, Dynamics, Imaging, Electron microscopy, Regeneration, Spectrin

1. Introduction

A critical step in neuronal regeneration after mechanical injury is the transformation of the cut axonal tip into a GC. While the axon is a stationary structure specialized for propagating action potentials and bidirectional transport of subcellular organelles and molecules, the GC is a motile, multitask compartment that orchestrates in time and space the use of cell resources for growth and regeneration. Competent GCs serve during development or in regenerating

Rolf Dermietzel (ed.), *The Cytoskeleton: Imaging, Isolation, and Interaction*, Neuromethods, vol. 79,
DOI 10.1007/978-1-62703-266-7_10,

modes as the site of cytoskeleton assembly, local insertion of lipid membrane in support of the elongation processes, and as a specialized compartment that reads the microenvironment and navigates the neurites to their appropriate targets. In addition, GCs can recognize postsynaptic elements and transform into presynaptic elements (for reviews see (1–4)).

The effective transformation of a differentiated nongrowing neuron into a regenerating mode involves the activation of a large number of mechanisms. This chapter describes the approaches used to document and analyze the remodeling cascades of the cytoskeleton underlying the transformation of a cut axonal end into a GC. As the functional organization of the cytoskeletal elements in axons and GCs differs substantially, it is obvious that the structural transformation of an axon into GC involves massive alterations.

To study these events we used Cultured *Aplysia*'*s* neurons. The use of this system proved to be very important for the analysis of synaptic transmission, the molecular, cellular, and biophysical mechanisms that underlie short- and long-term memory acquisition processes (see the Nobel prize lecture of Dr. E.R. Kandel (5) and (6, 7)). The use of *Aplysia* neurons was also very important in the elucidation of basic cell biological concepts related to the structural and functional organization of GCs (8–10) and in the discovery of injury signal transport (11–14).

We developed the use of cultured *Aplysia* neurons as a cell biological platform to analyze the mechanisms underlying GC assembly after axotomy (15–29) and study the role of cytoskeleton in cellular models of neurodegenerative diseases (30–33). The large size of identifiable cultured neurons and the ability to simultaneously express a number of exogenous genes and live image their fluorescently tagged translational products by confocal imaging over timescales of seconds to weeks provide a significant technical advantage. The correlation of live confocal imaging with electron microscopic observations, retrospective immunolabeling, and electrophysiological measurements provided the unique opportunity to study the cytoskeleton of GCs and the restructuring events following axotomy. For the studies we developed methodologies to reliably express and live image EGFP, EYFP, ECFP, and cherry-fusion proteins (17, 18, 23, 24, 30, 34–36) and advanced photo-switchable and photo-activatable fluorescent tools (35). These include probes to image microtubules, microtubule polarity, actin, membrane retrieval, exocytosis, and probes to image real-time proteolytic activities (19, 20), the free intracellular calcium, and other ions (27–29, 34, 37). In addition exogenous human genes such as mutated tau can be expressed by these cells (30–33). These, together with retrospective immunolabeling and electron microscopy, provide a comprehensive picture of the subject. It should be noted that where observations have been made on other invertebrate systems such as the *C. elegance* and vertebrate neurons, the cellular events that underlie the transformation of a cut *Aplysia's* axon into a GC are similar. The common cascades allow to formulate general cell biological principles for GC assembly.

Neurons typically possess two types of distinct processes, a single axon and several dendrites. This chapter describes step by step the sequences of subcellular events that transform the main neurite of cultured *Aplysia* neuron (referred to as the axon) into a GC and the corresponding methods that have been used to document it. We focus the discussion on three cytoskeletal elements: the microtubules, actin filaments, and the submembrane skeleton—spectrin. To provide a comprehensive scenario of the cell biological events and methodologies we also illustrate the contribution of anterogradely transported material to the restructuring of the cytoskeleton and how fusion of vesicles with the GC's plasma membrane is directed by the spatial reorganization of the cytoskeleton. The chapter deals with the cytoskeleton at the cellular level.

2. Materials and Methods

2.1. Culturing Procedures and Axotomy

Identifiable neurons from *Aplysia californica* are isolated and maintained in culture as described by Schacher and Proshansky (38) and Spira et al. (25, 39). Briefly, juvenile *Aplysia californica* (1–10 g) are anesthetized by injection of isotonic $MgCl_2$ solution (380 mM) into the animal's body cavity. The ganglia are then dissected out and incubated in marine supplemented L15 containing 0.1% protease (type XIV bacterial, from *Streptomyces griseus*, Sigma) at 34°C for 1.5–2.5 h. Following the protease treatment the ganglia are washed with L15 supplemented for marine species according to Schacher and Proshansky (38). The ganglia are then desheathed. The identified neurons are manually pulled out along with their original axon with the aid of a sharp glass microelectrode and plated in glass-bottom dishes coated with poly-L-lysine containing culture medium. All experiments are performed at room temperature (21–25°C) after replacing the culture medium with artificial sea water (ASW). Axonal transection is conducted by applying pressure on the axon with the thin shaft of a micropipette under visual control (Fig. 1 and (16, 25, 39)).

2.2. Organization of the Cytoskeleton in Intact Axon Microtubules

In differentiated neurons the MTs serve a number of functions: in concert with other skeletal elements they confer the polar nature of neurons and their mechanical stability contributes to the maintenance of the neuron morphology. MTs serve as polar tracks along which cell resources and signaling molecules are transported anterogradely and retrogradely (40). Three complementary methods

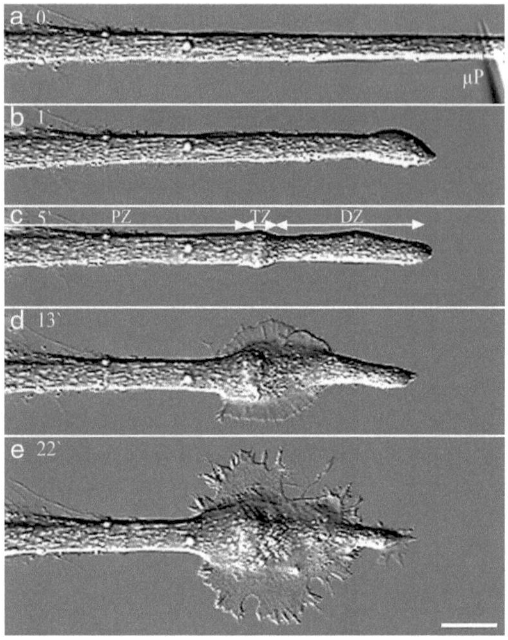

Fig. 1. The morphological sequence of axonal transformation into a growth cone after axotomy. A Buccal neuron was cultured for 24 h and then transected with a sharp micropipette (μP, in **a**). Following axonal transection a membrane seal is formed over the cut axonal end (**b**). Within 5 min of axotomy, a compartment located 75 μm proximally to the cut end of the axon swells (**c**). This swollen zone, referred to as the transition zone (TZ), subdivides the axon into a distal zone (DZ) and a proximal zone (PZ). The transition zone becomes the growth cone-organizing center (**d**), as indicated by the extension of a growth cone's lamellipodium initially from it (**d**). With time the lamellipodium also extends from the distal zone (**e**). Scale bar = 25 μm (from (24)).

are used to provide a comprehensive view of the MT distribution and dynamics: classical electron microscopy, retrospective immunolabeling, and live confocal imaging of fluorescently labeled tubulin or MT-associated proteins. The use of these methods is described below.

2.2.1. Classical Electron Microscopy

The spatial distribution of MTs in axons and dendrites has been extensively documented using classical electron microscopic methods. Cross or longitudinal sections of neurites also provide detailed information as to the organelle distribution along the MTs (Fig. 2). The nanometric resolution of the EM allows quantification of the MTs, their spacing, their spatial distribution, and even the documentation of MT-associated proteins and organelles (41). Electron microscopic methods can even disclose the critical feature of MT polar orientation. Mapping the polar orientation of individual MTs by EM is possible by the use of the so-called hook method (42–46). This method involves elaborate procedures and extensive examination of MT cross sections to reveal the directional curvature of hook-like structures formed by exogenous tubulin dimers that

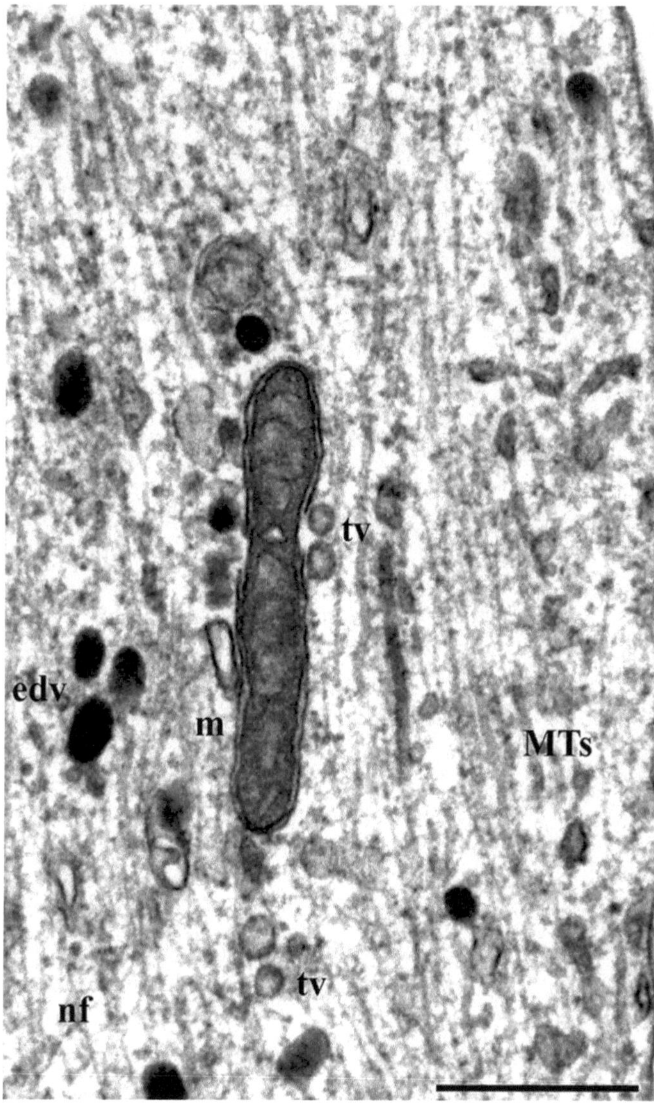

Fig. 2. The distribution of MTs and organelles in cultured *Aplysia* axon. Axons of cultured *Aplysia* neurons are characterized by linear arrays of microtubules and neurofilaments (nf). The MTs align in parallel to the long axis of the axon and are homogeneously distributed within the axoplasm. Mitochondria (m), translucent vesicles (tv), electron dense vesicles (edv), vacuoles, and the endoplasmic reticulum are evenly distributed among the MTs. Scale bar = 0.5 μm.

polymerize to generate curving sheets attached to the endogenous MTs (46). Careful observation of the "hook" directional curvature, nevertheless, allows mapping the polar orientation of individual MTs in a given EM section at a given point in time. Besides for the elaborate technical procedure it should be remembered that an EM image provides a structural snapshot. Thus, the use of the EM hook-method to study the dynamics of MT polar orientation is limited to slow processes (see for example (47)).

2.2.2. Retrospective Immunolabeling of Microtubules

We used classical retrospective immunolabeling methods to image MTs. As this method is extensively used we refer to (26).

2.2.3. Live Confocal Imaging of Fluorescently Labeled Tubulin or MT-Associated Proteins

The green fluorescent fusion protein and its isoforms are extensively used to monitor gene expression, protein localization, and their dynamics in relation to fundamental cellular processes (48). Dr. BK Kaang laboratory introduced in 1996 the use of pNEX derivative vector to transfect *Aplysia* neurons (49). Since then, the PNEX transfection vector was successfully used by a number of laboratories. We found that intracellular injection of mRNA encoding tagged proteins is a very effective procedure for the expression of fluorescently tagged proteins in *Aplysia* neurons. For most genes that we used, mRNA injection permits visualization of the fluorescent proteins within 4 h of injection. The method that we use consists of in vitro transcription of mRNA encoding tagged-fusion proteins and its subsequent intracellular pressure-microinjection into the cytoplasm (23). An example of a neuron injected by a solution containing mRNA encoding EGFP-α tubulin fusion protein is depicted in Fig. 5 (23, 24).

Another method to label tubulin is by microinjection of tetramethyl-rhodamine-labeled tubulin monomers (Fig. 10, red, and Fig. 12a, (50)). To that end we used 4 mg/ml tetramethyl-rhodamine tubulin, from bovine brain (Molecular Probes). The probe is loaded into a glass micropipette and pressure-injected into the cytoplasm of the cell body.

2.2.4. Labeling MTs by Plus-End Tracking Proteins for Live Imaging

To study the polar orientation of MTs, the microtubules are labeled by GFP-tagged plus-end-tracking protein (51–53). These fusion proteins transiently bind to the plus end of MTs and move with the growing MT's tips forming a comet taillike image, and thereby allowing to repeatedly live image the polar orientation of polymerizing MTs (54–57). In vertebrate this is done by transfection (55, 58) and in cultures *Aplysia* neurons by microinjection of mRNA encoding the plus-end-tracking protein (17, 18, 21, 30–32).

In cultured *Aplysia* neurons these observations revealed that within the cell body the MTs point their plus end in various directions, in the axon hillock the MTs form a typical microtubule organizing center (MTOC) (Fig. 3a), and in the main axon approximately 70% of the EB3-GFP-labeled MTs orient their plus ends distally, away from the cell body towards the tip of the neurites (Fig. 3b). In the axon approximately 15% of the MTs point their plus end proximally and the rest 15% perpendicular to the long axis of the axon (31, 32). It should be noted however that whereas the use of plus-end-tracking proteins allows to map the polar orientation of MTs, this tracking protein family only labels dynamically polymerizing MTs, excluding stable or depolymerizing MTs (59, 60). Thus, mapping of MT polar orientation by live imaging of plus-end-tracking proteins provides incomplete information.

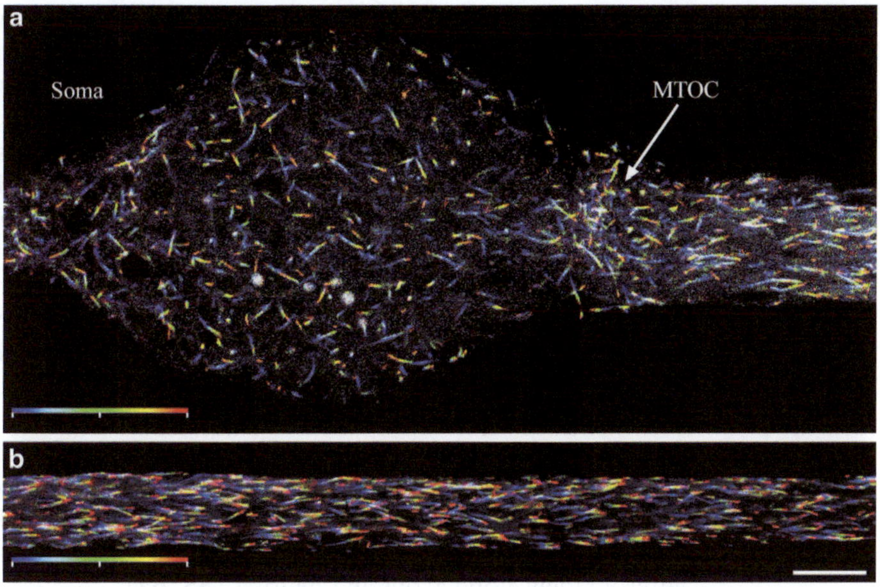

Fig. 3. Imaging of microtubules by the plus-end-tracking protein EB3. A B neuron was cultured for 4 h, and then microinjected with EB3-GFP mRNA. EB3-GFP comet tails were live imaged 16 h post mRNA injection. Images of the EB3 comet tails were taken 4 s and 6 s apart (**a** and **b**, respectively). Four frames of the movies were color coded and displayed as a single image (see color scales). The *blue-red* stretches indicate the directional orientation of the MT plus ends. In the soma the MT plus ends point in various directions, in the microtubule-organizing center (MTOC) they appear to radiate from a virtual center (**a**). In the axon (**b**), the majority of the MT plus ends point away from the cell body (the *right-hand side* of the figure). Scale bar = 10 μm.

Combining quantitative analysis of MTs from EM and live confocal images of EB3-GFP comet tails in cultured *Aplysia* neurons (33) revealed that EB3 labels only 10% of the MTs in intact axon. This observation raised the question as to the level to which MT-polar orientation analysis by EB3-GFP comet tails represents the remaining 90% of the MT population. This question can be addressed by complementary studies of the very same neurons using the EM Hook method. Nevertheless, to the best of our knowledge this was not done. Partial insight into this question comes from experiments in which the polar orientation of MTs is assessed by observing vesicle transport along an axon. In *Aplysia* neurons retrogradely transported endocytotic vesicles can be labeled by the fluid-phase pinocytotic marker sulforhodamine 101 (SR101) (17, 18, 22, 30). Live imaging of SR101-labeled vesicles revealed that under normal conditions the labeled vesicles translocate retrogradely along linear tracks across the entire cross section of the axon (17, 22). Under experimental conditions that lead to reversal of the MT polar orientation the retrograde transport of these vesicles is impaired (30, 33). In cultured *Aplysia* neurons these observations supported the view that the EB3-GFP comet tails, which report on only 10% of the MT population, represent well the polar orientation of the MTs within the sampled compartments.

2.3. Actin

In differentiated cells approximately 50% of the actin is maintained in its globular form (G actin) and the rest in the form of actin filaments (F actin). These filaments form actin bundles and weblike networks (Chapters 1, 3 and 8). In both forms the actin filaments face their barbed ends towards the membrane. Continuous incorporation of G-actin takes place at the barbed end, whereas the subunits dissociate at the pointed ends leading to treadmilling of the actin filaments (61). In concert with other skeletal elements actin filaments contribute in differentiated neurons to the mechanical stability and structural plasticity of various neuronal-compartments such as the dendritic spines and filopodia (62), and contribute to the assembly of subcellular organelles in compartments such as the assembly of synaptic vesicles in presynaptic terminals (63). Actin serves as part of the mechanical machinery underlying cytokinesis, endocytosis, and phagocytosis and serves as tracks for myosin-driven cargo transport. Interestingly, in both *Aplysia* and vertebrate neurons packages of actin are seen to be transported anterogradely along axons and neurites (35, 64).

Unlike MTs classical electron microscopy cannot be used to unequivocally identify actin filaments and their spatial distribution. Therefore, retrospective immunolabeling and live confocal imagings of fluorescent probes (such as phalloidin) are the tools to visualize actin filaments. Subcellular actin filament dynamics can be best studied by live fluorescent speckle microscopy (65). In cultured *Aplysia* neurons this method was extensively used by the laboratory of Dr. P. Forscher to study actin filament dynamics in GCs (66, 67). This was done by injection of low amounts of Alexa-594 phalloidin into a neuron followed by total internal reflection fluorescence microscope (TIRF) observations. TIRF imaging of growth cones was selected to provide high signal-to-noise imaging and low phototoxicity.

In intact cultured *Aplysia* neurons actin labeled by fluorescently tagged phalloidin or EGFP appears to be distributed within the bulk of the axoplasm. EGFP-actin hot spots are seen along the plasma membrane facing the glass substrate (Fig. 9, control). These actin puncta most likely represent adhesion complexes (68, 69).

2.4. Spectrin

Spectrin-based membrane skeleton is found in most cells, including neurons. A major function of the submembrane spectrin skeleton is to mechanically support the membrane bilayer permitting erythrocytes, for example, to survive as they are deformed during circulation through small capillaries. In neurons spectrin is considered important for stabilization of the axonal membrane and its elasticity and integrity, and in the localization of integral membrane proteins. Spectrin serves as a barrier restricting the diffusion of cytoplasmic and membrane proteins and links the plasma membrane to the submembrane actin skeleton (70, 71). The submembrane spectrin skeleton cannot be visualized by classical electron microscopy. Since we have not been able to express fluorescent spectrin and to live image spectrin cleavage product we turned to the use of retrospective spectrin immunolabeling. For spectrin immunolabling we use polyclonal rabbit anti-human spectrin as a primary antibody and sheep anti-rabbit IgG, Cy3 conjugated as a secondary antibody (19–21, 25).

Examination of the distribution of spectrin, by this method, shows that in the cell body and axons spectrin is found predominantly beneath the plasma (Fig. 11a).

2.5. The Transformation of Typical Axonal Cytoskeleton into that of a Growth Cone

The advantageous and limitations of the different methods described above to image in time and space cytoskeleton remodeling processes following axotomy are illustrated in the next section. It is important to note that a general problem in designing electron microscope- and immunolabeling-based experiments to study the cascade of events that underlie the transformation of a cut axon-tip into a GC is in defining the time at which to "freeze" the process by chemical fixation. In general the unfolding of the response to axotomy greatly varies depending on the neurons' phenotype, their age, physiological state, and microenvironment. Thus, time post injury cannot serve as a proper parameter for comparison of experimental results obtained by the use of different neurons, under different experimental conditions, or of different ages. Parameters such as membrane seal formation and down regulation of the free intracellular calcium concentration at the cut end provide better reference points for comparing the responses of axons to transection.

2.5.1. Electron Microscopy

When transected in normal physiological solution, the axon's tip subdivides within approximately 5 min into three distinct zones: the proximal zone (PZ), the transition zone (TZ), and the distal zone (DZ) (Figs. 1 and 4). The proximal zone is that part of the axon which is closest to the cell body and which is not affected by the transection, retaining its pre-axotomy MT and organelle ultrastructure. The distal zone is a segment that stretches from the transection point and proximally up to the TZ (Figs. 1 and 4a). The characteristic ultrastructure of the DZ corresponds to the axonal segment in which the $(Ca^{2+})_i$ exceeds ~500 μM. Within minutes of axotomy, the MTs undergo fragmentation and depolymerization along the DZ. Amorphous electron dense aggregates are formed at the distal tip of the DZ (Fig. 4a), and longitudinal electron dense deposits are seen along its length. The longitudinal electron dense deposits are composed of bundles of fragmented microtubules decorated by amorphous material. Surrounding the axoplasmic core of the DZ there exists an area of cytoplasmic material containing vesicles and endoplasmic reticulum, but lacking MTs and neurofilaments. This region interposes between the plasma membrane and the axoplasmic core (Fig. 4a see (26)).

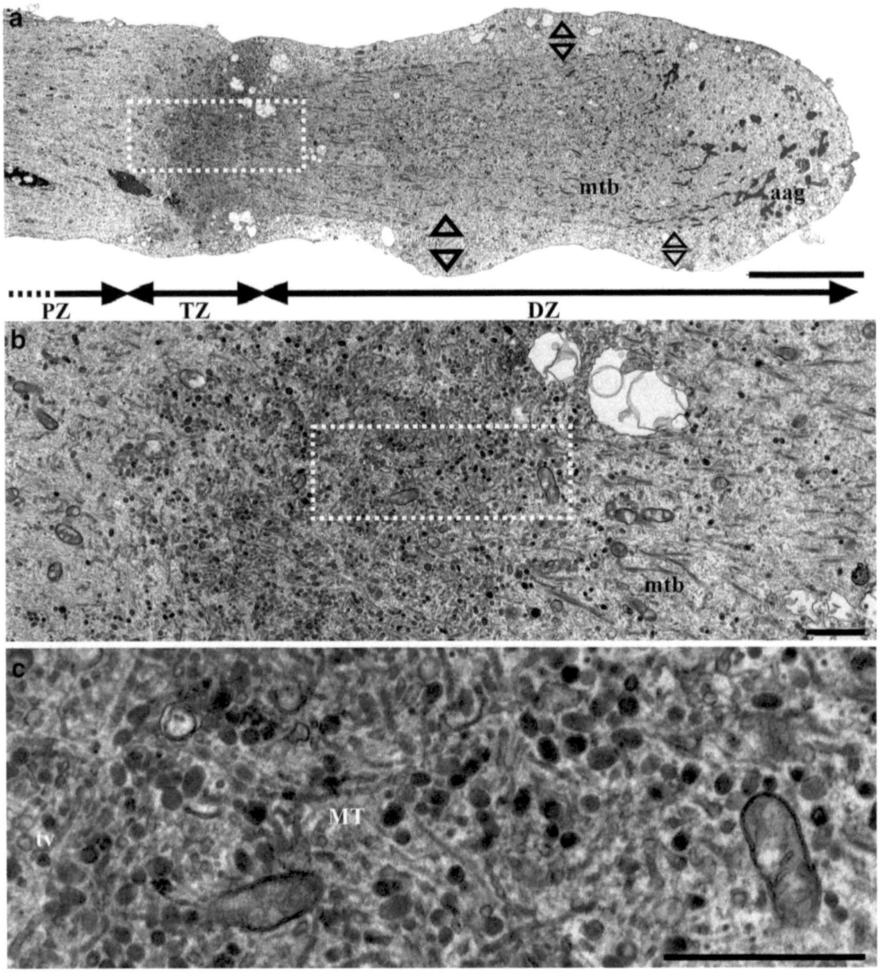

Fig. 4. Axotomy leads to rapid subdivision of the cut end into three distinct zones and the accumulation of vesicles in the transition zone. A B1 neuron was cultured for 24 h, transected, and fixed 3 min after axotomy. (**a**) A low-magnification image reveals the subdivision of the cut end into a distal zone (DZ), transition zone (TZ), and a proximal zone (PZ). Notice the detachment of the plasma membrane from the axonal core (*arrowheads*) along the distal zone, and the typical formation of amorphous aggregates (aag) and microtubular bundles (mtb). (**b**) High magnification of the transition zone (*dashed rectangle* in **a**) reveals that within minutes of axotomy, a large number of electron-dense vesicles and clear vesicles populate the transition zone. Note that the vesicles occupy the region in which the longitudinal microtubular bundles disappear. (**c**) Enlargement of the area marked by the *dashed rectangle* in (**b**) reveals that the transition zone contains a high density of vesicles and short fragments of microtubules. Scale bars = 10 μm in (**a**) and 1 μm in (**b**) and (**c**) (from (26)).

The "detachment" of the plasma membrane from the axoplasmic core suggests that in the DZ the anchoring of the plasma membrane to the cytoskeleton is severed (see Section 2.4 this chapter). This characteristic ultrastructure of the DZ already exists in neurons fixed less than 5 min after axotomy. As time progresses, the morphology of the cut end changes, most notably within the TZ. The TZ is the compartment found between the PZ and the DZ. It is the site from which the nascent GC emerges 10–30 min after axotomy. It is initially recognizable as an abrupt interface between the DZ and PZ

(Figs. 1 and 4). The interface contains fragments of NFs and MTs that orient in various directions in respect to the long axis of the axon (Fig. 4b, c). Initially, the TZ does not contain a large population of vesicles (Fig. 6). However, within minutes of axotomy, vesicles accumulate within this restricted zone (Figs. 4c and 6).

The extension of the GC's lamellipodia from the TZ is detectable approximately 15 min after axotomy (Figs. 1, 9, and 10). At a later stages (30–50 min after axotomy), MTs polymerize to extend radially towards the margins of the GC's lamellipodium (Fig. 10). In axotomy-induced GCs this event coincides with the subdivision of the lamella into branches, each led by a smaller GC. At this stage, the GC's center still contains a dense population of vesicles, and is enriched with mitochondria (17, 24, 26).

2.5.2. Microtubule Depolymerization and Restructuring After Axotomy: Use and Limitations of Molecular Probes for Confocal Imaging

To study the dynamics of MT remodeling after axotomy, we labeled the MTs either by intracellular microinjection of tetramethylrhodamine-labeled tubulin monomers (Fig. 10—red and Fig. 12 column I) or by intracellular microinjection of EGFP-α-tubulin mRNA (Fig. 5). The injection of the probe did not alter the behavior of the neurons or their response to axotomy in any noticeable way.

Axotomy in normal physiological solution (control conditions) induces a retrograde wave of MT disassembly as revealed by the dissipation of the linearly oriented structures and the ensuing MT reorganization after axotomy.

Such images revealed that MT dissipation starts at the point of axonal transection and progresses along an axonal segment 50–150 μm in length (Fig. 5). The retrograde wave of MT fragmentation or depolymerization propagates at a rate corresponding to the rate of propagation of the calcium concentration-front that originates at the ruptured end of the axon (28, 29). The termination of the MT disassembly retrograde wave after axotomy corresponds to the formation of a membrane seal over the cut end and the subsequent decline in calcium ion levels (25, 28, 29).

The termination of microtubule disassembly is followed within minutes of axotomy by a phase of MT bundle reformation. MT bundles reform at the edge of the proximal zone and progress distally (PZ, arrows in Fig. 5—24 s and 3 min). In the distal zone, microtubule bundles reform along the cortical region of the axoplasm while the core of the axoplasm along the distal zone becomes progressively depleted of fluorescent tubulin (Figs. 5 and 6). As time progresses, a transition zone is formed. This zone takes the form of an axoplasmic band devoid of microtubules which is confined by microtubules of the PZ on one side and by low-density microtubule bundles on the DZ (Figs. 5 and 6). The transition zone is the site in which vesicles accumulate and in which the growth cone's center is formed (Figs. 5, 6, and 10). We termed this compartment the growth cone-organizing center (GCOC).

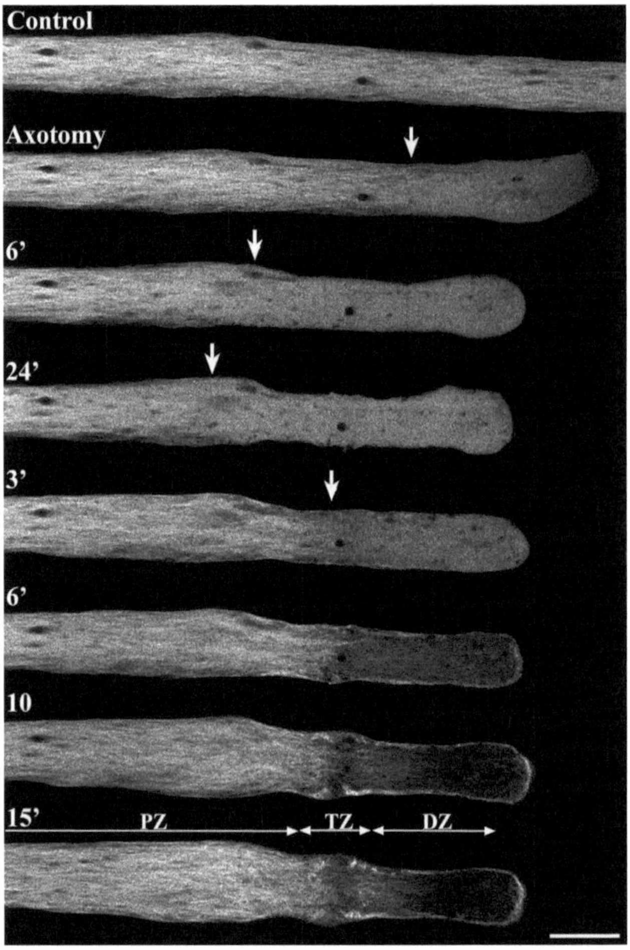

Fig. 5. Microtubules restructuring after axotomy. A Buccal neuron was cultured for 4 h, then microinjected with EGFP-α-tubulin mRNA. Twelve hours after the injection of the mRNAs the axon was transected and the distribution of microtubules was imaged 3 μm above the substrate. (Control) EGFP-tubulin fluorescence prior to axotomy. Axotomy (Axotomy) leads to a retrograde wave of microtubule disassembly (*arrows*, immediately after axotomy, 6 and 24 s after axotomy). MT repolymerization is first observed to occur in the proximo-distal direction 3 min after axotomy. Thereafter, repolymerization of MTs proceeds along the submembrane domains of the DZ. This is temporally associated with reduced tubulin fluorescence intensity in the core of the axoplasm (6–15 min after axotomy). In parallel, a band of axoplasm proximally and distally confined by MTs is formed (TZ). With time the subdivision of the axoplasm to PZ, TZ, and DZ becomes more pronounced. Scale bar = 15 μm (from (24)).

2.5.3. The Use of Plus-End-Tracking Proteins to Reveal the Formation of MT-Based Vesicle Traps by Reorientation of the MT Polar Orientation

To get complementary insight into the mechanisms that underlie the restructuring of the cytoskeleton at the cut end and in particular to understand the relationships between the MT restructuring and the accumulation of vesicles in the GCOC, we labeled the dynamic MTs by intracellular microinjection of mRNA encoding GFP-tagged EB3. Axotomy (conducted 12 h after the injection) leads within seconds to (a) a retrograde wave of EB3-GFP "comet

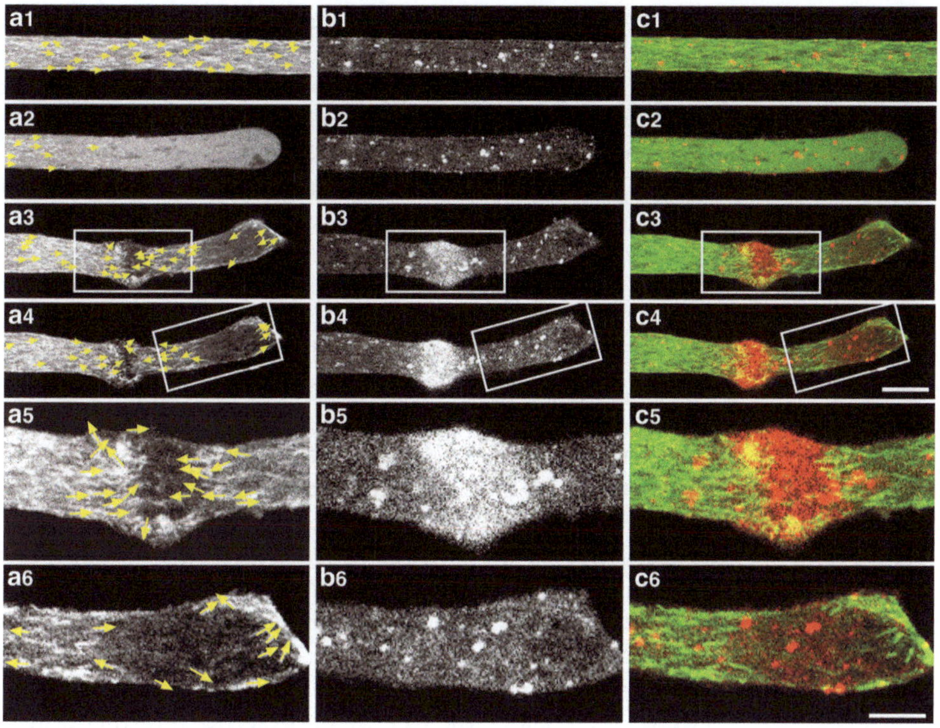

Fig. 6. Formation of MT-based vesicle traps after axotomy. A Buccal neuron was cultured for 4 h, and then microinjected with EB3-GFP mRNA. One hour later, the neuron was bathed in RH237 for 30 min. The experiment began 16 h later by online imaging of the distribution of the EB3-GFP signal (**a**), and RH237-labeled vesicles (**b**), 3 μm above the substrate. Merged images of EB3-GFP (*green*) and RH237 (*red*) fluorescent signals are shown in (**c**). *Arrows* indicate the plus end of the MTs as revealed by online imaging of the EB3-GFP signals. Before axonal transection (control), the majority of MT plus ends point towards the tip of the axon (A1). The RH237-labeled vesicles are distributed in the axoplasm (B1). The merged image of A1 and B1 is shown in (C1). Axotomy leads within seconds to the depolymerization of the MTs as indicated by the dissipation of the EB3-GFP "comet tail" structures (A2). The distribution of the RH237 fluorescence was unaltered (B2 and merged image, C2). Within minutes of axotomy the MTs repolymerize and the EB3-GFP reappears. Within the next 10 min the MTs reorient to form a plus-end trap (square in A3, and enlarged in A5) and a minus-end trap (square in A4 and enlarged in A6). The formation of the traps is associated with accumulation of the RH237 fluorescent signal mainly in the plus-end trap (square B3, and B5) and the minus-end trap (square B4, and B6). The corresponding merged images are shown in C3, C4, C5, and C6. The number of MTs that forms the traps increases with time and so is the accumulation of RH237-labeled vesicles. A1–C1 control, A2–C2 6 s after axotomy, A3–C3 7 min, A4–C4 9 min, post axotomy. Scale bar = 10 μm in C4 (applies to 1–4) and 5 μm in C6 (applies to 5–6) (from (17)).

tail" dissipation from the distal axonal zone (Fig. 6A2). This wave, temporally and spatially, corresponds to MT depolymerization (Fig. 4) ends when the cut axonal end reseals, and the free intra-axonal calcium concentration recovers to its resting level. (b) The termination of the EB3-GFP "comet tail" dissipation wave is followed within minutes of axotomy by an anterograde wave of EB3-GFP "comet tail" reformation. Within minutes of axotomy the polarity of the MTs at the cut axonal end is restructured in the following manner. In the proximal zone all MTs point their plus ends anterogradely (Fig. 6A3, A5). Distally, the polarity of the MTs reverses such that the plus ends point retrogradely (Fig. 6A4, A6).

This reversal of MT polarity creates the narrow GCOC, which is bordered both proximally and distally by the MT plus ends ("plus-end trap," Fig. 6A3, A5). Distal to the GCOC, the MT plus ends point towards the plasma membrane at the tip of the cut axon, or are aligned in parallel and close to the plasma membrane (Fig. 6A4, A6). This MT orientation forms a second zone, which is bordered by the minus ends of the MTs at the very tip of the cut axon and the minus ends of the MTs that border also the GCOC ("minus-end trap," Fig. 6A4, A6).

2.5.4. Accumulation of Fluorescently Labeled Vesicles Within the Cut End of the Axon

The MT polar reorientation at the cut axonal end is reflected immediately in the transport regime of anterogradely and retrogradely oriented molecular motors and their cargo. To examine the distribution kinetics of vesicles after axotomy in relation to the restructuring of the MT cytoskeleton the MTs were labeled by EB3-GFP, and membrane-bound organelles by the styryl dye RH237 (24, 72). The labeling is performed as follows: mRNA encoding EB3-GFP was injected into neurons and then the neurons are immersed for 30 min in RH237. Initially, RH237 partitions into the plasma membrane (24, 34); thereafter the dye is internalized by membrane retrieval and labels endocytotic vesicles. Twelve to 24 h later, the RH237 fluorescence is further distributed among membrane-bound organelles (24). Thus, using this protocol, RH237 indiscriminately labels lipid membranes of both anterogradely and retrogradely transported organelles.

As soon as the MT polarity is reorganized after axotomy (Fig. 6a), the RH237 fluorescence accumulates within the plus-end trap and to a lesser extent and more slowly along the minus-end trap (Fig. 6B1–B6, C1–C6). To establish that the accumulated RH237 signal represents the accumulation of vesicles, we compared the distribution of RH237 fluorescence with the ultrastructural composition of the axon at various times after axotomy, and confirmed that the RH237 fluorescence signal localizes at regions at which the vesicles and tubular structures concentrate (24).

2.5.5. Microtubule-Based Vesicle Traps Sort Anterogradely Transported Golgi-Derived Vesicles from Retrogradely Transported Pinocytotic Vesicles

Labeling of subcellular organelles by RH237 revealed the presence of two distinct zones at which vesicles accumulate (the GCOC and the DZ). However, as RH237 labels lipids in general, it did not allow differentiating between anterogradely and retrogradely transported vesicles. We therefore specifically labeled anterogradely transported vesicles using intracellular injections of mRNA encoding super-ecliptic-synaptopHluorin (73), EYFP-SNAP-25 (25 kDa synaptosome-associated protein (74, 75)), or cherry (76)-SNAP-25 4–6 h before axotomy and imaging. Retrogradely transported pinocytotic vesicles were labeled by bath application of the fluid-phase pinocytotic maker SR101 (77). In a series of control experiments we established that (a) injection of mRNA encoding EGFP, EYFP, or cherry leads to evenly distributed fluorescent signal that

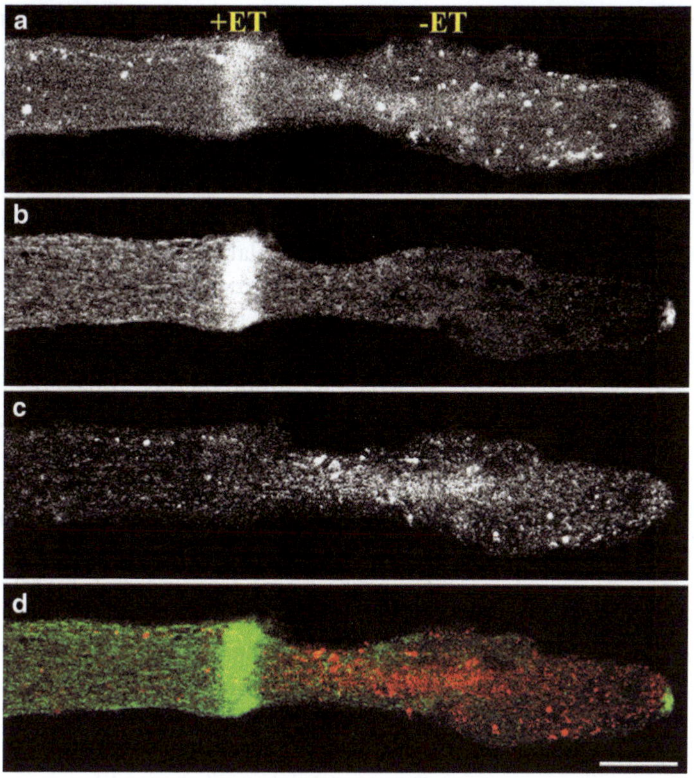

Fig. 7. Differential accumulation of anterogradely transported vesicles labeled by SNAP-25 and retrogradely transported vesicles labeled by the pinocytotic marker SR101. A Buccal neuron was cultured for 4 h. To label in a nonspecific manner membrane-bound organelles, the neuron was bathed in RH237 for 30 min. 24 h later and 4 h before the axon was transected EYFP-SNAP-25 mRNA was microinjected into the neuron. To label endocytotic vesicles, the neuron was incubated for 20 min in SR101. After washing with artificial sea water the axon was transected. Shown are images taken 20 min after axotomy. (**a**) The spatial distribution of RH237-labeled vesicles corresponds to the plus- (+ET) and minus-end (−ET) vesicle traps. (**b**) Anterogradely transported EYFP-SNAP-25 fluorescence concentrates in the region of the plus-end trap and at the very tip of the axon. (**c**) The fluid-phase pinocytotic marker SR101 concentrates in the distal zone, which corresponds to the minus-end trap. (**d**) The merged image of EYFP-SNAP-25 (*green*) and the SR101 fluorescence (*red*) demonstrates their differential accumulation in the 0020plus- and minus-end traps, respectively. Scale bar = 15 μm in (**d**) (applies to **a**–**d**) (from (17)).

does not concentrate in any of the traps after axonal transection. (b) Vesicles labeled by synaptopHluorins, EYFP-SNAP-25, or cherry-SNAP-25, 4–6 h before axotomy, are almost exclusively transported anterogradely, and that vesicles labeled by SR101 are transported retrogradely.

To illustrate the differential accumulation of the anterogradely and retrogradely transported vesicles into the plus-end and minus-end traps, we show in Fig. 7 the simultaneous imaging of the anterogradely transported EYFP-SNAP-25 (Fig. 7b), the retrogradely transported SR101 (Fig. 7c), and the distribution of RH237 (Fig. 7a). For the experiment, we first labeled the entire population of vesicles by 30-min incubation of the neuron in RH237; 20 h later, i.e., 4 h before axotomy, the neuron was injected with mRNA encoding EYFP-SNAP-25; and subsequently 2 h before transection the neuron was exposed for 20 min to the pinocytotic marker SR101. After axotomy under such conditions, anterogradely transported EYFP-SNAP-25-labeled vesicles specifically concentrated within the GCOC (Fig. 7b) and the retrogradely transported pinocytotic vesicles labeled by SR101 were retained at the distal zone (Fig. 7c); the merged image of both labeling patterns closely matches that of the distribution of the RH237 (compare Fig. 7a–d). As a complementary approach, we also performed experiments in which we simultaneously imaged EB3-GFP and cherry-SNAP-25 or EB3-GFP and SR101. In these experiments we also observed that anterogradely and retrogradely transported vesicles concentrate in the "plus-end trap" and "minus-end trap," respectively. Together, these observations demonstrate that the traps formed by MTs lead to sorting and concentration of vesicles in accordance to their directional movement along microtubules.

2.5.6. Mechanisms that Underlie the Reversal of MT Polarity and the Formation of the Vesicle Traps

What mechanisms underlie the highly orchestrated cytoskeleton-restructuring event, i.e., the formation of the plus- and minus-end vesicle traps? This subject is extensively discussed in the literature in the context of basic cell biological events for example (78–81). One mechanism that could account for the formation of the MT-based traps after axotomy is that following the massive depolymerization of the MTs at the DZ, when the calcium level is down regulated, MTs repolymerize in an altered intracellular microenvironment. The plus end of the newly assembled MTs is stabilized by sites exposed by the inner face of the plasma membrane along the DZ and organelles that accumulate at the TZ (similar to the "selective stabilization model" or "search-and-capture model" of (82)).

Another (or a complementary) mechanism that could facilitate the formation of the traps is the accumulation of molecular motors and factors arriving to the site of transection from the cell body or the cut end (17, 18, 21, 80, 81, 83, 84).

We began to examine whether molecular factors associated with anterograde or retrograde transport of vesicles are needed to generate the reverse polarity of the repolymerizing MTs. Since specific molecular and genetic tools that directly interfere with the normal functions of the molecular motors of *Aplysia* neurons are presently not available, and as vertebrate kinesin and dynein antibodies did not alter the axoplasmic transport, we utilized an indirect approach to gain some insights into these processes.

We reduced the supply of the molecular motors by mechanical removal of the cell body from the axon (Fig. 8). Isolated distal axonal segments of cultured *Aplysia* neurons maintain their structure and fire action potentials and release neurotransmitters for several days (16, 85–88). The experiments described below were

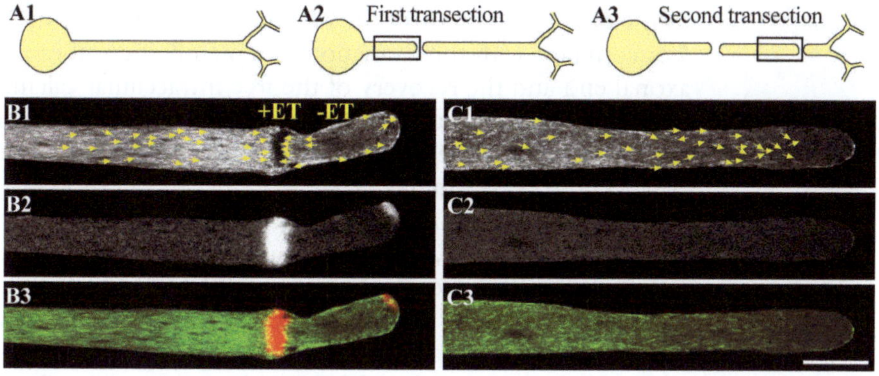

Fig. 8. The formation of the MT-based plus-end and minus-end traps depends on anterograde transport of component from the cell body to the axon. A (Buccal) neuron was cultured for 4 h, and then microinjected with EB3-GFP. Approximately 17 h later the neuron was injected with cherry-SNAP-25 mRNAs. Five hours later the axon was transected (A1 and A2, first transection) and the distribution of EB3-GFP and cherry-SNAP-25 was imaged. (B1–B3) Shown are images taken 5 min after axonal transection, 3 μm above the substrate. The *arrows* in B1 indicate the plus end of the MTs as revealed by EB3-GFP imaging. Note the formation of the plus-end (+ET) and the minus-end traps (−ET). (B2) The cherry-SNAP-25 fluorescent signal concentrated within the plus-end trap and at the tip of the axon. (B3) A merged image of B1 (*green*) and B2 (*red*). (C1–C3) 3 h after the first transection, the isolated axon was transected again (A3, second transection). The shown images were taken 18 min after the transection. Note that even 1 h after axotomy the MT polarity was unchanged (*arrows* in C1 indicate the plus ends), and the cherry-SNAP-25 fluorescent signal did not accumulate (C2). (C3) A merged image of C1 (*green*) and C2 (*red*). Scale bar = 20 μm in C3 (applies to **b–c**) (from (17)).

conducted between 5 min and 4.5 h after axonal isolation, well within a window of time in which many of the cellular functions of the isolated axon are operating quite normally.

For the experiments, the cell body, together with a short axonal segment, was mechanically removed, leaving a long isolated axon in the culture (first axotomy, Fig. 8A2, b). After various time intervals, which we believe allowed for depletion of the anterogradely transported motors from the main axon, the isolated axon was transected again distally to the first transection (second axotomy, Fig. 8A3, c). The restructuring of the MTs and accumulation of anterogradely or retrogradely transported vesicles under these conditions were then imaged. MTs were imaged by EB3-GFP, anterogradely transported vesicles by Cherry-SNAP-25 or synaptopHluorins, and retrogradely transported vesicles by SR101. We found that in most experiments, the cut end of the isolated axons (second axotomy) did not form the vesicles traps, although as evidenced by EB3 labeling, the MTs underwent a cycle of depolymerization and repolymerization in response to the second axotomy (Fig. 8c). These results are consistent with the hypothesis that anterogradely transported plus-end oriented components, driven from the cell body to the axon, participate in the formation of the traps; nevertheless this subject waits further analysis.

2.5.7. Actin Restructuring After Axotomy

Live confocal imaging of EGFP-actin revealed that following axotomy the actin puncta facing the glass substrate dissipate in parallel to the dissipation phase of the MT bundles after axotomy (Fig. 9). Concomitant with the formation of a membrane seal over the cut axonal end and the recovery of the free intracellular calcium level, large EGFP-actin puncta assemble in an anterograde direction along the plasma membrane facing the substratum of the distal zone (Fig. 9, 1.5–7 min).

In addition, within minutes of axotomy, and prior to the extension of a nascent GC lamellipodium, intense actin fluorescence accumulates along the axoplasmic cortex of the transition and distal zones (Fig. 9, 9.5 min). Close to the surface of the substratum actin assembles into radially oriented bundles in what is to become the leading edge of the GC's lamellipodium (Fig. 9, 9.5 and 15 min, and Fig. 10—green). Initially this domain is not invaded by microtubules (Fig. 10c—red). As time progresses, the GC's lamellipodium extends further and a delicate actin meshwork forms between the GC's center and the actin bundles at its leading edge. This actin meshwork is invaded by radially oriented microtubules (Fig. 10d–g). Throughout the process the GC's lamellipodium expansion actin-rich puncta are formed along the plasma membrane facing the substratum (Fig. 10). When the radius of the primary lamellipodium reaches 30–60 μm, the lamellipodium breaks up into branches, each led by a secondary motile GC (24).

2.5.8. Imaging of Spectrin Removal and Proteolysis at the Cut Axonal End

The submembrane spectrin skeleton is proteolyzed in the transected end of *Aplysia* neurons after axotomy in a calpeptin-dependent manner (19–21, 26). If the proteolysis of spectrin is inhibited by the calpain inhibitor calpeptin, a growth cone is not formed after axotomy. Spectrin immune-fluorescent intensity is attenuated sharply in the transected tip within less than a minute after transection ((19, 21), Fig. 11a, b), and spectrin immunofluorescence was reduced in a region encompassing the DZ and TZ of the cut end. In axons transected in the presence of calpeptin (a calpain inhibitor), this effect was not observed (Fig. 11c). These observations are consistent with the hypothesis that cleavage of spectrin by calpain is one of the initial steps that enable the neuron to form a competent GC (21).

Whereas the cleavage of spectrin was not documented by live imaging, axotomy-induced proteolytic activity can be imaged. To that end, neurons are continuously incubated in physiological solution containing 10 μM bis(CBZ-Alanyl-Alanine amine) Rhodamine 110 (bCAA-R110, Molecular Probes, Eugene, OR) and are imaged for the production of fluorescent Rhodamine 110 (R110). Ratio imaging in respect to an intracellular inert flurophore should be used to correct for volumetric changes. Since bCCAA-R110 is not fluorescent and membrane permeable and its cleavage product R110 is fluorescent and membrane permeable the appearance of the fluorescent product indicates that proteolytic activity is taking place and the site at which the cleavage product concentrate (19–21).

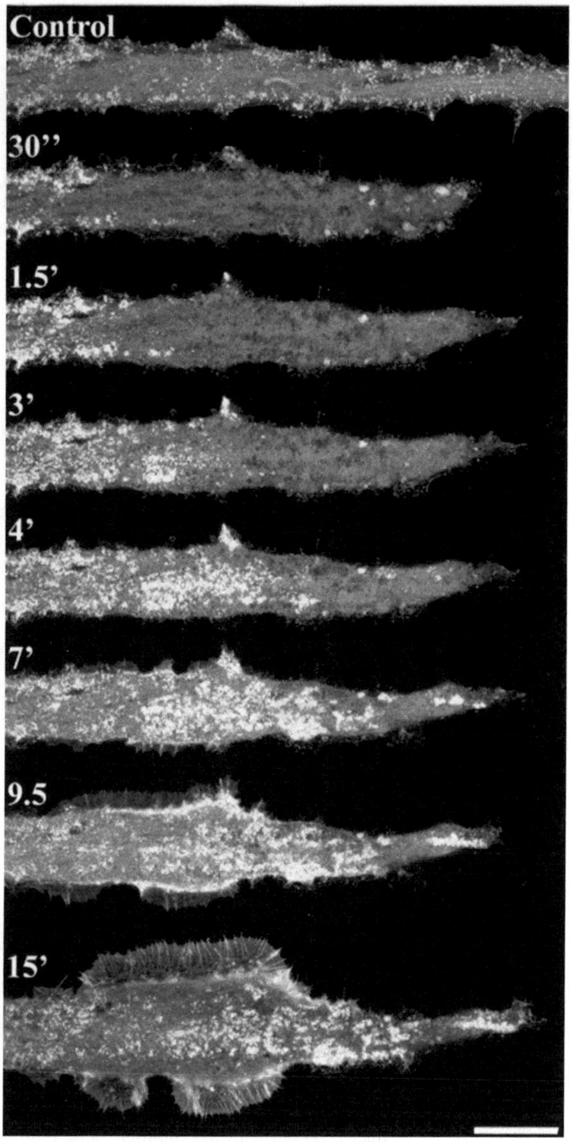

Fig. 9. Actin network restructuring after axotomy. Buccal neurons were cultured for 4 h, then microinjected with EGFP-actin mRNA. Twelve hours after the injections of the mRNA the axon was transected and the distribution of actin was imaged at the substrate level. Prior to axotomy the EGFP-actin fluorescence is homogeneously distributed in the bulk of the axoplasm. Fluorescent puncta are observed along the plasma membrane surface facing the glass substrate (control). Axotomy leads to dissipation of the actin puncta (30 s). Within minutes after axotomy, actin-rich hot spots gradually reappear from proximal region towards the most distal tip of the cut axon (0.5–7 min). With time, laterally oriented actin bundles formed along the edges of the plasma membrane lead to the extension of a nascent GC lamellipodium (9.5–15 min after axotomy). Scale bar = 20 μm (from (24)).

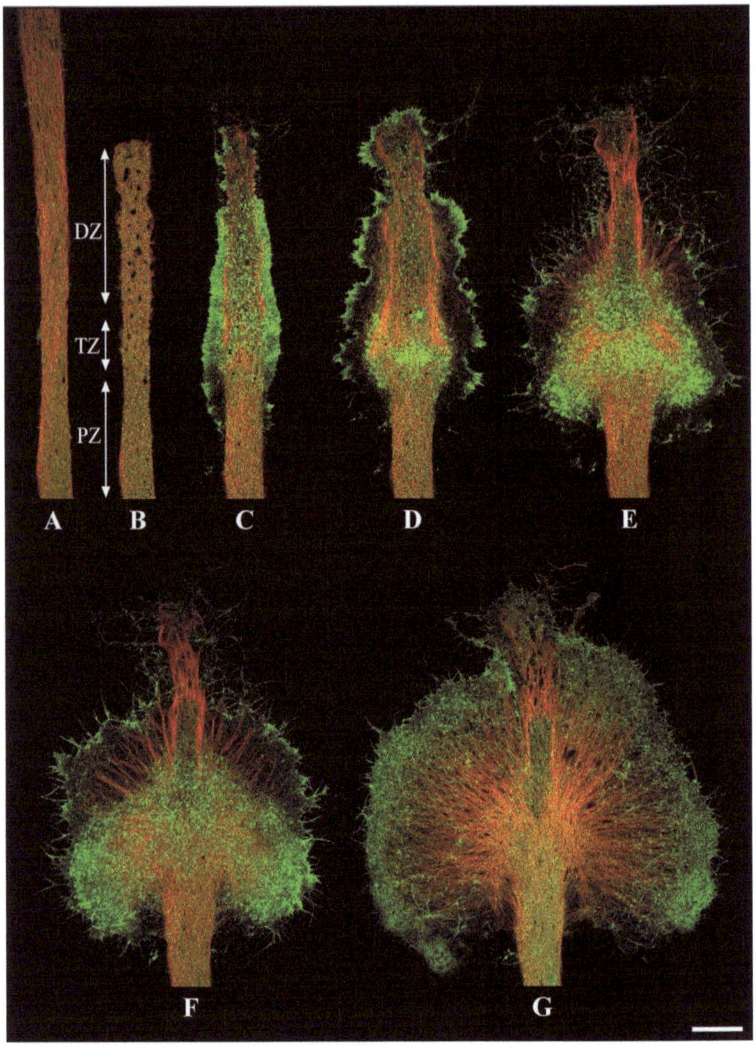

Fig. 10. Concomitant spatiotemporal mapping of microtubules and actin filaments during the transformation of a cut axon into a growth cone. A B neuron was cultured for 4 h, then microinjected with tetramethyl-rhodamine tubulin and EGFP-actin mRNA. Twelve hours later, the neuron was transected, and the distributions of microtubules (*red*) and actin network (*green*) were imaged at the substrate level. Prior to axotomy, the microtubules are oriented in parallel to the longitudinal axis of the axon (**A**—red). The actin is evenly distributed within the axoplasm (**A**—green). Axotomy leads within minutes to subdivision of the cut end into proximal (PZ), transition (TZ), and distal (DZ) zones (**B**). EGFP-actin assembles within minutes of axotomy at the perimeters of the transition and the distal zones (**c**). At a later stage, 30–50 min after axotomy, when the lamellipodium's diameter increases microtubules invade into the lamellipodium (**C**–**G**) and support its branching into neurites led by secondary GCs. A—before, **B** to **G**—1, 12, 25, 60, 90, and 220 min after axotomy. Scale bar = 20 μm (applies to **A**–**G**).

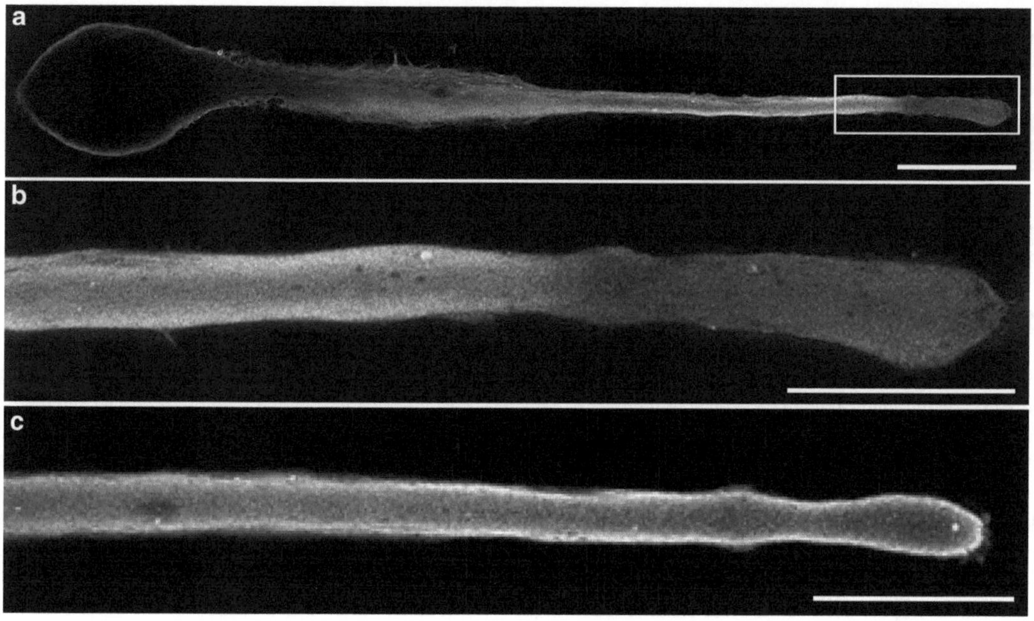

Fig. 11. Spectrin distribution after axotomy. A Buccal neuron was transected in physiological solution (**a** and **b**) or in the presence of 100 μM calpeptin (calpain inhibitor) (**c**). Minutes later the neurons were fixed and immunolabeled for spectrin. The distribution of spectrin within the neurons was examined by confocal microscopy. (**a**) Spectrin distribution in the cell body, the axon, and at the transected tip of a neuron. Spectrin staining within the cell body is confined to the plasma membrane. The density of spectrin is more pronounced at the axon's cortex. (**b**) Spectrin distribution at the tips of transected axon (enlargement of the *rectangle* seen in **a**). In the control experiment, axotomy leads to cleavage of the submembrane spectrin from the DZ but not along the axon. (**c**) When a neuron was transected in the presence of the calpeptin, the submembrane spectrin skeleton was not cleaved and a growth cone was not formed. Scale bar = 50 μm in (**a**), 10 μm in (**b**), and 20 μm in (**c**).

2.6. Combined Use of Imaging and Pharmacological Tools to Study the Relations Between MTs and Actin Dynamics in Growth Cone Formation

The observations reported above demonstrate that within minutes of transection, the cut axon undergoes robust structural remodeling to form a GC. Two alternative control-mechanisms could account for this highly coordinated cascade of events: (a) the restructuring of the microtubules and of the actin network and the localization of vesicles require continuous feedback among the participating elements. If this was the case, it would be expected that perturbation of the kinetics of any one of the participants would disrupt the kinetics of the others. Alternatively, (b) axotomy may set into motion in parallel and in an independent manner the cascades that lead to the restructuring of microtubules and of actin networks. In this case, the perturbation of one element would not be expected to affect the others immediately. We examine these hypotheses below.

To test the role of MT repolymerization on the assembly of the GC after axotomy we prevented MT repolymerization by the microtubule-depolymerizing reagent nocodazole. For the experiments, neurons were incubated for 15–40 min in 5 μM nocodazole and were then transected in the presence of the drug. Incubation of

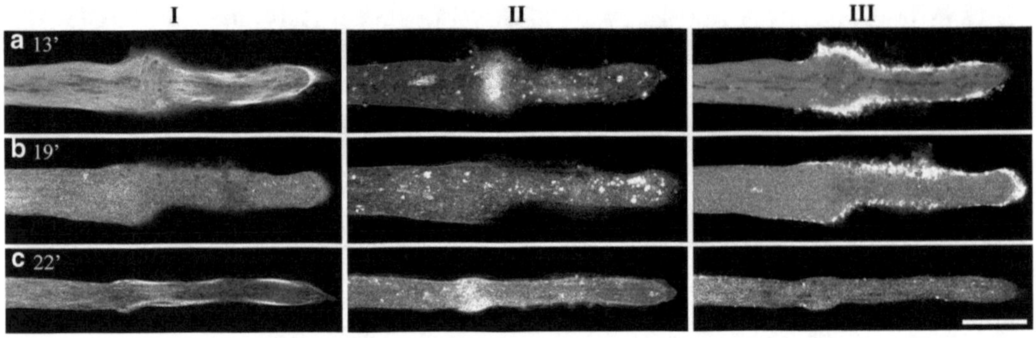

Fig. 12. Combined use of imaging and pharmacological tools to study the relations between MTs and actin dynamics in growth cone formation. B neurons were cultured for 4 h, then microinjected with tetramethyl-rhodamine tubulin and EGFP-actin mRNA. Thereafter, the neurons were bathed for 30 min in the styryl dye RH237, and the excess dye was thoroughly washed away. Twelve hours later, the neurons were transected in normal physiological solution (**a**), in physiological solution with 5 μM nocodazole (**b**), or in physiological solution with 10 μM cytochalasin D (**c**). The distributions of microtubules (I), RH237-labeled vesicles (II), and actin network (III) were imaged 3 μm above the substrate level. (**a**) In control condition axotomy leads within minutes to subdivision of the cut end into proximal (PZ), transition (TZ), and distal (DZ) zones (aI). RH237-labeled vesicles concentrate within the transition zone (aII). EGFP-actin assembles within minutes of axotomy at the perimeters of the transition and the distal zones (aIII). (**b**) In nocodazole, the MTs fail to subdivide the cut axonal end into PZ, TZ, and DZ (bI) and RH237-labeled vesicles do not accumulate (bII). Nevertheless, axotomy is followed by accumulation of actin at the perimeters of the cut end (bIII). (**c**) In cytochalasin D, the MTs subdivide the cut axonal end into PZ, TZ, and DZ (cI) and RH237-labeled vesicles accumulate at the TZ (CII). However, axotomy is not followed by the polymerization of actin at the TZ and DZ perimeters (cIII). Scale bar = 20 μm for (**a** and **b**) and 15.5 μm for (**c**).

neurons in 5 μM nocodazole for an hour does not affect the structural integrity of preexisting MTs.

In the presence of nocodazole, the cut end reseals and the neuron down regulated the increase in $(Ca^{2+})_i$ at rates similar to those observed in control experiments. Microtubules depolymerize and calpain is activated as evidenced by imaging of bis(CBZ-Alanyl-Alanin amine) Rhodamine 110 (19). Nevertheless, the characteristic restructuring of the cut axonal end into three distinct zones and the accumulation of RH237-labeled vesicles at the tip of the cut axon does not occur (Fig. 12bI, bII).

In the presence of nocodazole, the initial restructuring of the actin network preceded normally. That is, EGFP-actin concentrated along the perimeter of the distal axonal segment while the distal zone flattened. In nocodazole, only a small lamellipodium extends laterally and fails to proceed any further. In addition, RH237 fluorescence does not concentrate at any specific location along the distal segment of the axon (24).

In conclusion, nocodazole inhibits the repolymerization and restructuring of the depolymerized microtubules. Nevertheless, the restructuring of the actin network initially proceeds in a normal manner. This observation suggests that the initial stage of actin network restructuring is independent of the restructuring of the microtubules. However, the nascent GC fails to expand further, suggesting that late phases of the expansion of the GC's lamellipodium require the participation of invading microtubules from the GC center (Fig. 12bIII) (24).

To evaluate the role of the restructuring of the actin network in the processes that coordinate growth cone formation after axotomy, we examined the effects of an actin-filament-stabilizing reagent.

Neurons were incubated for 15–35 min in 10 μM cytochalasin D (CD) before axotomy. The axon was then transected in the presence of the drug. In such experiments the cut end reseals, and the transient increase in the $(Ca^{2+})_i$ is down regulated at rates similar to those of control experiments. In the CD, axotomy leads to the restructuring of the microtubules into the three axonal zones described previously, and vesicles labeled with RH237 accumulate in the transition zone, as occurs in control experiments (Fig. 12cI, cII). Nevertheless, the characteristic assembly of actin filaments in the perimeter of the transition and distal zones does not take place (Fig. 12bIII). Instead, large aggregates of EGFP-actin fluorescence form within the axoplasm. The actin hot spots imaged at the level of plasma membrane facing the substratum dissipate upon axotomy as in the control, but new hot spots do not reappear after the resealing of the axon. Under these conditions, a lamellipodium is not formed and the microtubules do not polymerize laterally for at least 12 h.

These observations suggest that the initial phases of axotomy-induced microtubular depolymerization and repolymerization are independent of actin restructuring. Nevertheless, the later stages of microtubule polymerization from the transition zone into the lamellipodium are not executed unless an actin network is assembled. Similar results were obtained when perturbing actin dynamics by 2.5 μM jasplakinolide.

3. Concluding Remarks

Through the use of cultured *Aplysia* neurons, live confocal imaging, retrospective immunolabeling, and classical electron microscopy this chapter illustrates that major cytoskeletal elements (microtubules, actin filaments, the submembrane spectrin skeleton) undergo robust disassembly, restructuring, and reorganization in response to axotomy. These events serve to transform the structure and functions of a differentiated stationary axon into a motile GC. Similar restructuring events may occur under pathological conditions that do not involve mechanical injury (30–33). The potential range of changes in cytoskeleton reorganization in differentiated neurons should be taken into account when evaluating mechanisms underlying neurodegenerative diseases as well as physiological events.

References

1. Hur EM, Saijilafu, Zhou FQ (2012) Growing the growth cone: remodeling the cytoskeleton to promote axon regeneration. Trends Neurosci 35:164–174
2. Lowery LA, Van Vactor D (2009) The trip of the tip: understanding the growth cone machinery. Nat Rev Mol Cell Biol 10:332–343
3. Dent EW, Gupton SL, Gertler FB (2011) The growth cone cytoskeleton in axon outgrowth and guidance. Cold Spring Harb Perspect Biol 3:1–40
4. McCormick AM, Leipzig ND (2012) Neural regenerative strategies incorporating biomolecular axon guidance signals. Ann Biomed Eng 40:578–597
5. Kandel ER (2001) The molecular biology of memory storage: a dialog between genes and synapses. Biosci Rep 21:565–611
6. Glanzman DL (2009) Habituation in Aplysia: the cheshire cat of neurobiology. Neurobiol Learn Mem 92:147–154
7. Baxter DA, Byrne JH (2006) Feeding behavior of Aplysia: a model system for comparing cellular mechanisms of classical and operant conditioning. Learn Mem 13:669–680
8. Forscher P, Smith SJ (1988) Actions of cytochalasins on the organization of actin filaments and microtubules in a neuronal growth cone. J Cell Biol 107:1505–1516
9. Lin CH, Forscher P (1993) Cytoskeletal remodeling during growth cone-target interactions. J Cell Biol 121:1369–1383
10. Suter DM, Forscher P (2000) Substrate-cytoskeletal coupling as a mechanism for the regulation of growth cone motility and guidance. J Neurobiol 44:97–113
11. Ambron RT, Zhang XP, Gunstream JD, Povelones M, Walters ET (1996) Intrinsic injury signals enhance growth, survival, and excitability of Aplysia neurons. J Neurosci 16:7469–7477
12. Ambron RT, Walters ET (1996) Priming events and retrograde injury signals. A new perspective on the cellular and molecular biology of nerve regeneration. Mol Neurobiol 13:61–79
13. Sung YJ, Ambron RT (2004) Pathways that elicit long-term changes in gene expression in nociceptive neurons following nerve injury: contributions to neuropathic pain. Neurol Res 26:195–203
14. Rishal I, Fainzilber M (2010) Retrograde signaling in axonal regeneration. Exp Neurol 223:5–10
15. Ashery U, Penner R, Spira ME (1996) Acceleration of membrane recycling by axotomy of cultured *Aplysia* neurons. Neuron 16:641–651
16. Benbassat D, Spira ME (1993) Survival of isolated axonal segments in culture: morphological, ultrastructural, and physiological analysis. Exp Neurol 122:295–310
17. Erez H, Malkinson G, Prager-Khoutorsky M, De Zeeuw CI, Hoogenraad CC, Spira ME (2007) Formation of microtubule-based traps controls the sorting and concentration of vesicles to restricted sites of regenerating neurons after axotomy. J Cell Biol 176:497–507
18. Erez H, Spira ME (2008) Local self-assembly mechanisms underlie the differential transformation of the proximal and distal cut axonal ends into functional and aberrant growth cones. J Comp Neurol 507:1019–1030
19. Gitler D, Spira ME (1998) Real time imaging of calcium-induced localized proteolytic activity after axotomy and its relation to growth cone formation. Neuron 20:1123–1135
20. Gitler D, Spira ME (2002) Short window of opportunity for calpain induced growth cone formation after axotomy of Aplysia neurons. J Neurobiol 52:267–279
21. Kamber D, Erez H, Spira ME (2009) Local calcium-dependent mechanisms determine whether a cut axonal end assembles a retarded endbulb or competent growth cone. Exp Neurol 219:112–125
22. Prager-Khoutorsky M, Spira ME (2009) Neurite retraction and regrowth regulated by membrane retrieval, membrane supply, and actin dynamics. Brain Res 1251:65–79
23. Sahly I, Erez H, Khoutorsky A, Shapira E, Spira ME (2003) Effective expression of the green fluorescent fusion proteins in cultured Aplysia neurons. J Neurosci Methods 126:111–117
24. Sahly I, Khoutorsky A, Erez H, Prager-Khoutorsky M, Spira ME (2006) On-line confocal imaging of the events leading to structural dedifferentiation of an axonal segment into a growth cone after axotomy. J Comp Neurol 494:705–720
25. Spira ME, Benbassat D, Dormann A (1993) Resealing of the proximal and distal cut ends of transected axons: electrophysiological and ultrastructural analysis. J Neurobiol 24:300–316
26. Spira ME, Oren R, Dormann A, Gitler D (2003) Critical calpain-dependent ultrastructural alterations underlie the transformation of an axonal segment into a growth cone after axotomy of cultured Aplysia neurons. J Comp Neurol 457:293–312
27. Ziv NE, Spira ME (1993) Spatiotemporal distribution of Ca2+ following axotomy and throughout the recovery process of cultured Aplysia neurons. Eur J Neurosci 5:657–668

28. Ziv NE, Spira ME (1995) Axotomy induces a transient and localized elevation of the free intracellular calcium concentration to the millimolar range. J Neurophysiol 74:2625–2637
29. Ziv NE, Spira ME (1997) Localized and transient elevations of intracellular Ca2+ induce the dedifferentiation of axonal segments into growth cones. J Neurosci 17:3568–3579
30. Shemesh OA, Erez H, Ginzburg I, Spira ME (2008) Tau-induced traffic jams reflect organelles accumulation at points of microtubule polar mismatching. Traffic 9:458–471
31. Shemesh OA, Spira ME (2010) Paclitaxel induces axonal microtubules polar reconfiguration and impaired organelle transport: implications for the pathogenesis of paclitaxel-induced polyneuropathy. Acta Neuropathol 119:235–248
32. Shemesh OA, Spira ME (2010) Hallmark cellular pathology of Alzheimer's disease induced by mutant human tau expression in cultured Aplysia neurons. Acta Neuropathol 120:209–222
33. Shemesh OA, Spira ME (2011) Rescue of neurons from undergoing hallmark tau-induced Alzheimer's disease cell pathologies by the antimitotic drug paclitaxel. Neurobiol Dis 43:163–175
34. Malkinson G, Spira ME (2006) Calcium concentration threshold and translocation kinetics of EGFP-DOC2B expressed in cultured Aplysia neurons. Cell Calcium 39:85–93
35. Malkinson G, Fridman ZM, Kamber D, Dormann A, Shapira E, Spira ME (2006) Calcium-induced exocytosis from actomyosin-driven, motile varicosities formed by dynamic clusters of organelles. Brain Cell Biol 35:57–73
36. Malkinson G, Spira ME (2010) Clustering of excess growth resources within leading growth cones underlies the recurrent "deposition" of varicosities along developing neurites. Exp Neurol 225:140–153
37. Gabso M, Neher E, Spira ME (1997) Low mobility of the Ca2+ buffers in axons of cultured Aplysia neurons. Neuron 18:473–481
38. Schacher S, Proshansky E (1983) Neurite regeneration by Aplysia neurons in dissociated cell culture: modulation by Aplysia hemolymph and the presence of the initial axonal segment. J Neurosci 3:2403–2413
39. Spira ME, Dormann A, Ashery U, Gabso M, Gitler D, Benbassat D, Oren R, Ziv NE (1996) Use of Aplysia neurons for the study of cellular alterations and the resealing of transected axons in vitro. J Neurosci Methods 69:91–102
40. Stiess M, Bradke F (2011) Neuronal polarization: the cytoskeleton leads the way. Dev Neurobiol 71:430–444
41. Tang-Schomer MD, Johnson VE, Baas PW, Stewart W, Smith DH (2012) Partial interruption of axonal transport due to microtubule breakage accounts for the formation of periodic varicosities after traumatic axonal injury. Exp Neurol 233:364–372
42. Heidemann SR, McIntosh JR (1980) Visualization of the structural polarity of microtubules. Nature 286:517–519
43. McIntosh JR, Euteneuer U (1984) Tubulin hooks as probes for microtubule polarity: an analysis of the method and an evaluation of data on microtubule polarity in the mitotic spindle. J Cell Biol 98:525–533
44. Baas PW, Black MM, Banker GA (1989) Changes in microtubule polarity orientation during the development of hippocampal neurons in culture. J Cell Biol 109:3085–3094
45. Heidemann SR (1991) Microtubule polarity determination based on formation of protofilament hooks. Methods Enzymol 196:469–477
46. Baas PW, Lin S (2011) Hooks and comets: the story of microtubule polarity orientation in the neuron. Dev Neurobiol 71:403–418
47. Takahashi D, Yu W, Baas PW, Kawai-Hirai R, Hayashi K (2007) Rearrangement of microtubule polarity orientation during conversion of dendrites to axons in cultured pyramidal neurons. Cell Motil Cytoskeleton 64:347–359
48. Tsien RY (2010) Nobel lecture: constructing and exploiting the fluorescent protein paintbox. Integr Biol (Camb) 2:77–93
49. Kaang BK (1996) Parameters influencing ectopic gene expression in Aplysia neurons. Neurosci Lett 221:29–32
50. Keating TJ, Peloquin JG, Rodionov VI, Momcilovic D, Borisy GG (1997) Microtubule release from the centrosome. Proc Natl Acad Sci U S A 94:5078–5083
51. Schroer TA (2001) Microtubules don and doff their caps: dynamic attachments at plus and minus ends. Curr Opin Cell Biol 13:92–96
52. Carvalho P, Tirnauer JS, Pellman D (2003) Surfing on microtubule ends. Trends Cell Biol 13:229–237
53. Mimori-Kiyosue Y, Tsukita S (2003) "Search-and-capture" of microtubules through plus-end-binding proteins (+TIPs). J Biochem 134: 321–326
54. Akhmanova A, Hoogenraad CC (2005) Microtubule plus-end-tracking proteins: mechanisms and functions. Curr Opin Cell Biol 17: 47–54
55. Stepanova T, Slemmer J, Hoogenraad CC, Lansbergen G, Dortland B, De Zeeuw CI, Grosveld F, van Cappellen G, Akhmanova A, Galjart N (2003) Visualization of microtubule growth in cultured neurons via the use of EB3-GFP (end-binding

protein 3-green fluorescent protein). J Neurosci 23:2655–2664
56. Ahmad FJ, He Y, Myers KA, Hasaka TP, Francis F, Black MM, Baas PW (2006) Effects of dynactin disruption and dynein depletion on axonal microtubules. Traffic 7:524–537
57. Kim T, Chang S (2006) Quantitative evaluation of the mode of microtubule transport in Xenopus neurons. Mol Cells 21:76–81
58. Nakagawa H, Koyama K, Murata Y, Morito M, Akiyama T, Nakamura Y (2000) EB3, a novel member of the EB1 family preferentially expressed in the central nervous system, binds to a CNS-specific APC homologue. Oncogene 19:210–216
59. Perez F, Diamantopoulos GS, Stalder R, Kreis TE (1999) CLIP-170 highlights growing microtubule ends in vivo. Cell 96:517–527
60. Jaworski J, Hoogenraad CC, Akhmanova A (2008) Microtubule plus-end tracking proteins in differentiated mammalian cells. Int J Biochem Cell Biol 40:619–637
61. Pak CW, Flynn KC, Bamburg JR (2008) Actin-binding proteins take the reins in growth cones. Nat Rev Neurosci 9:136–147
62. Bosch M, Hayashi Y (2012) Structural plasticity of dendritic spines. Curr Opin Neurobiol 22:383–388
63. Shupliakov O, Haucke V, Pechstein A (2011) How synapsin I may cluster synaptic vesicles. Semin Cell Dev Biol 22:393–399
64. Flynn KC, Pak CW, Shaw AE, Bradke F, Bamburg JR (2009) Growth cone-like waves transport actin and promote axonogenesis and neurite branching. Dev Neurobiol 69:761–779
65. Waterman-Storer CM, Desai A, Bulinski JC, Salmon ED (1998) Fluorescent speckle microscopy, a method to visualize the dynamics of protein assemblies in living cells. Curr Biol 8:1227–1230
66. Schaefer AW, Kabir N, Forscher P (2002) Filopodia and actin arcs guide the assembly and transport of two populations of microtubules with unique dynamic parameters in neuronal growth cones. J Cell Biol 158:139–152
67. Schaefer AW, Schoonderwoert VT, Ji L, Mederios N, Danuser G, Forscher P (2008) Coordination of actin filament and microtubule dynamics during neurite outgrowth. Dev Cell 15:146–162
68. Hynes RO, Lander AD (1992) Contact and adhesive specificities in the associations, migrations, and targeting of cells and axons. Cell 68: 303–322
69. Schoenwaelder SM, Burridge K (1999) Bidirectional signaling between the cytoskeleton and integrins. Curr Opin Cell Biol 11: 274–286
70. Machnicka B, Grochowalska R, Boguslawska DM, Sikorski AF, Lecomte MC (2012) Spectrin-based skeleton as an actor in cell signaling. Cell Mol Life Sci 69:191–201
71. Li J, Li XY, Feng DF, Pan DC (2010) Biomarkers associated with diffuse traumatic axonal injury: exploring pathogenesis, early diagnosis, and prognosis. J Trauma 69: 1610–1618
72. Grinvald A, Hildesheim R, Farber IC, Anglister L (1982) Improved fluorescent probes for the measurement of rapid changes in membrane potential. Biophys J 39:301–308
73. Sankaranarayanan S, De Angelis D, Rothman JE, Ryan TA (2000) The use of pHluorins for optical measurements of presynaptic activity. Biophys J 79:2199–2208
74. Oyler GA, Higgins GA, Hart RA, Battenberg E, Billingsley M, Bloom FE, Wilson MC (1989) The identification of a novel synaptosomal-associated protein, SNAP-25, differentially expressed by neuronal subpopulations. J Cell Biol 109:3039–3052
75. Kimura K, Mizoguchi A, Ide C (2003) Regulation of growth cone extension by SNARE proteins. J Histochem Cytochem 51: 429–433
76. Shaner NC, Campbell RE, Steinbach PA, Giepmans BN, Palmer AE, Tsien RY (2004) Improved monomeric red, orange and yellow fluorescent proteins derived from Discosoma sp. red fluorescent protein. Nat Biotechnol 22:1567–1572
77. Teng H, Cole JC, Roberts RL, Wilkinson RS (1999) Endocytic active zones: hot spots for endocytosis in vertebrate neuromuscular terminals. J Neurosci 19:4855–4866
78. Mimori-Kiyosue Y (2011) Shaping microtubules into diverse patterns: molecular connections for setting up both ends. Cytoskeleton (Hoboken) 68:603–618
79. McNiven MA, Wang M, Porter KR (1984) Microtubule polarity and the direction of pigment transport reverse simultaneously in surgically severed melanophore arms. Cell 37: 753–765
80. Cytrynbaum EN, Rodionov V, Mogilner A (2004) Computational model of dynein-dependent self-organization of microtubule asters. J Cell Sci 117:1381–1397
81. Serbus LR, Cha BJ, Theurkauf WE, Saxton WM (2005) Dynein and the actin cytoskeleton control kinesin-driven cytoplasmic streaming in Drosophila oocytes. Development 132: 3743–3752

82. Kirschner M, Mitchison T (1986) Beyond self-assembly: from microtubules to morphogenesis. Cell 45:329–342
83. Vorobjev I, Malikov V, Rodionov V (2001) Self-organization of a radial microtubule array by dynein-dependent nucleation of microtubules. Proc Natl Acad Sci U S A 98: 10160–10165
84. Malikov V, Cytrynbaum EN, Kashina A, Mogilner A, Rodionov V (2005) Centering of a radial microtubule array by translocation along microtubules spontaneously nucleated in the cytoplasm. Nat Cell Biol 7:1113–1118
85. Oren R, Dormann A, Benbassat D, Spira ME (1997) In: Teelken A, Korf J (eds) Neurochemistry: cellular, molecular and clinical aspects. Plenum, New York, pp 647–653
86. Benbassat D, Spira ME (1994) The survival of transected axonal segments of cultured *Aplysia* neurons is prolonged by contact with intact nerve cells. Eur J Neurosci 6:1605–1614
87. Schacher S, Wu F (2002) Synapse formation in the absence of cell bodies requires protein synthesis. J Neurosci 22:1831–1839
88. Martin KC, Casadio A, Zhu H, Yaping E, Rose JC, Chen M, Bailey CH, Kandel ER (1997) Synapse-specific, long-term facilitation of aplysia sensory to motor synapses: a function for local protein synthesis in memory storage. Cell 91:927–938

Chapter 11

Isolation of the Postsynaptic Density: A Specialization of the Subsynaptic Cytoskeleton

Karl-Heinz Smalla, Patricia Klemmer, and Ursula Wyneken

Abstract

Postsynaptic densities (PSDs) are cytoskeletal specializations present in synaptic contacts in the central nervous system. PSDs have been considered to be subcellular organelles that maintain and cluster the synaptic signal transduction apparatus in direct contact with the actin cytoskeleton and its regulators. Synapses and PSDs are highly dynamic structures that in addition of subserving transmission of information participate in its processing and storage. Therefore, the molecular components of PSDs reveal their signal transduction capacities in health and disease. Here, we present the experimental protocol we have been using for years to isolate PSDs from other cell components, including or excluding detergent-resistant membranes (or lipid rafts). The protocol can be applied to brain samples of different areas, ages, and mammalian species and is useful to obtain this subcellular organelle in a highly reproducible manner.

Key words: Subcellular fractionation, Synaptosome, Synaptic junctions, Postsynaptic densities, Detergent-resistant membranes, Cytoskeleton

1. Introduction

In the central nervous system (CNS), chemical synapses are highly specialized cell–cell contacts to accomplish local communication between neurons. The main excitatory neurotransmitter is L-glutamate. Glutamatergic synapses, with a typical size of up to 1 μm, are highly dynamic structures and undergo different forms of plasticity, thus being the major morphological substrate for learning and memory. They are composed of a pre- and postsynaptic element separated by the synaptic cleft. The characteristic feature of the presynaptic side is the presence of neurotransmitter-containing vesicles and an electron-dense structure associated with the presynaptic plasma membrane, the cytomatrix at the active zone (CAZ). The CAZ is a scaffolding structure organizing the machinery for controlled neurotransmitter exocytosis and

Rolf Dermietzel (ed.), *The Cytoskeleton: Imaging, Isolation, and Interaction*, Neuromethods, vol. 79,
DOI 10.1007/978-1-62703-266-7_11, © Springer Science+Business Media, LLC 2013

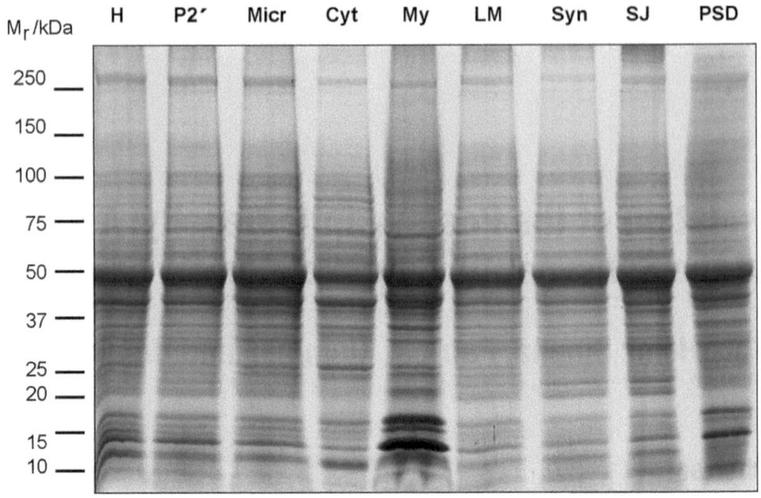

Fig. 1. SDS-PAGE of different intermediate stages during isolation of postsynaptic densities on a 5–20% gradient mini-gel indicates some characteristic changes in the protein patterns corresponding to enrichment and depletion of distinct proteins. Also the complexity of samples is decreasing comparing homogenate, crude membrane fraction, synaptosomes, synaptic junctions, and postsynaptic densities. *H* homogenate, *P2* crude membrane fraction, *Micr* microsomes (resulting from S2′), *Cyt* cytosolic proteins (resulting from combined S2 and S2′), *My* myelin, *LM* light membranes, *Syn* synaptosomes, *SJ* synaptic junctions, *PSD* postsynaptic densities (P5 according to (**b**)).

recycling of used synaptic vesicles. The postsynaptic side of the glutamatergic synapse typically contains a larger electron-dense subsynaptic structure, the postsynaptic density (PSD). The current view is that the PSD provides a molecular framework which assembles neurotransmitter receptors, intracellular signaling complexes, and cell adhesion/cell recognition molecules in close vicinity and connects on the other hand these complexes to the postsynaptic cytoskeleton (for reviews see, e.g., (1, 2)).

The PSD is an actin-rich meshwork of proteins which undergoes dynamic changes in response to neuronal activity (3, 4). In general synaptic activity is linked via different signaling pathways to the dynamics of the actin cytoskeleton (5). Thus, cytoskeleton remodeling is important for information processing and storage. Furthermore, alterations of actin dynamics have been linked to diverse neurological disorders like mental retardation (6, 7) and Morbus Alzheimer (8). Therefore the analysis of PSD constituents, especially actin regulators, increasingly attracts attention in order to understand molecular mechanisms underlying learning and memory as well as neurological disorders. Several experimental strategies have been used to identify components of the PSD fraction, and with the advent of proteomics even a full proteome of this fraction has been reported (4, 9–17). In these studies, PSDs basically have been isolated biochemically based on the procedure introduced by Carlin et al. (18) (Fig. 1). From these proteomic studies it became clear that the PSD fraction contains not only proteins from the postsynaptic compartment but also components of the presynaptic CAZ. In addition, components of glial end-feets associated with synaptic structures and extracellular matrix molecules can be found in biochemically prepared PSDs. Currently, approximately 1,600 different proteins have been detected in PSD preparations (37). A compilation of all proteins from the published proteomic analysis data of PSDs can be found in the SynProt database (http://www.synprot.de).

The term "postsynaptic density" has been introduced by Akert et al. in 1969 (19) after the discovery of PSDs in various electron microscopic studies in the 1950/1960s (20–22, 35). Until the early 1980s different protocols for the isolation of synaptic junctions and finally PSDs have been developed (18, 23–28), but the most commonly used procedure was introduced by the lab of Philip Siekevitz (18, 27) which constitutes the basis for the instructions in the protocol stated below. The method has been developed for the isolation of PSDs from adult brains/brain regions; it is based on differential centrifugation steps and the characteristic insolubility of PSDs in nonionic detergents such as Triton X-100. Furthermore, as detergent insolubility is shared by lipid rafts, they co-purify with PSDs. Both components can be separated by an additional centrifugation step that is also presented here. Typical yields of this protocol are 100–200 μg PSD proteins when starting from 1 g of wet forebrain tissue.

The abundance of presynaptic elements in biochemically isolated PSDs strongly depends on the pH of the buffers used in the final preparation steps. It had been shown that the content of presynaptic structures at pH 6 is remarkably higher compared to pH 8 and that presynaptic components can be removed from PSD fractions prepared at pH 6 when using a pH 8 stripping buffer (29).

2. Materials

2.1. Equipment

1. Homogenizer: Potter S (Sartorius AG, http://www.sartorius.com) including homogenizer vessels and appropriate pistils made of PTFE; available vessel volumes: 2 ml, 5 ml, 15 ml, 30 ml, 60 ml, Note 1.
2. Eppendorf Centrifuge 5810 R (Eppendorf AG, http://www.eppendorf.com), Note 2.
3. Sorvall® RC6 Centrifuge (Thermo Scientific, http://www.thermofisher.com/global/en/home.asp) with SS-34 rotor and appropriate centrifuge tubes, Note 2.
4. Optima™ L-70K Ultracentrifuge (Beckman Coulter, http://www.beckmancoulter.com) with suitable swing-out rotor (SW 28, SW 32 Ti or SW 40 Ti) and appropriate tubes, Note 2.

5. Accessory materials:
   Plastic Pasteur pipettes (suitable ranges 2–5 ml), Eppendorf pipettes (total range 20–5,000 μl), Eppendorf and Falcon tubes (1.5 ml and 15 ml or 50 ml resp.; standard), cooling box for crashed ice to store and transport samples and buffers, beaker glasses (different size), balance to adjust equal loading of centrifuge tubes for high-speed centrifugations (accuracy ≤ 0.02 g).

   Isolated synaptic structures resulting from step 6 in Sect. 3.3 (synaptosomes) onwards (pellet P3, synaptic junctions, PSDs) until step 23 in Sect. 3.3 have a strong tendency for aggregation especially after extraction with Triton X-100 and to stick to certain types of plasticware (cellulose nitrate) but also to glass. Whenever pellets of synaptic membranes (P3), synaptic junctions, or PSDs should be collected use preferably polypropylene- or polyallomer-based plasticware (centrifuge tubes, Pasteur pipettes, plastic tubes, pipette tips) and try to transfer these pellets in one or only a few pieces for resuspension; do not try to resuspend such pellets by means of pipetting but better use dounce-type homogenizers for this.

   The association of several components of membrane-containing subcellular fractions and particularly of PSDs depends strongly on ion concentrations (e.g., Ca^{2+}, Mg^{2+}). Therefore, water quality should consistently be, e.g., either distilled or double distilled (dd), or of adequate purity throughout a preparation, especially if a comparison of different preparations is planned.

2.2. Chemicals/Solutions

All chemicals should be at least of analytical grade and could be purchased from different companies like Sigma-Aldrich (http://www.sigmaaldrich.com) or Merck (http://www.merck-chemicals.com).

2.2.1. Chemicals

- 4-(2-hydroxyethyl)-1-piperazine ethanesulfonic acid (HEPES).
- 2-Amino-2-hydroxymethyl-propane-1,3-diol (Tris).
- Sucrose.
- Protease Inhibitor Cocktail Complete™ (Roche, Mannheim/Germany; http://www.roche-applied-science.com/sis/complete/index.jsp), Note 3.
- For analysis of phosphoproteins: Phosphatase inhibitor cocktails, e.g., PhosSTOP (Roche, Mannheim/Germany; http://www.roche-applied-science.com/sis/complete/index.jsp?id=com_020200), Note 3.
- Triton X-100 in best available quality.
- 1 N NaOH.
- Concentrated hydrochloric acid (HCl, 37%).

2.2.2. Solutions

Stock Solutions

The following stock solutions (1) and (2) are sufficient for many subcellular fractionation experiments, but solution (3) is adequate for the fractionation of approximately 12 g tissue (about 12 adult rat forebrains).

1. 500 mM HEPES/NaOH pH 7.4 (50 ml); dissolve 5.96 g HEPES in approximately 40 ml dd water, titrate with 1 N NaOH to pH 7.4, and bring it to the final volume with dd water, Note 4.
2. 500 mM Tris/HCl pH 8.1 (100 ml); dissolve 6.06 g Tris in approximately 90 ml dd water, titrate with concentrated HCl to pH 8.1, and bring it to the final volume with dd water, Note 4.
3. 2 M sucrose (600 ml); dissolve 410.76 g sucrose in dd water to the final volume.
4. 10% Triton X-100 (100 ml); mix 10 ml Triton X-100 with 90 ml dd water, Note 4.

Buffers and Working Solutions

The following buffers and solutions are sufficient to perform subcellular fractionation of 10–12 g brain tissue:

1. Buffer A (500 ml): 0.32 M sucrose in 5 mM HEPES buffer pH 7.4
   (80 ml 2 M sucrose, 5 ml 500 mM HEPES buffer pH 7.4, fill up with dd water to 500 ml).
2. Buffer A + PI (200 ml): Dissolve 4 Complete™ protease inhibitor pills in 200 ml buffer A; prepare immediately before starting the fractionation, Note 5.
3. Buffer B (200 ml): 0.32 M sucrose in 5 mM Tris/HCl pH 8.1
   (32 ml 2 M sucrose, 2 ml 500 mM Tris/HCl pH 8.1, fill up with dd water to 200 ml).
4. Buffer C (200 ml): 0.32 M sucrose in 12 mM Tris/HCl pH 8.1, 1% Triton X-100
   (32 ml 2 M sucrose, 4.8 ml 500 mM Tris/HCl pH 8.1, fill up with dd water to 200 ml).
5. 0.85 M sucrose/5 mM Tris/HCl pH 8.1 (200 ml)
   (85 ml 2 M sucrose, 2 ml 500 mM Tris/HCl pH 8.1, fill up to 200 ml with dd water).
6. 1 M sucrose/5 mM Tris/HCl pH 8.1 (200 ml)
   (100 ml 2 M sucrose, 2 ml 500 mM Tris/HCl pH 8.1, fill up to 200 ml with dd water).
7. 1.2 M sucrose/5 mM Tris/HCl pH 8.1 (200 ml)
   (120 ml 2 M sucrose, 2 ml 500 mM Tris/HCl pH 8.1, fill up to 200 ml with dd water).

8. 1.5 M sucrose/5 mM Tris/HCl pH 8.1 (50 ml)
   (37.5 ml 2 M sucrose, 0.5 ml 500 mM Tris/HCl pH 8.1, fill up to 50 ml with dd water).
9. 2 M sucrose/5 mM Tris/HCl pH 8.1 (15 ml)
   (15 ml 2 M Sucrose, 150 μl 500 mM Tris/HCl pH 8.1).

3. Methods

3.1. Purification of Postsynaptic Densities (Fig. 2)—General Considerations

Postmortem, the brain undergoes pronounced biochemical changes in a time-dependent manner (30). Thus, the dissection of brain tissue should be performed in the shortest time. Specific brain regions should be dissected on ice and stored in ice-cold buffer A until further use. Although best purification quality of PSDs is achieved using fresh brain tissue, sometimes the experimental design in comparative studies requires to freeze the tissues at −70°C or in liquid nitrogen (−196°C). The workflow for the fractionation is schematically depicted in Fig. 2 while the details are described in the Sects. 3.2–3.4. In general, samples and buffers are kept at 0–4°C during the fractionation procedure.

3.2. Preparation of a Crude Membrane Fraction

1. Homogenize brain tissue with 10 ml homogenization buffer (buffer A) per gram wet weight using the Potter S™ homogenizer with appropriate vessel and pistil. Use 900 rpm and perform 12 strokes (12 times down and up).
2. Transfer homogenates to Falcon tubes, Note 6, and balance tubes.
3. Spin samples at 1,000 × *g* for 10 min in (preferably) a swing-out rotor.
4. Collect supernatants S1 using Pasteur pipettes or by careful decantation, Note 7.
5. Rehomogenize pellets P1 in the same volume of buffer A as in step 1 with Potter S (900 rpm, 12 strokes).
6. Transfer homogenates to Falcon tubes, Note 6, and balance tubes.
7. "Wash" pellets by spinning samples at 1,000 × *g* for 10 min in (preferably) a swing-out rotor as in step 3.
8. Collect supernatants S1′ using Pasteur pipettes or by careful decantation, and combine S1 and S1′; the remaining pellets constitute P1, Note 8.
9. Transfer the combined S1 + S1′, Note 9, to centrifuge tubes suitable for Sorvall SS-34 rotors, and balance tubes which will be opposite to each other in the centrifuge (differences less than ±0.1 g).

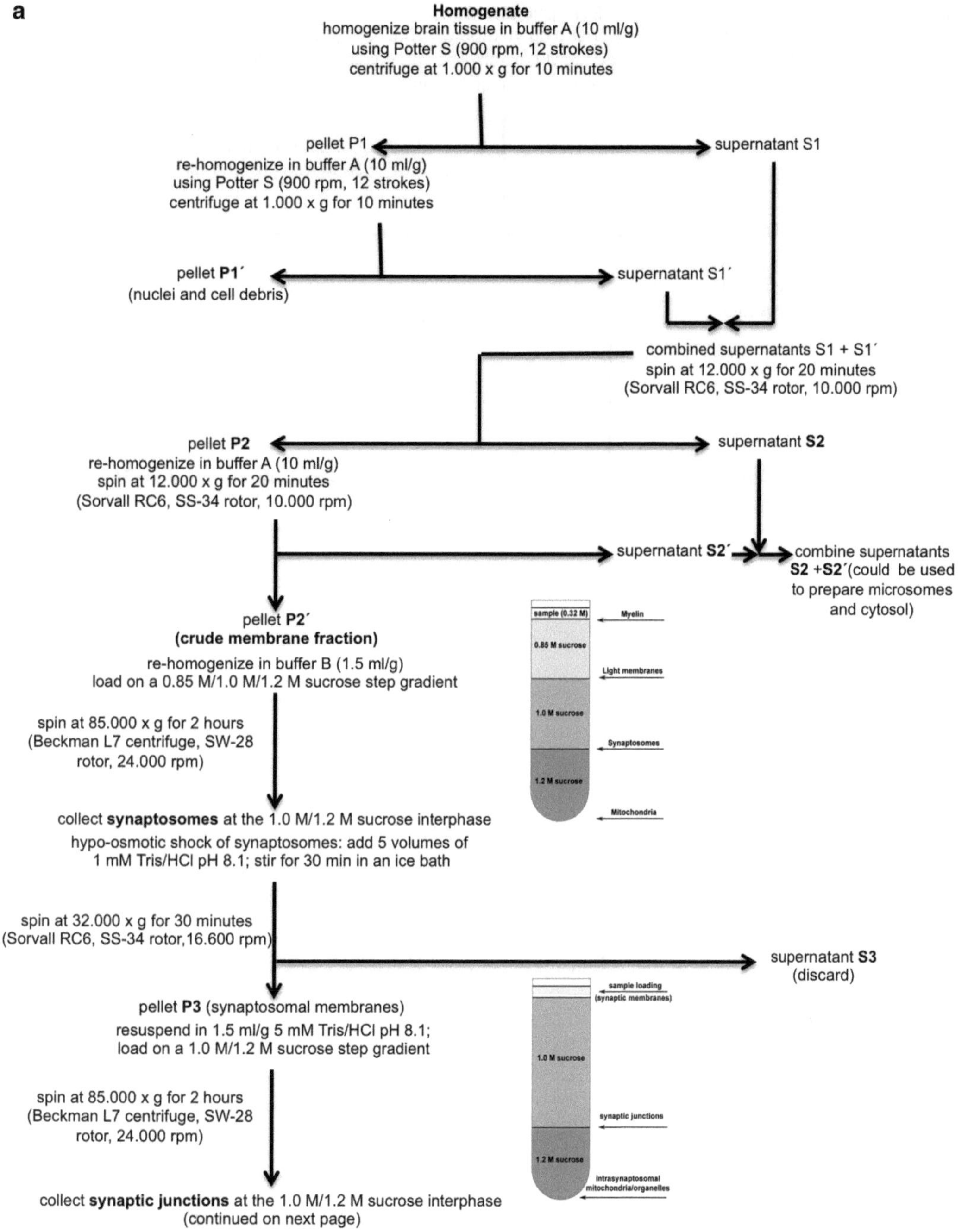

Fig. 2. (**a**) First part of the workflow for the isolation of postsynaptic densities: purification of synaptic junctions. (**b**) Workflow for the isolation of postsynaptic densities from synaptic junctions.

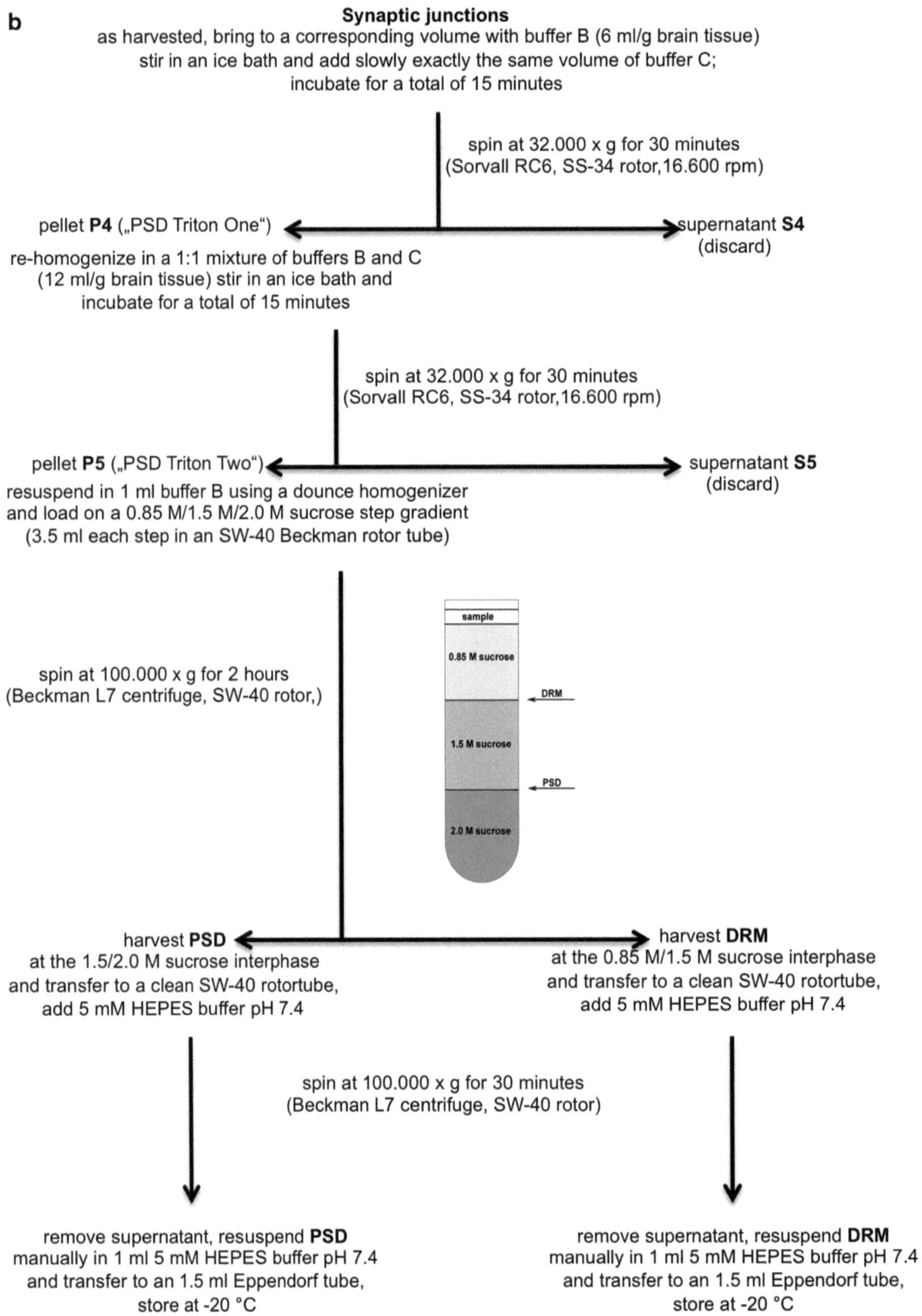

Fig. 2. (continued)

10. Spin samples at 12,000 × *g* for 20 min in a Sorvall high-speed centrifuge (Sorvall SW-34, 10,000 rpm), Note 10.
11. Collect supernatants S2 by decantation, Note 11.
12. Collect the pellets P2, Note 12, and resuspend pellets in 10 ml buffer A per (original) gram wet weight of tissue using a Potter S homogenizer with 900 rpm and six strokes in order to remove residual S2 by the next step.
13. Transfer to tubes for Sorvall SS-34 rotors and balance tubes which will be opposite to each other in the centrifuge (differences less than ±0.1 g).
14. Spin samples at 12,000 × *g* for 20 min in a Sorvall high-speed centrifuge (Sorvall SW-34, 10,000 rpm), Notes 10 and 13.
15. Collect supernatants S2′ by decantation and combine with corresponding S2, Note 14.

The remaining pellet P2′ is a crude membrane fraction, Note 6.

3.3. Preparation of Postsynaptic Densities

1. The starting material for further isolation of PSDs is the crude membrane fraction P2′.
2. Resuspend P2′ in 1.5 ml buffer B per gram (original) wet weight using either a dounce homogenizer or a Potter S (900 rpm, 6 strokes), Note 15.
3. Prepare a sucrose density step gradient consisting of 0.85 M/1.0 M/1.2 M sucrose solution in 5 mM Tris/HCl pH 8.1 with 9.1 ml of each solution in Beckman ultracentrifuge tubes for SW 28/SW 32 rotors (total tube volume is approximately 32 ml), Note 16.
4. Load 2.5–4 ml P2′ sample on top of a single gradient and balance tubes which will be opposite to each other in the centrifuge (weight difference ≤0.02 g).
5. Spin samples at 85,000 × *g* for 2 h (Beckman L7, SW 28 rotor at 24,000 rpm).
6. Harvest subcellular fraction with a Pasteur pipette.

   Interphase 0.32 M/0.85 M: Myelin, Note 6.

   Interphase 0.85 M/1.0 M: Light membranes, Notes 6 and 17.

   Interphase 1.0 M/1.2 M: Synaptosomes, Note 6.

   Pellet beneath 1.2 M: Mitochondria, Note 18 (Fig. 3).
7. In order to prepare synaptic junctions, synaptosomes are shocked hypo-osmotically. This can be accomplished by adding 5 volumes of freshly prepared 1 mM Tris/HCl, pH 8.1 to the original synaptosomal fraction volume followed by stirring gently for 30 min at 4°C.
8. Transfer solution of shocked synaptosomes to tubes for Sorvall SS-34 rotor and balance (weight difference ≤0.02 g).

Fig. 3. Separation of myelin, light membranes, synaptosomes, and mitochondria from a crude membrane fraction. (**a**) P2′ fraction loaded on top of the 0.85 M/1.0 M/1.2 M sucrose step gradient. (**b**) After 2 h of ultracentrifugation myelin is layered on top of the 0.85 M sucrose solution, while light membranes are found at the 0.85 M/1.0 M interface and synaptosomes are at the 1.0 M/1.2 M interface. The pellet consists mainly of mitochondria.

9. Spin at 32,000 × *g* for 30 min to settle a pellet of synaptic membranes (Sorvall SS-34, 16,600 rpm), Notes 19 and 20.
10. Prepare a sucrose density step gradient consisting of a 1.0 M/1.2 M sucrose solution in 5 mM Tris/HCl pH 8.1 with 18.2 ml of 1.0 M sucrose solution and 9.1 ml of 1.2 M sucrose solution in Beckman ultracentrifuge tubes for SW 28/SW 32 rotors (total tube volume is approximately 32 ml).
11. Discard the supernatant resulting from step 9 and rehomogenize the pellet P3 in 1.5 ml of freshly prepared 5 mM Tris/HCl pH 8.1 per (original) wet weight of tissue.
12. Load the resuspended synaptic membrane pellet P3 on top of the gradient as prepared in step 10.
13. Spin samples at 85,000 × *g* for 2 h (Beckman L7, SW 28 rotor at 24,000 rpm).
14. Harvest synaptic junctions at the 1.0 M/1.2 M interphase, Notes 6 and 21.
15. Bring the synaptic junction fraction with buffer B to a final volume corresponding to 6 ml per gram (original) wet weight.
16. Stir this suspension gently on a magnetic stirrer in an ice bath and add slowly exactly the same volume of buffer C giving a final Triton X-100 concentration of 0.5% within 15 min. Stick to time! See Note 22.
17. Transfer samples to tubes for Sorvall SS-34 rotor and balance (weight difference ≤0.02 g).

18. Spin at 32,000 × *g* for 30 min to settle a pellet of PSDs (Sorvall SS-34, 16,600 rpm). Meanwhile prepare a 1:1 (v/v) mixture of buffer B and buffer C with exactly the same volumes as used in step 16.
19. Discard the supernatant and transfer the remaining pellet into a hand-homogenizer to resuspend PSDs in the previously prepared 1:1 mixture of buffer B and buffer C. Bring this suspension to a final volume of 12 ml solution per gram of (original) wet weight. The pellet is essentially comparable to the fraction "PSD Triton One" from the Kennedy laboratory (31).
20. Stir this suspension gently on a magnetic stirrer in an ice bath for 15 min. Stick to time!
21. Transfer samples to tubes for Sorvall SS-34 rotor and balance (weight difference ≤0.02 g).
22. Spin at 32,000 × *g* for 30 min to settle the PSD pellet (Sorvall SS-34, 16,600 rpm). The pellet is essentially comparable to the fraction "PSD Triton Two" from the Kennedy laboratory (31).
23. Discard supernatant and transfer the PSD pellet to an Eppendorf tube, Note 23.

3.4. Separation of PSDs and Lipid Rafts (DRMs)

Triton X-100-insoluble parts of the synaptic junctions include both lipid rafts (detergent-resistant membranes, DRMs) and PSDs. Both subcellular compartments are characterized by distinct different densities and can therefore be separated by an additional sucrose step gradient as described below. Recently, the proteome of synapse-associated DRMs has been analyzed by the laboratory of T. Suzuki (36).

1. Prepare a sucrose step gradient composed of 0.85/1.5/2.0 M sucrose with 3.5 ml for each step in an ultraclear tube for SW-40 Beckman rotor, Note 16.
2. Resuspend PSDs in a small homogenizer manually in 1 ml buffer B and load on top of the 0.85/1.5/2.0 M sucrose gradient, Note 24.
3. Balance centrifugation tubes which will be opposite to each other in the centrifuge (weight difference ≤0.02 g).
4. Centrifuge the tubes at 100,000 × *g* for 2 h.
5. Remove first step until 0.5 cm above 0.85 M sucrose step.
6. Harvest the DRMs at the interphase to 0.85 M sucrose and transfer into a clean centrifugation tube kept on ice.
7. Remove all liquid until 0.5 cm above 1.5 M sucrose step.
8. Harvest PSDs at the 1.5/2.0 interphase and transfer into a clean centrifugation tube kept on ice.
9. Fill up both centrifugation tubes containing DRMs or PSDs with 5 mM HEPES buffer. Mix the solutions and adjust weight of both centrifugation tubes (weight difference ≤0.02 g).

10. Centrifuge the tubes at 100,000 × *g* for 30 min.
11. Remove supernatant and resuspend subcellular fractions in a small volume of 5 mM HEPES and transfer sample in an Eppendorf tube (1.5 ml).
12. Store samples at −80°C until further use, Note 25.

4. Notes

1. It is important to use homogenizers of dounce-type or teflon-glas-homogenizers to achieve a type of "gentle" homogenization by shear forces. The clearance of such homogenizers should be best ≤0.25 mm. Blenders as for instance Ultraturrax are not suitable for PSD preparations.
2. Corresponding equipment from other companies can also be used without restriction in the quality of the fractionation.
3. We used reagents from the indicated companies. Adequate reagents are also available from other companies and may also work very well.
4. At the day before the isolation of PSD both stock and working solutions should be prepared except for those which have been explicitly stated as "freshly prepared" which means earliest 2 h before use. Stock solutions (1) and (2) can be kept for up to 6 months at 4°C and (3) can be stored at −20°C.
5. Since some of the inhibitors are not stable for longer times the Complete™ pills should be dissolved in buffer A not earlier than 30 min before homogenization is started.
6. Retain an aliquot for determination of protein content and electrophoretic control; notice the total volume of the fraction in order to calculate later total protein contents.
7. S1 and S1′ should be collected with caution since the pellets P1 and P1′are very soft.
8. P1′ looks quiet inhomogeneous and mainly comprises nuclei, cell debris, fragments of blood vessels, and also erythrocytes.
9. This combined supernatant is still a turbid fluid.
10. Suitable Sorvall centrifuge types are RC-5B, RC-6, and RC-6 Plus.
11. The pellet P2 is still very heterogeneous. Therefore, stop decanting the supernatant when the white top layer of the pellet reaches the end of the tube.
12. Usually there is a red dot in the center of P2 consisting of erythrocytes. If the approach is a rather "qualitative" test and later quantitation is not intended, the purity of the subsequently isolated PSDs can be increased, when this red dot is mainly

excluded from P2 for further purification. This can be achieved using a Pasteur pipette with small amounts of buffer A for the collection of P2 excluding the very tight red "sub-pellet."

13. Latest at this point of fractionation start to prepare the sucrose density gradients required at step 3 in Sect. 3.3.

14. Spinning the combined S2 + S2′ at 100,000 × g for 1 h in a Beckman ultracentrifuge (SW 28, SW 32, or SW 40 rotor) leads to the separation of the cytosol in the supernatant and microsomes (fragments of intracellular membranous structures like endoplasmic reticulum or golgi apparatus) in the pellet.

15. It is important to resuspend P2′ for far better separation of subcellular organelles in the subsequent sucrose density gradient centrifugation.

16. Gradients can be assembled in 2 different ways, either by overlaying high-density solutions with less dense solutions or by sublayering low-density solutions with high-density solutions. Overlaying can be performed using a 10 ml pipette letting out the next denser solution very slowly from top (let the drops run down the tube walls and keep the tube tilted). Sublayering can be achieved using a long canula to deliver a heavier solution below a less dense solution (in that case keep the tube upright).

17. Light membranes contain fragments of intracellular membranous structures as, e.g., endoplasmic reticulum and Golgi apparatus. If the focus of the experiment is only the isolation of synaptosomes, synaptic junctions, or PSDs, the step gradient could be designed as 9.1 ml of 1.2 M sucrose/Tris/HCl pH 8.1 and 18.2 ml of 0.85 M sucrose/Tris/HCl pH 8.1.

18. This fraction is a rather crude mitochondrial fraction and contains also other not well-definable structural elements. Therefore, electrophoretic and immunoblot analysis of this fraction will never be very conclusive. Mitochondria enriched in such a way are also altered in function. If the focus of the study is on isolation of functionally intact mitochondria we strongly suggest to follow other protocols (32).

19. At this point latest the gradients for the next purification step should be prepared.

20. This pellet contains still very different membranous structures including synaptic vesicles, intrasynaptosomal mitochondria, and synaptic junctional membranes.

21. The resulting pellet contains mainly synaptosomal mitochondria but also residual undefined insoluble material.

22. Using 0.5% Triton X-100 several lipids and transmembrane proteins will be solubilized while the core of the synaptic structures, especially cytomatrix components from the postsynaptic and to a certain degree also from the presynaptic side remains insoluble.

23. A useful way to store PSDs is to resuspend the PSD pellet in a dounce homogenizer in a small volume (for instance 1 ml) of 5 mM Tris/HCl pH 8.1 and to store 100 μl aliquots at −20°C. Determination of protein concentration should be started immediately after resuspension of PSDs because PSDs begin to form very soon larger aggregates. This should also be considered when thawing stored aliquots: if they are not used as a single sample they must be resuspended to avoid sample heterogeneity. On the other hand this property of PSDs can be used to transfer them into a new buffer just by leaving them for 1–2 h at 4°C, then spinning them down at 20,000 × *g* for 30 min, and resuspending the pellet in the buffer of interest.
24. Isolated PSDs by the described procedure will be highly pure. As an alternative, PSDs can be enriched from the crude membrane fraction. Therefore extract PSDs and DRMs from crude membranes and sediment the insoluble particles by centrifugation (20,000 × *g* for 30 min). The pellet will be resuspended in 2 M sucrose. Then the sample will be put under a sucrose step gradient (0/0.85/1.5 M sucrose). Following centrifugation DRMs can be collected at the 0/0.85 M sucrose interphase and PSDs at the 1.5/2.0 M sucrose interphase. Keep in mind that the obtained DRM fraction might also contain insoluble Myelin sheets.
25. DRMs and PSDs can be subjected to immuno-blot or proteomics analysis.

Acknowledgements

This work was supported to KHS and UW by bilateral programs of the Deutsche Forschungsgemeinschaft (DFG SM38/8-1) and Bundesministerium für Bildung und Forschung (BMBF CHL 06/027) with Conicyt (Chile), to UW by Fondecyt(1100322) and Proyecto Anillo 09–06 (PBCT, Conicyt Chile), and to KHS by the European Structural Funds 2007–2013 (CBBS/ZVOH). PK was supported by DFG as a guest scientist within the SFB 779.

References

1. Boeckers TM (2006) The postsynaptic density. Cell Tissue Res 326(2):409–422
2. Ziff EB (1997) Enlightening the postsynaptic density. Neuron 19(6):1163–1174
3. Wyneken U et al (2001) Kainate-induced seizures alter protein composition and N-methyl-D-aspartate receptor function of rat forebrain postsynaptic densities. Neuroscience 102(1): 65–74
4. Ehlers MD (2003) Activity level controls postsynaptic composition and signaling via the ubiquitin-proteasome system. Nat Neurosci 6(3):231–242
5. Dillon C, Goda Y (2005) The actin cytoskeleton: integrating form and function at the synapse. Annu Rev Neurosci 28:25–55
6. van Spronsen M, Hoogenraad CC (2010) Synapse pathology in psychiatric and neurologic

disease. Curr Neurol Neurosci Rep 10(3): 207–214
7. Boda B, Dubos A, Muller D (2010) Signaling mechanisms regulating synapse formation and function in mental retardation. Curr Opin Neurobiol 20(4):519–527
8. Bamburg JR, Bloom GS (2009) Cytoskeletal pathologies of Alzheimer disease. Cell Motil Cytoskeleton 66(8):635–649
9. Cheng D et al (2006) Relative and absolute quantification of postsynaptic density proteome isolated from rat forebrain and cerebellum. Mol Cell Proteomics 5(6):1158–1170
10. Collins MO et al (2006) Molecular characterization and comparison of the components and multiprotein complexes in the postsynaptic proteome. J Neurochem 97(Suppl 1):16–23
11. Collins MO et al (2005) Proteomic analysis of in vivo phosphorylated synaptic proteins. J Biol Chem 280(7):5972–5982
12. Dosemeci A, Tao-Cheng JH, Vinade L, Jaffe H (2006) Preparation of postsynaptic density fraction from hippocampal slices and proteomic analysis. Biochem Biophys Res Commun 339(2):687–694
13. Jordan BA et al (2004) Identification and verification of novel rodent postsynaptic density proteins. Mol Cell Proteomics 3(9):857–871
14. Li K et al (2005) Organelle proteomics of rat synaptic proteins: correlation-profiling by isotope-coded affinity tagging in conjunction with liquid chromatography-tandem mass spectrometry to reveal post-synaptic density specific proteins. J Proteome Res 4(3):725–733
15. Li KW et al (2004) Proteomics analysis of rat brain postsynaptic density. Implications of the diverse protein functional groups for the integration of synaptic physiology. J Biol Chem 279(2):987–1002
16. Peng J et al (2004) Semiquantitative proteomic analysis of rat forebrain postsynaptic density fractions by mass spectrometry. J Biol Chem 279(20):21003–21011
17. Yoshimura Y et al (2004) Molecular constituents of the postsynaptic density fraction revealed by proteomic analysis using multidimensional liquid chromatography-tandem mass spectrometry. J Neurochem 88(3):759–768
18. Carlin RK, Grab DJ, Cohen RS, Siekevitz P (1980) Isolation and characterization of postsynaptic densities from various brain regions: enrichment of different types of postsynaptic densities. J Cell Biol 86(3):831–845
19. Akert K, Moor H, Pfenninger K, Sandri C (1969) Contributions of new impregnation methods and freeze etching to the problems of synaptic fine structure. Prog Brain Res 31:223–240
20. Palay SL, Palade GE (1955) The fine structure of neurons. J Biophys Biochem Cytol 1(1):69–88
21. De Robertis E (1959) Submicroscopic morphology of the synapse. Int Rev Cytol 8:61–96
22. Gray EG (1959) Electron microscopy of synaptic contacts on dendrite spines of the cerebral cortex. Nature 183(4675):1592–1593
23. Gray EG, Whittaker VP (1962) The isolation of nerve endings from brain: an electron-microscopic study of cell fragments derived by homogenization and centrifugation. J Anat 96:79–88
24. Cotman CW, Taylor D (1974) Localization and characterization of concanavalin A receptors in the synaptic cleft. J Cell Biol 62(1):236–242
25. Cotman CW, Banker G, Churchill L, Taylor D (1974) Isolation of postsynaptic densities from rat brain. J Cell Biol 63(2 Pt 1):441–455
26. Matus AI, Walters BB (1975) Ultrastructure of the synaptic junctional lattice isolated from mammalian brain. J Neurocytol 4(3):369–375
27. Cohen RS, Blomberg F, Berzins K, Siekevitz P (1977) The structure of postsynaptic densities isolated from dog cerebral cortex. I. Overall morphology and protein composition. J Cell Biol 74(1):181–203
28. Matus A, Ackermann M, Pehling G (1981) Regularity and differentiation within the structure of brain postsynaptic densities. J Neurocytol 10(6):889–896
29. Phillips GR et al (2001) The presynaptic particle web: ultrastructure, composition, dissolution, and reconstitution. Neuron 32(1):63–77
30. Hunsucker SW et al (2008) Assessment of postmortem-induced changes to the mouse brain proteome. J Neurochem 105(3):725–737
31. Walikonis RS et al (2000) Identification of proteins in the postsynaptic density fraction by mass spectrometry. J Neurosci 20(11):4069–4080
32. Frezza C, Cipolat S, Scorrano L (2007) Organelle isolation: functional mitochondria from mouse liver, muscle and cultured fibroblasts. Nat Protoc 2(2):287–295
33. De Robertis ED (1965) Some new electron microscopical contributions to the biology of neuroglia. Prog Brain Res 15:1–11
34. Trinidad JC, Thalhammer A, Specht CG, Schoepfer R, Burlingame AL (2005) Phosphorylation state of postsynaptic density proteins. J Neurochem 92(6):1306–1316
35. Whittaker VP, Gray EG (1962) The synapse: biology and morphology. Br Med Bull 18:223–228

36. Suzuki T et al (2011) Association of membrane rafts and postsynaptic density: proteomics, biochemical, and ultrastructural analyses. J Neurochem 119(1):64–77
36. Pielot R, Smalla KH, Müller A, Landgraf P, Lehmann AC, Eisenschmidt E, Haus UU, Weismantel R, Gundelfinger ED, Dieterich DC (2012) SynProt: A Database for Proteins of Detergent-resistant Synaptic Protein Preparations. Front Syn Neurosci 4:1. doi: 10.3389/fnsyn.2012.00001.

Part IV

Astroglia and the Cytoskeleton

Chapter 12

Cell-to-Cell Communication in Astroglia and the Cytoskeletal Impact

Georg Zoidl, Stefan Olk, Andrey Tuchinowitz, and Rolf Dermietzel

Abstract

The cytoskeleton constitutes a dynamic intracellular framework for inbound and outbound trafficking of cell organelles and is responsible for cell shaping including cell division. While transport in neuronal compartments has extensively been conquered in particular the shipping of cargo vesicles along microtubules for support of chemical synapses, the supply of molecular constituents to electrical synapses is still enigmatic. Here we describe protocols for proteomic analysis of cytoskeletal elements involved in the trafficking of connexin43, the major gap junction protein expressed in astrocytes. Methods described here offer hands-on instructions for further successful exploration of the cytoskeletal and its interaction with gap junction proteins.

Key words: Astroglia, Connexin43, Gap junctions, Cytoskeletal components, Proteomic analysis, Dynamics

1. Introduction

1.1. Connexin43 and the Astrocytic Syncytium

While glial cells were initially considered to be merely connective cells or "neuronal glue," today a multitude of functions of these cells is acknowledged (1, 2). Among glia cells astrocytes constitute the major cell type in the brain forming structural barriers at vascular surfaces, ensheathing neuronal elements and separating regions of fluctuating or dissimilar ionic composition, thereby physically and metabolically compartmentalizing various neuronal components of the central nervous system (CNS) (3). This supports CNS homeostasis by regulating extracellular pH, K^+ concentration, and glutamate levels through astrocytic ion channels and membrane transporters (4, 5). Other functions involve intercellular signaling between astrocytes, where they propagate Ca^{2+} waves over considerable distances from the focus of initiation (6). Since Ca^{2+}-dependent glutamate release from astrocytes profoundly affects

Rolf Dermietzel (ed.), *The Cytoskeleton: Imaging, Isolation, and Interaction*, Neuromethods, vol. 79,
DOI 10.1007/978-1-62703-266-7_12, © Springer Science+Business Media, LLC 2013

synaptic plasticity, such Ca^{2+} waves enable distant modulation of synaptic activity (7).

All these functions are influenced by gap junction communication (8–10) coupling neighboring astrocytes through gap junctions (GJ) composed of connexins to establish a glial syncytium. Such coupling was found at a variety of sites in brain parenchyma. Individual astrocytes can form, on average, a staggering 30,000 gap junctions per cell (11), connecting the processes of different astrocytes as well as different processes of the same astrocyte (autologous junctions). Several connexins have been reported to be expressed by astrocytes, including connexin43 (Cx43), connexin26 (Cx26), and connexin30 (Cx30) (3, 12–14). In addition, astrocytes form numerous gap junctions with oligodendrocytes, thus incorporating them into the pan-glial syncytium (15–17). The gap junctional coupling of astrocytes into multicellular networks may play a role in both normal function and CNS disorders in the developing and mature brain (18, 19). In summary, the physiological functions served by gap junctions connecting the glial syncytium provides a conduit for removal of extracellular K^{+}, taking up neurotransmitters (e.g., glutamate) from the local environment of neurons, serving as a route for delivery of metabolites (20) throughout the network and increasing buffer capacity for dilution of toxic or metabolically active compounds (21).

1.2. Cx43 and the Cytoskeleton

The predominant gap junction protein expressed by astrocytes is Cx43 (12, 22). Recent studies imply a role of Cx43 in the regulation of important cellular processes, including migration, proliferation and apoptosis. Mutations in this gene have been associated with oculodentodigital dysplasia and heart malformations (23). In addition, several types of CNS injury and various diseases are associated with the development of a glial scar. Astrocytes play a decisive role in scare formation and Cx43 mRNA and protein expression as well as a high degree of proliferation being hallmarks of this process (24, 25). Since all these events involve cytoskeletal dynamics, it is not surprising that most of the work related to connexin/cytoskeletal interaction has focused on the dominant astrocytic connexins (Cx43). While interactions and regulatory events might actually differ in details from other connexins (Cx), particularly those with a short carboxy-terminus like Cx26, it is likely that in case of interactions with the cytoskeleton common mechanisms may be at work.

In general, Cx43 is a typical member of the connexin gene family. Among shared structural and biological properties is a short half-life of only a few hours (26), and regulation of turnover rates comprises an important means of controlling of gap junction-mediated intercellular communication (GJIC). Interaction with microfilaments and microtubules is involved in the turnover mechanisms and enables the trafficking of Cx43 to, within and from the cell membrane. In a more general perspective, Cxs are cotranslationally inserted into the endoplasmic reticulum (ER) membrane. Depending on the Cx subtype, oligomerization can occur either in the ER-Golgi intermediate compartment (e.g., Cx32) or the trans-Golgi network (e.g., Cx43) (27–29). For transport to the cell membrane, connexons are released into cargo vesicles that travel along microtubules or actin filaments (30–33). Upon insertion in the cell membrane, hemichannels between adjacent cells can dock to each other and form functional GJ channels. Degradation of connexons happens through their internalization, giving rise to cytoplasmic annular GJs, which are degraded via lysosomal and proteasomal pathways (for a detailed review see (34)).

Both, shuttling to the plasma membrane and intramembrane movements are not well understood. Among plausible scenarios how newly assembled connexons are recruited to GJ plaques, it is suspected that connexon-containing cargo vesicles may fuse with the plasma membrane in the center of plaques or at their marginal boundaries, followed by lateral diffusion in the plasma membrane to reach the plaque site. According to Gaietta et al. (35), the latter pathway seems to be the preferred route for plaque renewal with removal of older connexons from the center of the plaques. Incorporation of new connexons preferentially at the periphery of plaques indicated that a directed transport of connexons via cargo vesicles is fundamental for GJ assembly. However, observations regarding GJ formation and involvement of cytoskeletal components as reported are highly controversial. Upon pharmacological disruption of microtubules with nocodazole, Cx43 assembly was hardly affected in transfected HeLa cells, whereas Cx26 assembly was inhibited (36). In contrast, nocodazole treatment affected Cx43 recruitment in GJs, whereas Cx26 transport and clustering remained nearly unchanged (37). Inhibition of Cx transport in the Golgi by brefeldin A and inhibition of microtubules by nocodazole in Cx43-transfected HeLa cells blocked GJ assembly after the pool of non-junctional hemichannels in the plasma membrane was depleted. The same effect of nocodazole was observed in Cx43-transfected MDCK cells, where lateral insertion of unopposed connexons into GJ plaques was still possible, whereas forskolin induced increase of GJ growth was blocked (38). In rat alveolar epithelial cells, inhibition of tubulin polymerization by colchicine initiated redistribution of immunopositive Cx43 from the membrane to the cytosol, which is in support of an association of microtubules with connexons during their transport to the plasma membrane (39).

Directed microtubule-mediated Cx43 transport to the plasma membrane was reported recently, where microtubule plus-end-tracking proteins (+TIPs), such as EB1 and interaction partners like p150 (40) were shown to contribute to connexon assembly. Fluorescence recovery after photobleaching (FRAP), total internal reflection fluorescence (TIRF) microscopy and time-lapse imaging demonstrated a rapid Cx43 delivery to GJ plaques and a preferential and prolonged association of microtubule plus ends with the plaques (41). Most strikingly, Shaw et al. (40) showed that GJ formation was disrupted by short interfering RNA (siRNA)-mediated downregulation of EB1 expression. These recent findings provide a new and unanticipated aspect to the mechanism for connexin delivery to GJs.

In this context, the question arises, whether pools of connexons prefer specific domains within the plasma membrane. An interesting observation made by Laing et al. (42), demonstrates that connexons transiently reside in lipid rafts, where they do not participate in GJIC. Rapid transfer of Cx hemichannels from such lipid rafts to actin-anchored GJ plaques was suggested in cases where sufficient amounts of the scaffolding protein ZO-1 were present. However, if there is reduced interaction between ZO-1 and the cytoskeleton, the Cx/ZO-1 complex is trapped in a lipid raft compartment (42). From the above studies it becomes obvious, that different cytoskeletal elements seem to be involved in connexon transport, and that their individual contribution is likely to depend on cell- and/or tissue-specific cues.

Looking at the other bookend of the connexin life cycle, connexons are internalized into cytoplasmic vesicles termed annular gap junctions during degeneration, with cytoplasmic GJ vesicles surrounded by actin-containing filaments (43). Consistent with these data were electron microscopical studies characterizing the plasma membrane and its cytoskeletal undercoat at sites of GJ invagination. Here, invaginating GJs revealed a network of actin-containing microfilaments (44). In deeper invaginations of withdrawing GJs, parallel microfilament arrays formed a submembrane sheath with an orthogonal arrangement toward the long axis of the invagination. These findings imply that the microfilament network and sheath of actin act in concert to generate a mechanical force driving the invagination-endocytosis of GJ in form of annular GJs (43, 44). In addition, many annular Cx43 GJs appear associated with myosin II. Disruption of microfilaments with cytochalasin B decreased the number, but increased the size of annular GJs, suggesting a role for myosin-containing cytoskeletal elements in annular GJ turnover (45).

More recent studies using epitope and fluorescent protein-tagged Cx43 support the mechanisms of GJ internalization by annular junctions (35, 46). Mechanistic analyses revealed that components of the clathrin-dependent endocytosis machinery are directly involved in the internalization, inward movement, and initial degradation of these intercellular double-membrane vesicles. The components described in this study include clathrin, the alternative clathrin-adaptor Dab2, the GTPase dynamin, the unconventional myosin-VI, and actin filaments (47). The mechanism by which Cx43 GJ plaques are internalized might involve Src-mediated dissociation of ZO-1 from one side of the GJ plaque (48, 49).

A further cytoskeletal candidate involved in maintaining GJs in their functional state at the plasma membrane is drebrin. This F-actin binding protein, when silenced with siRNA, displayed impaired junctional coupling, increased GJ internalization, and targeting of Cx43 to degradative pathways (50). The group of Linda Musil has shown elegantly the dependence of GJIC on internalization and subsequent degradation (51, 52). Their data demonstrate that degradation from the cell surface can be downregulated by physiologically relevant forms of stress. For connexins, this may serve to enhance or preserve GJ-mediated intercellular communication even under conditions in which protein synthesis and/or intracellular transport are compromised. A diagrammatic representation of the intricate cytoskeletal nexus involved in gap junction assembly and turnover is indicated in Fig. 1.

With accruing evidence demonstrating that the role of Cx43 in the regulation of important cellular processes, including migration, proliferation, and shape formation is mediated and even dependent on interaction with subsets of the cytoskeleton analysis tools are needed to identify the relevant components and track physiological and pathological dynamics. Using astrocytes isolated from newborn wild type treated with (si)RNA or astrocytes from genetically modified mice has been proven as an excellent starting point to address quantitative changes imposed on cytoskeletal dynamics due to alterations of GJIC. Technically, high-throughput, fluorescence or calorimetric cell-based assays that can be performed in conventional 96-well formats compatible with standard ELISA-based instrumentation available to many laboratories. They are suitable to obtain robust data sets discriminating differences between cell populations even in circumstances when regulation at the protein level is small (53). Finally, quantitative Western blot (qWB) and quantitative immunocytochemistry (qICC) substantially benefit from use of dual infrared laser fluorescence imaging technology with an expanded dynamic range and simultaneous detection of reference and test protein in the same sample as major assets.

2. Protocols: Isolation and Culture of Mouse Astrocytes

2.1. Animals

Astrocytic cell cultures as described in this protocol are obtained from 2-day old C57 Black mice according to Dermietzel et al. (54).

2.2. Method for Isolation of Astrocytes

- For preparation of astrocytes, mice are decapitated, the cortices dissected and transferred to ice-cold PBS.
- After removal of the meninges, the cortices are incubated in PBS containing 0.1% trypsin.

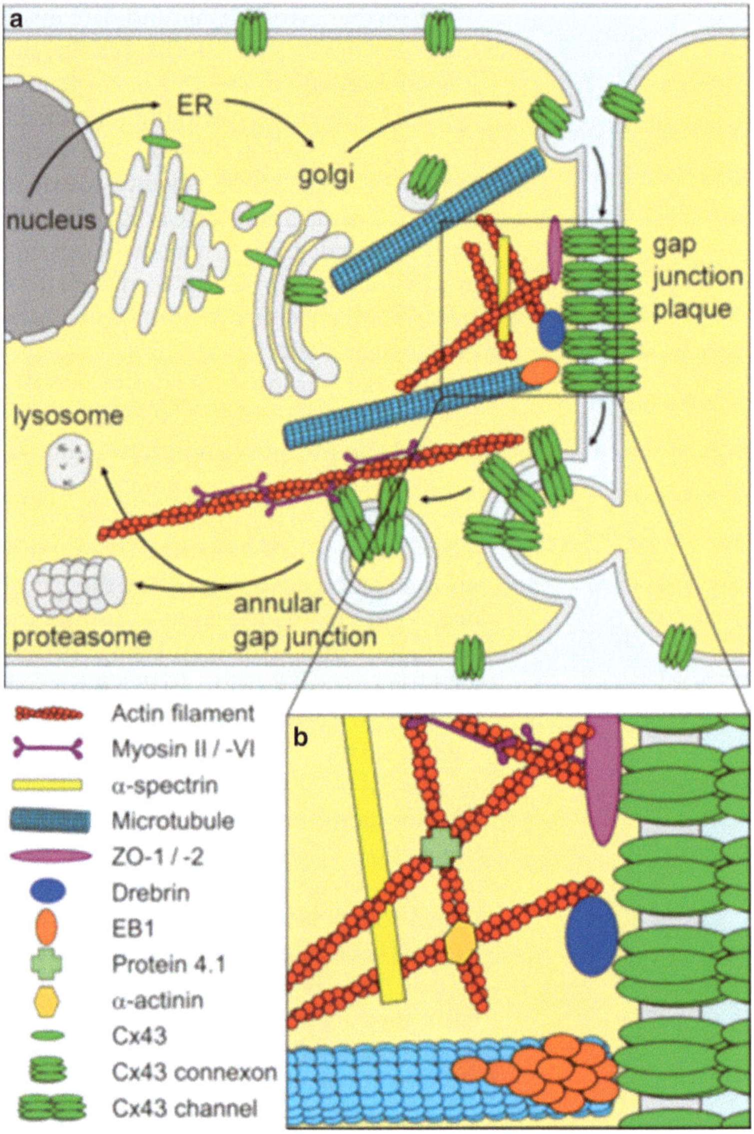

Fig. 1. Multiprotein complex around Cx43, the "nexus." (**a**) The cytoskeleton is involved in different steps of Cx turnover. Trafficking of Cxs to, within and from the cell membrane is enabled by cytoskeletal filaments. After their translation in the endoplasmic reticulum and oligomerization in the Golgi, connexon containing vesicles are shuttled to the membrane along microtubules and actin filaments. Upon insertion into the cell membrane, connexon mobility depends on its anchorage to cytoplasmic protein scaffolds. For lysosomal and proteasomal degradation, the actin cytoskeleton and associated proteins are likely to mediate internalization of connexons in cytoplasmic annular gap junction vesicles. (**b**) Depicts the subjunctional "nexus" of the cytoskeletal scaffold. The *symbols* on the *left side* indicate the different cytoskeletal and junctional elements as shown in (**a** and **b**) (Adapted from Olk et al. 2009).

- DNA digestion is done by incubating the cells in DMEM containing 0.1% DNase I and terminated by addition of two volumes DMEM with 10% fetal calf serum (FCS).
- The cells are then washed two times and plated in 25 cm^2 culture flasks.
- Medium is changed twice a week and astrocytes are used after 3–4 weeks of maturation in culture.

2.2.1. Note

When preparing astrocytes from Cx43 conditional knockout or Cx43/Cx30 double knockout animals, astrocytic cell cultures are prepared from 2-day mice as described above. Each mouse brain is prepared separately and cultured in separate flasks, because the genotype is not known at the time of preparation. To analyze the genotype, a tail biopsy is taken simultaneously, and used for PCR genotyping, while astrocytes are maintained in culture. This workflow ensures that neither animal welfare nor molecular and cellular analysis is compromised.

2.3. Protocol: Transfection with siRNA

2.3.1. Materials and Instruments

- Wild-type astrocytes (3–4 weeks in culture).
- 24-well cell culture plates.
- siGENOME or siGENOME ON-TARGETplus (Dharmacon, Lafayette, USA).
- siControl and siControlPlus pools (Dharmacon, Lafayette, USA).
- Transfection reagents, e.g., *siLentFect* (BioRad, Hercules USA), Transit-TKO (Mirus, Madison, USA), RNAiFect (Qiagene) or DharmaFECT (Dharmacon, Lafayette, USA).

2.3.2. Protocol

- Passage astrocytes 3 days before the transfection on 24-well cell culture plates either untreated or containing coverslips treated with poly-L-lysine at a density of 40,000 cells/mL.
- Transfect when astrocytes achieve 70–90% confluency.
- On the day of transfection mix 0.75 μL of siLentFect with 3 pmol of siRNA in 50 μL of serum-free DMEM to obtain a final concentration of 60 nM siRNA.
- Add transfection cocktail drop wise to the well containing 500 μL of culture media.
- Shake gently for 3 min.
- Incubate for 24 h.
- Change media replacing with DMEM containing 10% FCS, 1% nonessential amino acids, penicillin/streptomycin (100 g/mL) and glutamine (2 mM). Use for downstream applications as outlined in protocols below. In general, analyses conducted 48 h after transfection will give best results.

2.3.3. Technical Note

Each experiment should include the following samples in triplicate: (1) untreated cells, (2) positive control siRNA (targeting an endogenous or reporter gene), (3) negative control siRNA (nontargeting) and (4) the desired test siRNA. To avoid contribution of "off-target" effects, use different mixtures of duplexes with alternative sequences. Among different suppliers for siRNA transfection lipofection with siLentFect (Bio-Rad) according to the manufacturer's instructions proved most efficient and reliable.

2.4. Protocol for Quantitative Western Blots

2.4.1. Materials and Instruments

- Protein lysates of siRNA transfected, wild-type or astrocytes from KO animals.
- Nitrocellulose or PVDF membrane (we prefer nitrocellulose).
- Odyssey® Blocking Buffer (LI-COR, Cat. #927-40000).
- Primary antibodies.
- Infrared (IR)-labeled secondary antibodies, e.g., anti-mouse and anti-rabbit IRDye 680 and 800 (1:15,000, LI-COR Biosciences, USA).
- Odyssey® CLx Infrared Imaging System (LI-COR Biosciences, USA).

2.4.2. Protocol

- Prepare samples for quantitative WB (qWB) by washing astrocytes with ice-cold PBS and scraping into standard Laemmli loading buffer.
- Separated by SDS-PAGE at least in triplicates and transfer to nitrocellulose membranes (GE Healthcare).
- After incubation with blocking reagent (for WB: Roche, Mannheim, Germany; for qWB: Li-Cor, Bad Homburg, Germany), probe membranes with polyclonal anti-Cx43 antibodies (1:1,000, Zytomed, Berlin, Germany), anti-EB1 (1:3,000), anti-tropomyosin (1:500), anti-transgelin (1:500), anti-cofilin (1:500), anti-Ser-3-phosphorylated cofilin (p-cofilin) (1:500, all Santa Cruz, Santa Cruz, CA, USA), anti-GFAP (1:3,000), anti-tubulin (1:15,000, both Sigma).
- Monoclonal anti-GAPDH (1:10,000, Santa Cruz) and anti-actin (1:30,000, Sigma) serve as reference proteins.
- Use anti-mouse and anti-rabbit IRDye 680 and 800 (1:15,000 LI-COR) as secondary antibodies.
- Acquire qWB images with a dual infrared laser fluorescence imager (Odyssey system, LI-COR).
- Perform quantitative analysis using image analysis software of the manufacturer (Li-Cor) and/or Image J (NIH, Bethesda, MD, USA).
- Perform statistical analysis, e.g., using Graph Pad Prism 4.03 (Graph Pad Software, San Diego, CA, USA) or equivalent software package.

- Note: Since intraexperimental variability is a critical issue in qWB, we suggest measuring single astrocytic cell lysates up to six times to evaluate the variance of the method. The calculated coefficients of variation should be small enough to reliably quantify small expression changes.

2.4.3. Technical Notes

For qWB the following guidelines to design two color experiments are highly recommended. You must use primary antibodies from different host species to permit discrimination by secondary antibodies of different specificities. Trail test preliminary tests with each primary antibody alone to determine possible background bands. Cross-reactivity may occur and obscure interpretation of data. Use secondary antibodies labeled with 700 or 800 nm channel dyes. A good start is the Li-Cor Web site, providing a list of fluorescent dyes and how they can be visualized. Use highly cross-adsorbed secondary antibodies for two color detection. Failure to use cross-adsorbed antibodies may result in cross-reactivity and obscure data.

2.5. Quantitative Immunocytochemistry

2.5.1. Materials and Instruments

- SiRNA transfected, wild type or astrocytes from KO animals.
- 96-well cell culture plates.
- Primary antibodies
  1. Infrared (IR)-labeled secondary antibodies, e.g., anti-mouse and anti-rabbit IRDye 680 and 800 (1:15,000, LI-COR Biosciences, USA).
- Odyssey® CLx Infrared Imaging System (LI-COR Biosciences, USA).

2.5.2. Protocol

- For quantitative ICC (qICC, described as "In-cell Western" (Kortum and Lewis 2004)), fix cells in 96-well plates with 4% PFA for 15 min.
- After permeabilization in PBS containing 1% Triton X-100, block cells in blocking solution (PBS containing 3% normal goat serum and 1% bovine serum albumin).
- Incubate primary antibodies over night at 4°C.
- A monoclonal anti-GAPDH antibody (1:100, Santa Cruz) can be used as a reference for standardization.
- Wash cells thoroughly in PBS-T (containing 0.1% Tween-20) before incubation with the secondary antibodies (dilution 1:15,000–1:30,000) for 1 h.
- Aspirate and wash at least three times with PBS before 96-well plate is scanned with the dual infrared laser fluorescence imager.
- Use the LI-COR software for primary quantification and Graph Pad Prism 4.03 for statistics. Equivalent software packages can be used. We recommend using the nonparametric Wilcox Mann–Whitney test as significance test.

2.5.3. Technical Notes

In general, the same guidelines apply for qWB and qICC for design two color experiments. A critical issue of qICC is the determination of background signals. Here, mixing astrocytes from wild-type and knockout mice in defined ratios allows establishing a standard curve, unspecific background subtraction and calculation of siRNA-mediated knockdown efficiency.

2.6. Cell Migration Assay

2.6.1. Materials and Instruments

- SiRNA transfected, wild type or astrocytes from KO animals.
- 96-well cell culture plates.
- Orbital shaker.
- Microplate reader (BioRad, Hercules, USA; Model 550).
- CellTiter CytoSelect 24-well Cell Migration Assay with 8 μm pore size (Cell Biolabs, San Diego, USA).
- Laminin, BSA (Sigma-Aldrich, St. Louis, USA; Cat. Num: 114956-81-9), FCS (Invitrogen).

2.6.2. Protocol

- Measure migration of astrocytes by colorimetric readout 96 h after siRNA transfection.
- Collect cells after trypsin/EDTA treatment and brief centrifugation at 500 × *g*.
- Wash with DMEM containing 10% FCS, centrifuge at 500 × *g*.
- Resuspend cell pellet in DMEM containing 0.2% BSA, and place into the inserts of the upper chamber (see Fig. 2) at a concentration of 10^5 cells/300 μL ($n = 3$).
- Add 500 μL of DMEM containing 10% FCS or 10 μg/mL of laminin as attractants for migration (55) into the lower compartment.
- Incubate at 37°C for 18 h.
- Stain the fraction of astrocytes, which had migrated into the lower chamber, with Cell Stain Solution (Cell Biolabs). Stained cells will turn blue! Wash intensively!
- Destain with extraction solution (Cell Biolabs).
- Transfer 100 μL from each sample to a 96-well microtiter plate.
- Measure colorimetric signals at 570 nm.

2.7. Cell Adhesion Assay

2.7.1. Materials and Instruments

- SiRNA transfected, wild type or astrocytes from KO animals.
- 96-well cell culture plates.
- Orbital shaker.
- Microplate reader (BioRad, Hercules, USA; Model 550).
- CellTiter 96 AQueous One Solution Cell Proliferation Assay (Promega GmbH, Mannheim, Germany).

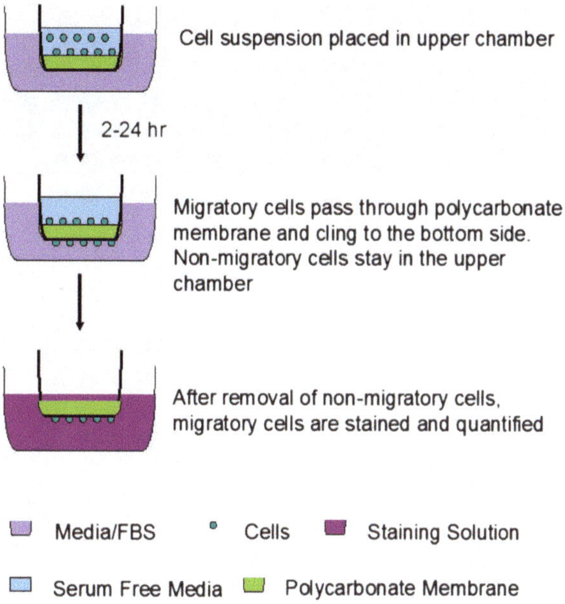

Fig. 2. The principle of migration assay (taken from CytoSelect 24-well Cell Migration Assay supporting literature). The assay Kit contains polycarbonate membrane inserts (8 μm pore size) in a 24-well plate. The membrane serves as a barrier to discriminate migratory cells from nonmigratory cells. Migratory cells are able to extend protrusions towards chemoattractants (via actin cytoskeleton reorganization) and ultimately pass through the pores of the polycarbonate membrane. Finally, the cells are removed from the top of the membrane and the migratory cells are stained and quantified.

- Laminin (Sigma-Aldrich, St. Louis, USA; Cat. Num: 114956-81-9), collagen G from bovine calf skin (Biochrom AG, Berlin, Germany; Cat. Num: L7213) poly-L-lysine (Sigma-Aldrich, St. Louis, USA; Cat. Num: P4832).

2.7.2. Protocol

- For studying cell–cell adhesion, transfect astrocytes with either 10 nM siControl or 10 nM Cx43 siRNA pools.
- After 120 h in culture, trypsinate cells and passage on 96-well plates (50,000 cells per well) containing either siControl or Cx43 siRNA-treated confluent astrocytes monolayers. A typical experimental layout is shown in (Fig. 3). We recommend for statistical reasons plating 12 wells for each test condition.
- For adhesion assay treat standard 96-well plates with 10 μg/mL of laminin, 40 μg/mL of collagen G from bovine calfskin or 10 μg/mL poly-L-lysine for 1 h at RT.
- Let passaged astrocytes adhere to cell monolayer for 4 h.
- Remove nonadherent or poorly adherent astrocytes on orbital shaker shaking for 5 s at 500 rpm.
- This step is repeated three times with changing the culture media between steps.

Fig. 3. Representation of an experimental design for the migration (**a**) and cell-to-cell-adhesion (**b**) assays. (**a**) 96-well plate treated with different indicated substrates for migration assessment. (**b**) 96 well with strategy for cell-to-cell adhesion. Please find the detailed protocol under Sect. 2.7.2 this chapter.

- Determine the relative number of remaining cells in each well (monolayer cells + adherent cells) using CellTiter 96 AQueous One Solution Cell Proliferation Assay (Promega GmbH, Mannheim, Germany) according to the manufacturer's protocol.
- Measure colorimetric signals at 490 nm using a standard microplate reader.
- Calculate background by subtracting the signal from wells containing only astrocytic monolayers.

3. Conclusions

Accruing evidence supports a view that concerted crosstalk and interaction between the major astrocytic GJP Cx43 and the cytoskeleton regulates directed migration. While regulatory differences of a cluster of several cytoskeletal proteins are discrete and small as discovered by proteomic analyses, quantitative Western blotting, and quantitative immunocytochemistry, as well as other cell-based assays described above aided in demonstrating these important players in the formation of cortical platforms that contribute to process formation, migration, and adhesion. The technologies described allow to obtained data providing mechanistic clues for an understanding of Cx43 interaction with cellular motor activities, by linking its expression to differentially expressed components of the cytoskeleton, which are involved in process formation, migration, and adhesion. Application of these methodologies provides the opportunity to gain further insights into the mechanisms by which astrocytes can achieve changes of their structural phenotype.

References

1. Dermietzel R, Spray DC (1998) From neuroglue ('Nervenkitt') to glia: a prologue. Glia 24(1):1–7
2. Allen NJ, Barres BA (2009) Neuroscience: glia—more than just brain glue. Nature 457(7230):675–677. doi:10.1038/457675a
3. Nagy JI, Rash JE (2000) Connexins and gap junctions of astrocytes and oligodendrocytes in the CNS. Brain Res Brain Res Rev 32(1): 29–44
4. Orkand RK, Nicholls JG, Kuffler SW (1966) Effect of nerve impulses on the membrane potential of glial cells in the central nervous system of amphibia. J Neurophysiol 29(4): 788–806
5. Anderson CM, Swanson RA (2000) Astrocyte glutamate transport: review of properties, regulation, and physiological functions. Glia 32(1):1–14
6. Scemes E, Dermietzel R, Spray DC (1998) Calcium waves between astrocytes from Cx43 knockout mice. Glia 24(1):65–73
7. Haydon PG (2001) GLIA: listening and talking to the synapse. Nat Rev Neurosci 2(3):185–193. doi:10.1038/35058528
8. Naus CC, Bani-Yaghoub M, Rushlow W, Bechberger JF (1999) Consequences of impaired gap junctional communication in glial cells. Adv Exp Med Biol 468:373–381
9. Spray DC, Duffy HS, Scemes E (1999) Gap junctions in glia. Types, roles, and plasticity. Adv Exp Med Biol 468:339–359
10. Rozental R, Giaume C, Spray DC (2000) Gap junctions in the nervous system. Brain Res Brain Res Rev 32(1):11–15
11. Wolburg H, Rohlmann A (1995) Structure–function relationships in gap junctions. Int Rev Cytol 157:315–373
12. Dermietzel R, Traub O, Hwang TK, Beyer E, Bennett MV, Spray DC, Willecke K (1989) Differential expression of three gap junction proteins in developing and mature brain tissues. Proc Natl Acad Sci U S A 86(24): 10148–10152
13. Dermietzel R, Gao Y, Scemes E, Vieira D, Urban M, Kremer M, Bennett MV, Spray DC (2000) Connexin43 null mice reveal that astrocytes express multiple connexins. Brain Res Brain Res Rev 32(1):45–56
14. Altevogt BM, Paul DL (2004) Four classes of intercellular channels between glial cells in the CNS. J Neurosci 24(18):4313–4323
15. Nagy JI, Rash JE (2003) Astrocyte and oligodendrocyte connexins of the glial syncytium in relation to astrocyte anatomical domains and spatial buffering. Cell Commun Adhes 10(4–6):401–406
16. Nagy JI, Ionescu AV, Lynn BD, Rash JE (2003) Coupling of astrocyte connexins Cx26, Cx30, Cx43 to oligodendrocyte Cx29, Cx32, Cx47: Implications from normal and connexin32 knockout mice. Glia 44(3):205–218
17. Nagy JI, Ionescu AV, Lynn BD, Rash JE (2003) Connexin29 and connexin32 at oligodendrocyte and astrocyte gap junctions and in myelin of the mouse central nervous system. J Comp Neurol 464(3):356–370
18. Nedergaard M, Ransom B, Goldman SA (2003) New roles for astrocytes: redefining the functional architecture of the brain. Trends Neurosci 26(10):523–530
19. Ransom B, Behar T, Nedergaard M (2003) New roles for astrocytes (stars at last). Trends Neurosci 26(10):520–522
20. Rouach N, Koulakoff A, Abudara V, Willecke K, Giaume C (2008) Astroglial metabolic networks sustain hippocampal synaptic transmission. Science 322(5907):1551–1555
21. Nagy JI, Dudek FE, Rash JE (2004) Update on connexins and gap junctions in neurons and glia in the mammalian nervous system. Brain Res Brain Res Rev 47(1–3):191–215
22. Giaume C, Fromaget C, el Aoumari A, Cordier J, Glowinski J, Gros D (1991) Gap junctions in cultured astrocytes: single-channel currents and characterization of channel-forming protein. Neuron 6(1):133–143
23. Zoidl G, Dermietzel R (2010) Gap junctions in inherited human disease. Pflugers Arch 460(2):451–466
24. Haupt C, Witte OW, Frahm C (2007) Up-regulation of Connexin43 in the glial scar

following photothrombotic ischemic injury. Mol Cell Neurosci 35(1):89–99
25. Haupt C, Witte OW, Frahm C (2007) Temporal profile of connexin 43 expression after photothrombotic lesion in rat brain. Neuroscience 144(2):562–570
26. Fallon RF, Goodenough DA (1981) Five-hour half-life of mouse liver gap-junction protein. J Cell Biol 90(2):521–526
27. Das Sarma J, Meyer RA, Wang F, Abraham V, Lo CW, Koval M (2001) Multimeric connexin interactions prior to the trans-Golgi network. J Cell Sci 114(Pt 22):4013–4024
28. Das Sarma J, Wang F, Koval M (2002) Targeted gap junction protein constructs reveal connexin-specific differences in oligomerization. J Biol Chem 277(23):20911–20918
29. Maza J, Mateescu M, Das Sarma J, Koval M (2003) Differential oligomerization of endoplasmic reticulum-retained connexin43/connexin32 chimeras. Cell Commun Adhes 10(4–6):319–322
30. Falk MM (2000) Biosynthesis and structural composition of gap junction intercellular membrane channels. Eur J Cell Biol 79(8):564–574
31. Saez JC, Berthoud VM, Branes MC, Martinez AD, Beyer EC (2003) Plasma membrane channels formed by connexins: their regulation and functions. Physiol Rev 83(4):1359–1400
32. Koval M (2006) Pathways and control of connexin oligomerization. Trends Cell Biol 16(3):159–166
33. Laird DW (2006) Life cycle of connexins in health and disease. Biochem J 394(Pt 3):527–543
34. Segretain D, Falk MM (2004) Regulation of connexin biosynthesis, assembly, gap junction formation, and removal. Biochim Biophys Acta 1662(1–2):3–21
35. Gaietta G, Deerinck TJ, Adams SR, Bouwer J, Tour O, Laird DW, Sosinsky GE, Tsien RY, Ellisman MH (2002) Multicolor and electron microscopic imaging of connexin trafficking. Science 296(5567):503–507
36. Martin PE, Blundell G, Ahmad S, Errington RJ, Evans WH (2001) Multiple pathways in the trafficking and assembly of connexin 26, 32 and 43 into gap junction intercellular communication channels. J Cell Sci 114(Pt 21): 3845–3855
37. Thomas T, Jordan K, Laird DW (2001) Role of cytoskeletal elements in the recruitment of Cx43-GFP and Cx26-YFP into gap junctions. Cell Commun Adhes 8(4–6):231–236
38. Johnson RG, Meyer RA, Li XR, Preus DM, Tan L, Grunenwald H, Paulson AF, Laird DW, Sheridan JD (2002) Gap junctions assemble in the presence of cytoskeletal inhibitors, but enhanced assembly requires microtubules. Exp Cell Res 275(1):67–80
39. Guo Y, Martinez-Williams C, Rannels DE (2003) Gap junction-microtubule associations in rat alveolar epithelial cells. Am J Physiol Lung Cell Mol Physiol 285(6):L1213–L1221
40. Shaw RM, Fay AJ, Puthenveedu MA, von Zastrow M, Jan YN, Jan LY (2007) Microtubule plus-end-tracking proteins target gap junctions directly from the cell interior to adherens junctions. Cell 128(3):547–560
41. Prochnow N, Dermietzel R (2008) Connexons and cell adhesion: a romantic phase. Histochem Cell Biol 130(1):71–77
42. Laing JG, Koval M, Steinberg TH (2005) Association with ZO-1 correlates with plasma membrane partitioning in truncated connexin45 mutants. J Membr Biol 207(1):45–53
43. Larsen WJ, Tung HN, Murray SA, Swenson CA (1979) Evidence for the participation of actin microfilaments and bristle coats in the internalization of gap junction membrane. J Cell Biol 83(3):576–587
44. Watanabe H, Washioka H, Tonosaki A (1988) Gap junction and its cytoskeletal undercoats as involved in invagination-endocytosis. Tohoku J Exp Med 156(2):175–190
45. Murray SA, Williams SY, Dillard CY, Narayanan SK, McCauley J (1997) Relationship of cytoskeletal filaments to annular gap junction expression in human adrenal cortical tumor cells in culture. Exp Cell Res 234(2): 398–404
46. Jordan K, Chodock R, Hand AR, Laird DW (2001) The origin of annular junctions: a mechanism of gap junction internalization. J Cell Sci 114(Pt 4):763–773
47. Piehl M, Lehmann C, Gumpert A, Denizot JP, Segretain D, Falk MM (2007) Internalization of large double-membrane intercellular vesicles by a clathrin-dependent endocytic process. Mol Biol Cell 18(2):337–347
48. Gilleron J, Fiorini C, Carette D, Avondet C, Falk MM, Segretain D, Pointis G (2008) Molecular reorganization of Cx43, Zo-1 and Src complexes during the endocytosis of gap junction plaques in response to a non-genomic carcinogen. J Cell Sci 121(Pt 24):4069–4078
49. Fiorini C, Gilleron J, Carette D, Valette A, Tilloy A, Chevalier S, Segretain D, Pointis G (2008) Accelerated internalization of junctional membrane proteins (connexin 43, N-cadherin and ZO-1) within endocytic vacuoles: an early event of DDT carcinogenicity. Biochim Biophys Acta 1778(1):56–67

50. Butkevich E, Hulsmann S, Wenzel D, Shirao T, Duden R, Majoul I (2004) Drebrin is a novel connexin-43 binding partner that links gap junctions to the submembrane cytoskeleton. Curr Biol 14(8):650–658
51. VanSlyke JK, Musil LS (2005) Cytosolic stress reduces degradation of connexin43 internalized from the cell surface and enhances gap junction formation and function. Mol Biol Cell 16(11):5247–5257
52. Kelly SM, Vanslyke JK, Musil LS (2007) Regulation of ubiquitin-proteasome system mediated degradation by cytosolic stress. Mol Biol Cell 18(11):4279–4291
53. Olk S, Turchinovich A, Grzendowski M, Stuhler K, Meyer HE, Zoidl G, Dermietzel R (2010) Proteomic analysis of astroglial connexin43 silencing uncovers a cytoskeletal platform involved in process formation and migration. Glia 58(4):494–505
54. Dermietzel R, Hertberg EL, Kessler JA, Spray DC (1991) Gap junctions between cultured astrocytes: immunocytochemical, molecular, and electrophysiological analysis. J Neurosci 11(5):1421–1432
55. Takenaga K, Kozlova EN (2006) Role of intracellular S100A4 for migration of rat astrocytes. Glia 53(3):313–321

Chapter 13

Reactive Astrocytes, Astrocyte Intermediate Filament Proteins, and Their Role in the Disease Pathogenesis

Milos Pekny, Michelle Porritt, Yolanda de Pablo, Marcela Pekna, and Ulrika Wilhelmsson

Abstract

Astrocyte activation and reactive gliosis are seen in many neuropathologies, e.g., neurotrauma, stroke, epilepsy, or neurodegenerative diseases. Astrocyte activation alters gene expression and leads to morphological and functional changes in astrocytes with important functional consequences for the central nervous system (Eddleston and Mucke, Neuroscience 54:15-36, 1993; Eng and Ghirnikar, Brain Pathol 4:229-237, 1994; Hernandez et al., Glia 38:45-64, 2002; Pekny and Nilsson, Glia 50:427-434, 2005; Wilhelmsson et al., Proc Natl Acad Sci U S A 103:17513-17518, 2006; Sofroniew, Trends Neurosci 32:638-647, 2009; Sofroniew and Vinters, Acta Neuropathol 119:7-35, 2010). The understanding of astrocyte activation and reactive gliosis in pathological situations remains incomplete but the increasing amount of experimental evidence points to its importance in disease pathogenesis (Wilhelmsson et al., J Neurosci 24:5016-5021, 2004; Sofroniew, Neuroscientist 11:400-407, 2005; Maragakis and Rothstein, Nat Clin Pract Neurol 2:679-689, 2006; Seifert et al., Nat Rev Neurosci 7:194-206, 2006; Correa-Cerro and Mandell, J Neuropathol Exp Neurol 66:169-76, 2007; Barres, Neuron 60:430-440, 2008; Li et al., J Cereb Blood Flow Metab 28:468-481, 2008; Macauley et al. J Neurosci 31:15575-15585, 2011). One of the principal hallmarks of astrocyte activation and reactive gliosis is the upregulation of astrocyte intermediate filament (nanofilament) proteins and reorganization of intermediate filaments that are part of the cytoskeleton. This review focuses on the role of the intermediate filament system of astrocytes in neuropathological context and presents some of the relevant model systems.

Key words: Astrocytes, Intermediate filaments, GFAP, Vimentin, Reactive gliosis, Entorhinal cortex lesion, Photothrombotic model, Hypoxia, Oxygen–glucose deprivation, Experimental models

1. Introduction

1.1. Astrocytes in the Normal Central Nervous System

Before focusing on the function of reactive astrocytes in the disease context, we should consider the function of astrocytes in the normal central nervous system (CNS). With certain degree of simplification, we can view most of the mammalian CNS as being divided among morphological and functional domains of

Rolf Dermietzel (ed.), *The Cytoskeleton: Imaging, Isolation, and Interaction*, Neuromethods, vol. 79,
DOI 10.1007/978-1-62703-266-7_13, © Springer Science+Business Media, LLC 2013

individual astrocytes with other cellular elements contained within these domains (16–20). There is an intimate contact between astrocyte processes and neurons or endothelial cells and pericytes of blood capillaries, and this contact is essential for an efficient cell–cell communication. Astrocytes recycle molecules involved in neurotransmission and play an important role in supplying sources of energy to neuronal cells (21–24), induce formation of neuronal synapses and control their function (25–28), and regulate blood–brain barrier properties (13, 29) as well as blood flow (30, 31). A single astrocyte in the human brain can have up to two million neuronal synapses within its domain (32). This vast complexity of interaction opportunities with other cells is further amplified by gap junctional coupling between astrocytes, which links them into a complex communicating network with the capacity to sense the environment and respond to it. Astrocyte population might well be as heterogenous as neuronal population (33, 34). Attempts to characterize astrocytes on a molecular level come from increasing number of laboratories (34–36). They will form the basis for functional classification of these cells, a prerequisite for understanding their roles in the disease pathogenesis (R1).

1.2. Activated Astrocytes and Reactive Gliosis

As stated above, activated astrocytes exhibit characteristic molecular and morphological features. Upregulation of intermediate filament proteins, in particular glial fibrillary acidic protein (GFAP), the main constituent of astrocyte intermediate filaments, has long served as a marker of astrocyte activation in human neuropathologies as well as in experimental neuroscience research (37). These pathological conditions include brain and spinal cord trauma and infections, stroke, epilepsy, brain tumors, or neurodegenerative diseases such as Alzheimer's disease, multiple sclerosis, or amyotrophic lateral sclerosis.

On a morphological level, reactive gliosis ranges from subtle to moderate to very prominent, the latter often being accompanied by glial scarring. With respect to the primary lesion (which ranges from local to generalized), reactive astrocytes can be present throughout the lesion or form a border that surrounds it. At least in some situations, e.g., in traumatic brain lesions or around amyloid plaques in Alzheimer's disease, reactive astrocytes seem to constitute a border between a focal pathology and the tissue around it. When such lesion demarcation was inhibited in animal models of neurotrauma, the pathology expanded and the functional impairment became more prominent (38, 39). It is conceivable that evolution selected for temporary or permanent demarcation of a lesion by reactive astrocytes and its isolation from relatively unaffected CNS (40, 41). The price for a quick clinical stabilization allowed by such a sequestering of a lesion, however, might be a compromised regenerative response at a later stage (42).

What happens if one eliminates astrocytes or reactive astrocytes or prevents astrocytes from becoming fully reactive? Perhaps not surprisingly, elimination of astrocytes in a mammal is not compatible with life (43). Elimination of the dividing subpopulation of reactive astrocytes is detrimental and exacerbates the negative consequences of brain or spinal cord trauma (44–47). Combined genetic deficiency of the intermediate filament (nanofilament) proteins GFAP and vimentin achieved by in vivo gene targeting in mice leads to attenuated reactive gliosis (48, 49). The $GFAP^{-/-}Vim^{-/-}$ mice show reduced glial scar formation, slower healing process, and increased loss of neuronal synapses after neurotrauma (8, 49), and increased baseline and injury-induced neurogenesis (R2, R3). When exposed to focal brain ischemia, the $GFAP^{-/-}Vim^{-/-}$ mice develop larger infarctions compared to wild-type mice (14). The astrocyte intermediate filament system has been linked to viscoelastic properties of astrocytes (50), astrocyte motility (51), intracellular trafficking of vesicles (52, 53, R4), or impaired response to hypoosmotic stress (54).

Are there any benefits of attenuated reactive gliosis? There is a growing body of evidence to support an affirmative answer to this question and provide a rationale for the modulation of reactive gliosis as a means to achieve a better regenerative response. Thus, attenuation of reactive gliosis after neurotrauma in $GFAP^{-/-}Vim^{-/-}$ mice led to better regeneration of neuronal synapses (8), neuronal axons (55), or improved regenerative response and functional recovery after spinal cord trauma (56). $GFAP^{-/-}Vim^{-/-}$ mice also show better integration of neural grafts in the retina (57) or neuronal and astrocyte differentiation of adult neural stem cells transplanted in the hippocampus (58), either due to attenuated reactive gliosis or as a result of altered interactions between the grafted cells and intermediate filament-free astrocytes. Thus, the benefits of reactive gliosis at the acute phase of the injury seem to be balanced against restricted regenerative potential at later stages.

The regeneration-inhibiting effects of prominent reactive gliosis and glial scarring were well described in various animal models (59–63) with a number of molecules other than intermediate filament proteins being involved. These regeneration-inhibiting effects might be viewed as the cost of the efficient handling of the acute cellular stress, relatively good neuroprotection, and effective isolation of the rest of the CNS from the toxic and potentially expanding lesion area (46, 64). It is likely that during evolution, such acute responses were selected since they allowed survival and a measure of repair, albeit at a price of a restricted regenerative capacity. Modifying such an equilibrium at the right time point might provide new opportunities for therapy of specific neuropathological situations.

1.3. Astrocytes and Reactive Astrocytes as a Therapeutic Target

Will astrocytes become the target for new therapeutic strategies for diseases such as neurotrauma, stroke, or neurodegenerative diseases? We can safely say that the question is not if but when (13, 65–67). The work currently pursued in several laboratories shifts attention from assessing individually selected potential molecular targets to high-throughput screening strategies with molecular candidates being evaluated in the context of the whole transcriptome or proteome (35, 68, 69 and Chaps 5, 7, 15 this volume). Some drugs currently used in clinical practice may well exert their therapeutic effects predominantly by modulating the function of astrocytes without us yet realizing their true mode of action. Future therapeutic strategies will likely aim at a gentle adjustment of multiple equilibria (61, 70) and will often be combined with non-pharmacological regeneration-promoting modalities, e.g., targeted rehabilitation, exercise, or multisensory stimulation (71).

2. Material and Methods: Selection of Experimental Models

2.1. Entorhinal Cortex Lesion as a Neurotrauma Model

Many CNS diseases such as trauma, stroke, or neurodegenerative diseases result in denervation and synaptic loss in specific areas of the brain and spinal cord. Entorhinal cortex lesion is a highly useful experimental paradigm for studying axonal degeneration and synaptic plasticity in the hippocampus (72–74). Entorhinal cortex lesion models in mouse and rat have been used since the 1970s for investigations on synaptic loss, synaptic plasticity, and reorganization of synaptic input after denervation. In this model, a lesion interrupts the axonal connections in the perforant path, which innervate the outer molecular layer in the dentate gyrus of the hippocampus. The axonal degeneration triggers synapse remodeling and astrocyte activation, in the molecular layer of the dentate gyrus, an area not directly affected by the lesion itself (Fig. 1).

During the first days after denervation, the loss of synaptic input into the molecular layer of the dentate gyrus results in the elimination of synapses on the granule neurons. However, these synapses are later on at least partially replaced by collateral sprouting of remaining axons within the perforant pathway. Astrocytes in the dentate gyrus of the hippocampus respond to the lesion within 24 h by upregulation of GFAP mRNA resulting in an increase in GFAP protein level at 2 days after lesion. This upregulation is most prominent around 4 days after lesion and persists for several weeks (75, 76). This is accompanied by hypertrophy of astrocyte processes (77). Upregulation of astrocyte intermediate filament proteins vimentin, nestin, and synemin is less prominent and is detectable by immunohistochemistry in astrocytes in the denervated dentate gyrus at one to 2 weeks after the lesion (78, 79 and our own observations). Microglial cells respond faster than astro-

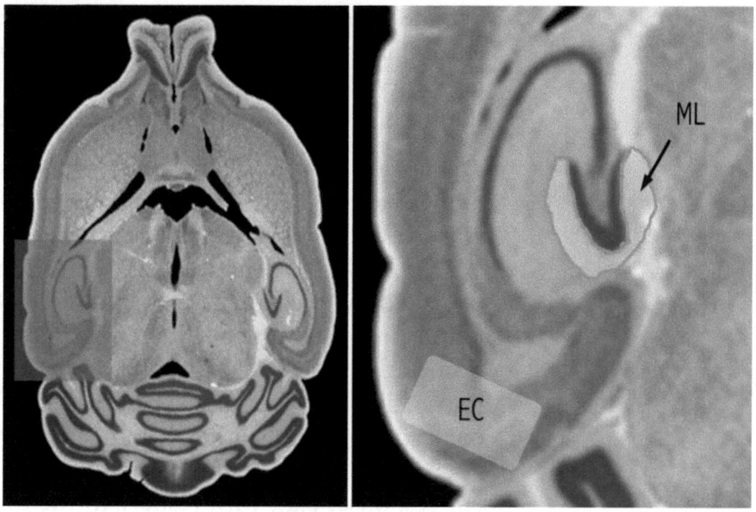

Fig. 1. Horizontal section of the mouse brain. Entorhinal cortex (EC) lesions partially denervate the molecular layer of the dentate gyrus of the hippocampus (ML).

cytes and exhibit changes in morphology and CD11b expression already 1 day after lesion (80). The population of NG2-positive glial cells shows increased expression of NG2 mRNA at 2 and 4 days after lesion (81). Astrocytes and microglia respond to the denervation by the production of cytokines and extracellular matrix molecules (82, 83) and are involved in the clearance of synapses and extracellular debris generated by the Wallerian degeneration of the perforant path (84). After clearance of synapse debris, new synapses are formed on the granule cell dendrites in the molecular layer of the dentate gyrus from sprouting of remaining unlesioned perforant path axons. The spine density on the granule cell dendrites increases rapidly and returns back to levels comparable to unlesioned mice during the third week after lesion, although structural alterations of the granule cell dendrites persist over several months or even longer (85). We assessed the role of reactive astrocytes and astrocyte intermediate filament proteins in this process using mice deficient in GFAP and vimentin that show attenuation of astrocyte activation and reactive gliosis. We found that in an early stage after the lesion, reactive astrocytes are beneficial by limiting the extent of synaptic loss and improving the removal of cell debris (8). Importantly, at a later stage, reactive astrocytes inhibit the formation of new synapses (8), suggesting that reactive astrocytes have both beneficial and detrimental roles after neurotrauma.

Advantages of the entorhinal cortex lesion model: A mechanically induced neurotrauma causes blood–brain barrier disruption with immunoglobulins, other plasma proteins, and blood-borne monocytes infiltrating the neural tissue and, at a later stage, contributing to the development of glial scar. As the area of degeneration in the dentate gyrus is located at a distance from the lesion itself (in the entorhinal cortex), this lesion model can be used to study astrocyte response to neurodegenerative processes in the absence of a direct trauma and with the blood–brain barrier remaining relatively intact.

Disadvantages of the entorhinal cortex lesion model: Several techniques have been used to sever the entorhinal afferents to the dentate gyrus. In rats, electrolytic lesions have been frequently used (77, 86, 87). Alternatively, the entorhinal cortex neurons were damaged by injection of ibotenic acid (88, 89). Since electrolytic lesions may have an epileptogenic effect (90), mechanical transection of the perforant path with a special wire knife or electrolytic transection has been used in rats (91, 92) and became the most commonly used entorhinal transection method in mice (93–95). Each of the de-afferentation method results in a somewhat different response, e.g., with respect to the immune and inflammatory parameters, and this will affect astrocyte activation and must be taken into account.

2.2. Unilateral Entorhinal Cortex Lesion of Adult Mice: Protocol

2.2.1. Material and Equipment Needed

Stereotactic frame with a wire knife assembly (Kopf Instruments) and a retractable wire knife (McHughMilieux.com)

Surgical tools: Forceps, scalpel, a small retractor, drill with a 1–2 mm size burr

Shaver

Cotton swabs

Suture material or surgical clips

2.2.2. The Procedure

1. Assemble the equipment, set the top and bottom stops for the knife, calibrate the knife length to 2.0 mm, and set the direction of the knife extension to pointing laterally to the left parallel to the ear bars on the frame (this is determined as 0° angle of the knife). Disinfect or sterilize surgical tools.
2. Anesthetize the mouse and administer buprenorphine to minimize post-surgery pain and to promote faster post-surgery recovery.
3. The mouse is shaved on the scalp and placed in the stereotactic frame. Tooth bar should be kept horizontal. Disinfect and remove shaved fur from the shaved area thoroughly with 70% ethanol and cotton swabs.
4. Make a 1 cm skin incision along the midline of the skull and retract the skin on the left side to visualize the Lambda landmark on the skull. Use a dry cotton swap to remove the connective tissue.
5. Set the anterior–posterior and medial–lateral zero coordinates at Lambda. Position the knife assembly 3.6 mm laterally and 0.2 mm posterior from Lambda and make a mark. Drill a hole through the skull. Drilling may sometimes cause the mouse

head to move slightly. It is therefore recommended to go back to Lambda after the drilling is done, check the zero coordinates, and if necessary adjust the lesion site coordinates accordingly.

6. Dry off excessive fluid and lower the knife assembly until it lightly touches the brain surface; set this as the dorsal–ventral zero coordinate. Lower the knife assembly for an additional 1.0 mm down into the brain cortex.
7. Extend the knife blade 2.0 mm at +30° angle. Lower the knife assembly 2.0 mm and withdraw the knife. Go up 2.0 mm and change the angle of the knife assembly to −135° angle. Extend the knife 2.0 mm, lower the knife assembly 2.0 mm, withdraw the knife, and elevate the knife assembly up and out from the mouse skull.
8. Suture the wound, apply bupivacaine to the wounded skin area, and administer saline intraperitoneally to decrease dehydration.
9. Place the mouse in a heated cage for observation until it is fully recovered.

2.3. Photothrombotic Stroke/Neurotrauma Model

By refining the model of Rosenblum, Watson and colleagues established the induction of reproducible thrombosis by photochemical reaction in vivo, which results in cerebral infarction in rodents (96, 97).

A photosensitive superoxide-forming dye such as rose Bengal (disodium tetraiodo-tetrachloro-fluorescein) or erythrosin B is injected directly into the bloodstream or peritoneum. Then illumination from a light source targeting a specific blood vessel or cortical region through the exposed skull is used to induce thrombus formation. The ischemic lesion is limited to the illuminated area, including all layers of the cortex. Because of limited light penetration, the subcortical tissue is undamaged by this method. Focal illumination by a light source activates the photosensitive dye to release a singlet oxygen with free radical generation and subsequent peroxidative damage of the endothelial membrane. The damaged endothelium activates platelet aggregation and the coagulation cascade leading to thrombotic occlusion of small vessels (98, 99). Abnormal blood perfusion through the vessels develops as early as 30 min after photothrombosis induction and is maximal by 4 h (98). The photothrombotic ischemia produces an irreversibly damaged core surrounded by a small penumbral region (100, 101) (Fig. 2). The presence of an ischemic penumbra has been observed surrounding the infarct core by MRI imaging technique (102).

The photothrombotic model has traditionally been used as a permanent model of cerebral ischemia. However, Yao and colleagues who specifically targeted the distal middle cerebral artery (MCA) have demonstrated reperfusion by use of ultraviolet light illumination at the same site (103).

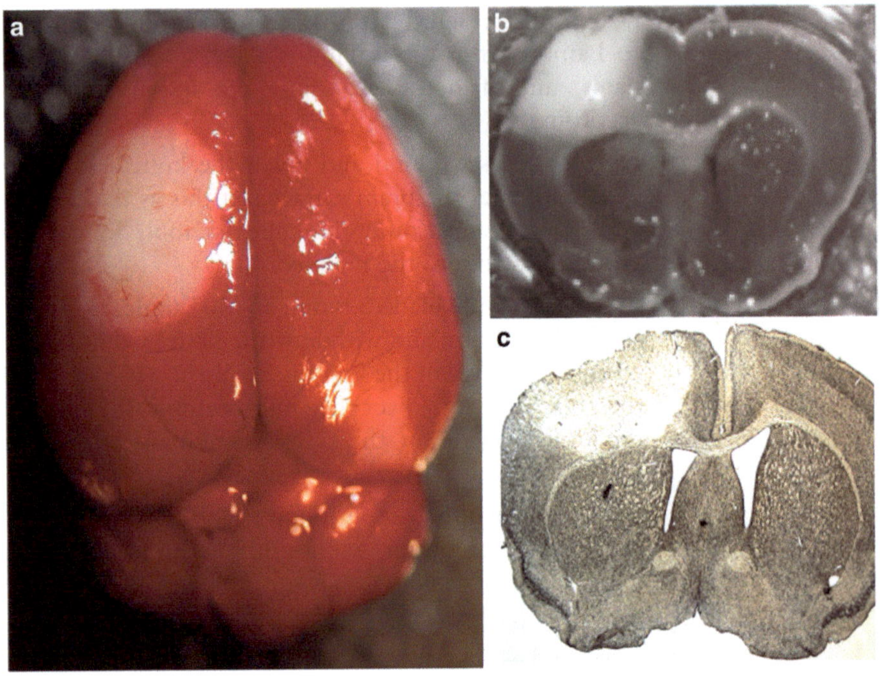

Fig. 2. Photothrombotic infarct within the somatosensory cortex of a mouse (C57BL/6 strain) (**a**) visualized by 2,3,5-Triphenyltetrazolium chloride (TTC) staining. TTC staining provides a measure of tissue viability. The infarct is unstained, i.e., white, whilst the remaining viable tissue is stained red because mitochondrial dehydrogenases convert the tetrazolium salt to a red formazin by-product. (**b**) A TTC-stained 1 mm thick coronal section in grayscale. (**c**) Thionine-stained 20 μm thick coronal section in grayscale.

After exposure to ischemia, astrocytes increase their glycolytic capacity, upregulate immediate early genes, and express intermediate filament proteins GFAP and vimentin (104, 105). Timing and distribution of reactive gliosis following photothrombotic ischemia have been reported in the literature (106, 107). In rats, a conical area of focal pannecrosis is visible 3 days after the cortical surface was subjected to photothrombotic ischemia. This area consists of a core that contains a few "ghost neurons" and transformed microglia with the tissue immediately adjacent containing numerous reactive astrocytes as identified by immunostaining for GFAP and vimentin. Reactive cortical astrocytes were detected over the entire ipsilateral hemisphere from 3 days after ischemia induction. By 7 days, transformed microglia and macrophages begin to infiltrate the ischemic core and the astrocytic rim is becoming clearly defined. Fourteen days following photothrombosis, the infarct core consists entirely of microglia/macrophages. The cortex atrophies and may form cysts. Staining for astrocytes 10 weeks after ischemia onset reveals that the band of reactive astrocytes constricts to a narrow band surrounding the core with GFAP immunoreactive astrocytes (106, 107). Reactive astrocytes are also reported in other neocortical areas such as the motor, sensory, and frontoparietal cortex 4 days after photothrombotic ischemia. GFAP immunoreactive cells have been observed in the hippocampus and amygdala but not in the contralateral hemisphere (108). Therapeutic agents have been evaluated for the ability to modulate glial activity after photothrombotic ischemia.

2.3.1. Neuroprotection

Photothrombotic ischemia has been used to evaluate numerous neuroprotective agents such as antioxidants, antidepressants, calcium channel blockers free radical scavengers, NMDA and $GABA_A$ receptor antagonists, and allosteric modulators (109–112). The successful neuroprotective agents in this model were administered prior to ischemic onset. Since the mechanism of action of these agents was antioxidative, it is thought that they affected singlet oxygen free radical damage rather than displayed true neuroprotection. Remarkably, therapeutic agents shown to be neuroprotective in other models of ischemia, such as MK801 and nimodipine, have failed to protect against the photothrombosis-induced ischemia (109–113). Jointly these results point to the limited usability of the photothrombotic model for neuroprotection studies.

2.3.2. Functional Assessment

A clear advantage of the photothrombotic model over other models of ischemia is that it can be used to target particular brain regions. Lesions over the motor cortex, usually isolated to only the forepaw representation, are used to evaluate the effectiveness of therapeutic agents in recovery. Lesions of the cortex are quite small, yet they do result in functional or use deficits. The early functional deficit may arise from the edema affecting the surrounding brain tissue (114). Shanina and co-workers demonstrated functional deficits in rat forelimb use and skilled walking after photothrombotic ischemia over the forelimb somatosensory cortex. They noted that this functional deficit was only transient and could be recorded for 2 weeks following infarction using the forelimb preference test and the coordinated limb placement tests (115).

Note 1: Advantages of the photothrombotic model: An apparent advantage of photothrombotic lesioning is direct access to the cerebral arteries through an intact skull; however, thrombosis can occur in any vessel illuminated by the light source with high concentration of photo-activating agent. Photothrombotic occlusion of cerebral microvessels induces small cerebral infarctions that are highly reproducible in size and location. Photothrombotic lesioning can be used to target specific blood vessels such as the distal MCA or large areas such as the MCA watershed area, i.e., the parietal cortex (116). There is also a ring model where the illumination source is circular, thus generating a small penumbral area encircled by an ischemic core (117). The pattern of glial activation and leukocyte infiltration is sequential (118). Inflammation in the post-ischemic cerebral tissue after photothrombosis is similar to that produced by the MCA occlusion model (119, 120). Photothrombotic ischemia is associated with strong and early microglial activation and proliferation within the lesioned cortex (121).

Note 2: Disadvantages of the photothrombotic model: A disadvantage of the photothrombotic model is that cerebral microvessels are also occluded; therefore it is not possible to investigate the effect of collateral perfusion (122). If the distal MCA is directly targeted, then reperfusion can be performed by use of ultraviolet light illumination at the same site (103). The ultraviolet light treatment is not successful at reperfusing the watershed area of the MCA over the cortex. A certain disadvantage is that the lesions are relatively small.

2.4. Photothrombotic Stroke/Neurotrauma Model of Adult Mice: Protocol

2.4.1. Material and Equipment Needed

Rose Bengal solution 20 mg/ml solution, sterile saline as diluent, syringe filter to remove any particulates (protect from the light)

Surgical Setup

Anesthesiology equipment including Isoflurane vaporizer, compressed O_2, and air

Heat mat and temperature control box

Stereotactic frame

Light source: Dual flexible arm cool light source/Laser 561 nm

Surgical Equipment

Scalpel

2 small retractors

A pair of Scissors for cutting sutures

A needle holder and hemostats for suturing

4-0 silk sutures to close the scalp incision

Gauze and cotton tips

2.4.2. The Procedure

1. Anesthetize the mouse by induction of Isoflurane in air and oxygen (1:1) initially at 2.5% until mouse is unresponsive, and then transfer it to the stereotaxic frame. Reduce Isoflurane to 1% for the duration of the surgical procedure.
2. Insert the rectal probe to monitor and maintain body temperature at 37°C using a homeothermic control unit (Harvard Apparatus, USA). Since mice have little thermal mass and hypothermia is known to be neuroprotective, it is important to control body temperature.
3. Pinch tail and paw to confirm the desired depth of anesthesia. Using the scalpel, make a 1 cm long incision in the scalp. Gently scrape the surface of the skull to disrupt the connective tissue and then retract the scalp. Dry the skull with a cotton tip.
4. Position the cold light source or the laser over Anterior 0.5 mm and Lateral 2.7 mm relative to Bregma (corresponds to the somatosensory cortex for a 20–30 g mouse). Inject rose Bengal

(100 μl, 20 mg/ml solution in sterile saline) intraperitoneally. After 5 min, turn on the laser (Cobolt, 50 mW, 561 nm) and illuminate the area for 10 min.

5. Close the scalp with 4/0 silk and let the mouse recover in a 22°C chamber for 1 h prior to returning it to the home cage. Mice should be provided with a heat mat and moist food in their home cage for the next 3–5 days of their post-stroke survival.

2.4.3. TTC Staining Protocol

1. Prepare a 1% 2,3,5-Triphenyltetrazolium chloride (TTC) solution in 0.9% NaCl. Always make fresh prior to use, protect from the light, and warm to 37°C.
2. Anesthetize the mouse with Isoflurane or barbiturate overdose until unresponsive.
3. Decapitate mouse, remove brain, and place in a brain matrix. For a mouse, 1 mm slice intervals are best.
4. Transfer sections to the TTC solution, incubate at 37°C for 5 min, and then flip sides. Six-well culture plates are ideal for this incubation.
5. Carefully aspirate TTC solution and add fresh 10% formalin solution. Sections can stay in this solution for up to 1 week. If measuring volume is important then ensure that the time in formalin is consistent, as tissue will shrink.
6. Photograph or scan sections with a scale for the infarct area and volume determination.

2.5. Oxygen and Glucose Deprivation as an In Vitro Stroke Model

In ischemic stroke, we can distinguish two areas: the core, where the blood flow is so low that cells undergo necrotic death, and the penumbra, where cells are metabolically compromised but still alive. At the time of reperfusion, the penumbra will be rescued or become part of the core. Understanding the mechanisms determining the outcome is of vital importance for finding a treatment to limit ischemic damage. Astrocytes in the penumbra become activated and among other changes, intermediate filament proteins are upregulated. The protocol described below represents a penumbra model with minimal astrocyte death and with the possibility to control the extent of neuronal damage.

The refinement of the conditions for tissue culture in the first half of the twentieth century led to the first in vitro models of stroke to analyze events on a cellular level. The effects of oxygen and glucose deprivation (OGD) in brain slices have been studied since the 1950s (123), and on neuronal cultures since three decades later (124) (for review of OGD in glia refer to (125)).

Note 1: Advantages of the OGD model: One advantage of the in vitro model is that the use of incubators or hypoxic chambers allows for the delivery of controlled oxygen concentration with

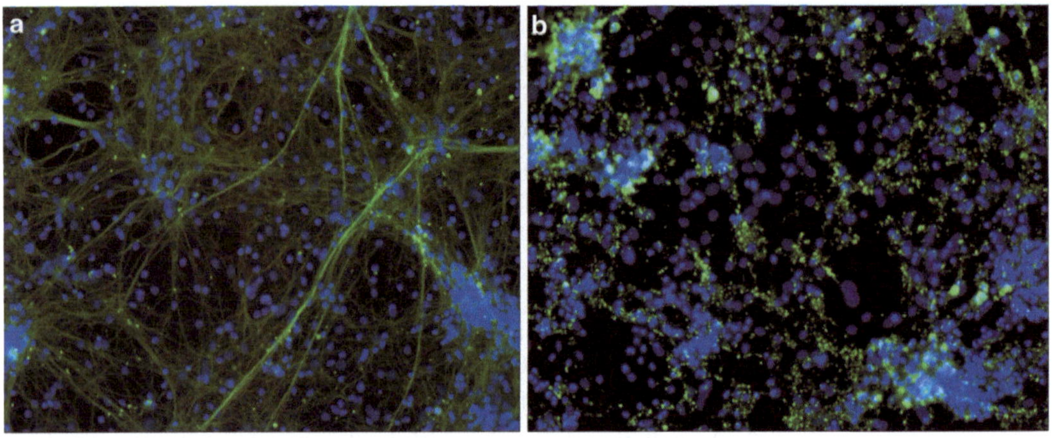

Fig. 3. Cocultures of astrocytes and neurons where neurons were plated on top of astrocytes for 10 days. (**a**) Numerous neurons forming a network are visible when cells are cultured under standard conditions (control). (**b**) After 12-h OGD and 24-h reperfusion, most of the neurons have died and the neurites appear fragmented and the nuclei condensed while at the same time, astrocyte nuclei are morphologically normal. Neurons are visualized with antibodies against beta-III-tubulin (*green*), all cell nuclei are stained with DAPI (*blue*).

fine regulation of the duration and severity of the insult. In the same way, oxygen and glucose can be reintroduced in a controlled manner mimicking some aspects of reperfusion. When measured after 24 h of reperfusion, OGD with total absence of oxygen requires the duration of around 4–6 h to cause significant death of astrocytes (126); 1% oxygen requires 12–20 h to induce measurable astrocyte death. Neurons, on the other hand, die after 1 h of OGD (127). Depending on the stimulus, reactive astrocytes upregulate many different cellular pathways and this protective response allows them to survive OGD for a longer time. For example, OGD increases the expression of HSP70 that protects mitochondria from oxidative stress (128). Notably, upregulation of HIF-1 protects astrocytes (129), but is detrimental for the cocultured neurons (130). Astrocyte death occurs more rapidly when hypoxia is combined with low pH and ionic shift, conditions present during acute brain damage (131). Under such conditions, astrocytes die in less than 1 h of OGD and reoxygenation.

Figure 3 shows the evolution of an astrocyte–neuron coculture after 12 h of OGD. Initially, the few surviving neurons lose most of the neurites. Another advantage of the in vitro model is the possibility of studying the response of different cell populations on their own and in combination. In the model described below, we omit the contribution of the immune response and the consequences of the blood–brain barrier disruption and we focus on the astrocyte and neuronal response alone or in combination, in the latter case by using astrocyte and neuron cocultures.

Note 2: Disadvantages of the OGD model: The main disadvantage of an in vitro OGD is the increase in the baseline astrocyte reactivity.

Besides the initial injury caused to cells in the process of dissociating the brain tissue into a single cell suspension, the artificial two-dimensional plastic surface on which they are grown, the presence of serum in the culture medium (that contains many cytokines and signaling molecules), and hyperoxia, all contribute to a pronounced cellular stress and affect subsequent adaptation. Several sources of additional stress have to be taken into account and minimized when studying the effect of OGD on astrocyte activation.

The key regulatory factor for the OGD model is the level of oxygen: ambient oxygen generates a partial pressure of 159 mmHg (dry air at sea level, 760 mmHg atmospheric pressure) while the arterial and tissue partial pressure are around 100 and 25 mmHg, respectively (132). The standard culture conditions are therefore hyperoxic and may cause preconditioning of the astrocytes. This could explain why astrocytes become more susceptible to OGD in aged cultures, when the preconditioning effect has faded due to adaptation. To study the effect of astrocyte intermediate filaments on OGD, cultures from genetically modified animals or siRNA transfection techniques are useful approaches.

2.6. Oxygen and Glucose Deprivation as an In Vitro Stroke Model: Protocol

2.6.1. Material and Equipment Needed

Dissection materials: Microscope, fine forceps, curved forceps, scissors, and fine scissors.

Media: Astrocyte medium (DMEM, 10% fetal bovine serum, 100 U/ml penicillin and 0.1 mg/ml streptomycin (P/S), and 2 mM L-glutamine), neuronal medium (Neurobasal (GIBCO), B27 (GIBCO), P/S, and 2 mM L-glutamine). Adjust the pH with CO_2 prior to use if required. Keep the B27, serum, 100× P/S, and glutamine stocks frozen at −20°C, avoid repeated freeze–thaw cycles.

Buffers: Phosphate-buffered saline (PBS), Hanks balanced salt solution (HBSS), OGD buffer: 51 mM NaCl, 65 mM K-gluconate, 0.13 mM $CaCl_2$, 1.5 mM $MgCl_2$, 10 mM HEPES pH 6.8.

Enzymes: Trypsin–EDTA 0.25% frozen stock, DNaseI 5 mg/ml frozen stock, TrypLE (GIBCO).

Plasticware: 35 mm plates for dissection, 15 ml centrifuge tubes, poly-D-lysine-coated T75 flasks, poly-D-lysine-coated 24-well plates for the cell cultures. Special care must be taken when growing neurons alone or on top of glass coverslips, coating the surface twice is recommended.

2.6.2. Astrocyte Culture Method

1. Kill 0–2-day-old mice by decapitation.
2. Sterilize the head by brief immersion in 70% ethanol and transfer to a 35 mm plate containing sterile PBS.
3. Under sterile conditions, cut the skin along the midline from the occipital area to the interocular line. Fold it to the side.

4. Cut the skull along the midline and to the sides in the occipital area. Fold the bone to the side to make enough space to remove the brain.
5. Dissect the desired brain area and remove the meninges. In vivo and in vitro, the astrocyte sensitivity to ischemia is different depending on the area of the brain (126). It is thus important to isolate the same brain region consistently between experiments.
6. Transfer the tissue to a sterile 15 ml tube with PBS. Continue under the laminar flow hood. Rinse with 5 ml PBS.
7. Incubate with 0.12% trypsin for 5–10 min at 37°C depending on the size of the tissue.
8. Aspirate the trypsin and rinse once with 2 ml astrocyte medium.
9. Triturate the tissue with a 1 ml pipette tip in 2 ml of astrocyte medium supplemented with DNAse1.
10. Plate in poly-D-lysine-coated flasks: 2 cortices into one T75 flask in 12 ml astrocyte medium.
11. Change the medium after 24 h. Shake the flasks vigorously to detach microglia and change the media every 2 days until confluency is reached, typically 7–10 days. High confluency and insufficient medium changes lead to an increase in the percentage of microglia; we have observed that this increases the culture's susceptibility to OGD.
12. Rinse with 6 ml PBS, add 3 ml TrypLE, and incubate for 10 min at 37°C. Detach the cells by shaking. We observe better survival rates when detaching the cells with TrypLE instead of trypsin.
13. Add 5 ml of astrocyte medium and transfer the cell suspension to a 15 ml tube. Spin at 300 × *g* for 4 min and resuspend the cells in fresh growing medium. We do not recommend freezing and replating primary astrocytes at this step. It can lead to a decreased response to OGD.
14. Seed the cells at an initial confluency of 40,000 cells/well in poly-D-lysine-coated 24-well cell culture plates. Incubate in 0.5 ml astrocyte media in a 5% CO_2 humidified incubator.
15. Change medium every 2 days.
16. At 90% confluency, typically after 5 days, proceed to coculture with neurons.

2.6.3. Neuronal Culture Method

1. Perform a caesarian section on a 16-day pregnant mice, killed or under deep anesthesia.
2. Extract the uterus and isolate the individual embryonic sacs. Decapitate each embryo and transfer the heads into a tube with sterile HBSS on ice.

3. Under a dissection microscope, cut the skull along the midline and to the sides. Remove the brain. Transfer it to a new 35 mm plate with cold HBSS.
4. Dissect the desired brain region and remove the meninges. Transfer it to a sterile 15 ml tube containing cold HBSS. Continue in the laminar flow hood.
5. Rinse the tissue pieces with 5 ml cold HBSS.
6. Incubate with 0.12 trypsin for 5–10 min at 37°C.
7. Aspirate the trypsin and rinse once with 2 ml astrocyte medium.
8. Add 1 ml of astrocyte medium containing DNAse1.
9. Dissociate the tissue with the help of a fire-polished Pasteur pipette. Do not over-triturate the tissue as it will result in neuronal death and a lot of cell debris. Instead, if large pieces of tissue remain, transfer the floating dissociated cells to a new tube and repeat from step 8.
10. Spin down at $250\times g$ for 5 min, discard the supernatant, and resuspend the pellet.
11. Pass through a 40 μm nylon mesh and count the cells.
12. Plate neurons on top of the astrocytes or in poly-D-lysine double-coated 24-well plates, 100,000 cells/well.
13. Next day change the medium to neuronal medium. It is useful to replace serum with a defined serum-free supplement that supports astrocytes; besides reducing the baseline activation it reduces variability between media batches and it does not interfere with downstream assays such as LDH.
14. Change half the medium every 2 days thereafter. Avoid imposing additional stress to the cells by using warm and CO_2-equilibrated medium. Replacing half of the medium in each well avoids drying.
15. Neuronal projections, or neurites, will start forming the day after plating. By the third day they will form an interconnected network.

2.6.4. OGD Protocol

1. Once the neuronal culture is mature, at least 7 days in vitro, initiate OGD by replacing the media with warm, nitrogen-bubbled OGD buffer. Use a minimum amount of medium so that the gas exchange in the incubator is fast.
2. Incubate in 1% oxygen, 5% CO_2, 94% nitrogen-humidified incubator for 16 h. Control cultures receive a media change and are incubated in the regular incubator. Extreme care has to be taken when aspirating the media from each well so as not to damage the neurites. At the same time, the medium has to be replaced immediately after so that the cells do not dry. It is advisable to change one well at a time and have the pipette ready with the new medium while aspirating the old one.

3. Initiate reperfusion by changing the medium to neuronal medium and transferring the plates to the regular incubator.
4. Morphological changes can be observed from time 0 of reperfusion as a loss of contact between adjacent astrocytes and deterioration of neurites.
5. Take a supernatant aliquot for cell viability and lyse or fix the cells at different time points of reperfusion. Cell death can be determined by the addition of propidium iodide before fixation. Cell lysates can be used for mRNA/protein quantification.

Acknowledgements

This work was supported by Swedish Research Council (11548 to MPy, 20116 to MPa), ALF Gothenburg (11267 to MPa and 146051 to MPy), STENA Foundation, AFA Insurance, the EU projects *EduGlia* (237956 to MPy), *TargetBraIn* (279017 to MPy) and *NanoNet* COST Action (BM1002), and R. and T. Söderberg's, E. Jacobson's, and R. and U. Amlöv's Foundations.

References

1. Eddleston M, Mucke L (1993) Molecular profile of reactive astrocytes–implications for their role in neurologic disease. Neuroscience 54:15–36
2. Eng LF, Ghirnikar RS (1994) GFAP and astrogliosis. Brain Pathol 4:229–237
3. Hernandez MR, Agapova OA, Yang P et al (2002) Differential gene expression in astrocytes from human normal and glaucomatous optic nerve head analyzed by cDNA microarray. Glia 38:45–64
4. Pekny M, Nilsson M (2005) Astrocyte activation and reactive gliosis. Glia 50:427–434
5. Wilhelmsson U, Bushong EA, Price DL et al (2006) Redefining the concept of reactive astrocytes as cells that remain within their unique domains upon reaction to injury. Proc Natl Acad Sci U S A 103:17513–17518
6. Sofroniew MV (2009) Molecular dissection of reactive astrogliosis and glial scar formation. Trends Neurosci 32:638–647
7. Sofroniew MV, Vinters HV (2010) Astrocytes: biology and pathology. Acta Neuropathol 119:7–35
8. Wilhelmsson U, Li L, Pekna M et al (2004) Absence of glial fibrillary acidic protein and vimentin prevents hypertrophy of astrocytic processes and improves post-traumatic regeneration. J Neurosci 24:5016–5021
9. Sofroniew MV (2005) Reactive astrocytes in neural repair and protection. Neuroscientist 11:400–407
10. Maragakis NJ, Rothstein JD (2006) Mechanisms of disease: astrocytes in neurodegenerative disease. Nat Clin Pract Neurol 2:679–689
11. Seifert G, Schilling K, Steinhauser C (2006) Astrocyte dysfunction in neurological disorders: a molecular perspective. Nat Rev Neurosci 7:194–206
12. Correa-Cerro LS, Mandell JW (2007) Molecular mechanisms of astrogliosis: new approaches with mouse genetics. J Neuropathol Exp Neurol 66:169–176
13. Barres BA (2008) The mystery and magic of glia: a perspective on their roles in health and disease. Neuron 60:430–440
14. Li L, Lundkvist A, Andersson D et al (2008) Protective role of reactive astrocytes in brain ischemia. J Cereb Blood Flow Metab 28:468–481
15. Macauley SL, Pekny M, Sands MS (2011) The role of attenuated astrocyte activation in infantile neuronal ceroid lipofuscinosis. J Neurosci 31:15575–15585
16. Kosaka T, Hama K (1986) Three-dimensional structure of astrocytes in the rat dentate gyrus. J Comp Neurol 249:242–260

17. Grosche J, Matyash V, Moller T et al (1999) Microdomains for neuron-glia interaction: parallel fiber signaling to Bergmann glial cells. Nat Neurosci 2:139–143
18. Bushong EA, Martone ME, Jones YZ et al (2002) Protoplasmic astrocytes in CA1 stratum radiatum occupy separate anatomical domains. J Neurosci 22:183–192
19. Bushong EA, Martone ME, Ellisman MH (2004) Maturation of astrocyte morphology and the establishment of astrocyte domains during postnatal hippocampal development. Int J Dev Neurosci 22:73–86
20. Haber M, Murai KK (2006) Reshaping neuron-glial communication at hippocampal synapses. Neuron Glia Biol 2:59–66
21. Simard M, Nedergaard M (2004) The neurobiology of glia in the context of water and ion homeostasis. Neuroscience 129:877–896
22. Pellerin L, Bouzier-Sore AK, Aubert A et al (2007) Activity-dependent regulation of energy metabolism by astrocytes: an update. Glia 55:1251–1262
23. Rouach N, Koulakoff A, Abudara V et al (2008) Astroglial metabolic networks sustain hippocampal synaptic transmission. Science 322:1551–1555
24. Suzuki A, Stern SA, Bozdagi O et al (2011) Astrocyte-neuron lactate transport is required for long-term memory formation. Cell 144: 810–823
25. Ullian EM, Sapperstein SK, Christopherson KS et al (2001) Control of synapse number by glia. Science 291:657–661
26. Ullian EM, Christopherson KS, Barres BA (2004) Role for glia in synaptogenesis. Glia 47:209–216
27. Christopherson KS, Ullian EM, Stokes CC et al (2005) Thrombospondins are astrocyte-secreted proteins that promote CNS synaptogenesis. Cell 120:421–433
28. Eroglu C, Barres BA (2010) Regulation of synaptic connectivity by glia. Nature 468: 223–231
29. Liebner S, Czupalla CJ, Wolburg H (2011) Current concepts of blood–brain barrier development. Int J Dev Biol 55:467–476
30. Zonta M, Angulo MC, Gobbo S et al (2003) Neuron-to-astrocyte signaling is central to the dynamic control of brain microcirculation. Nat Neurosci 6:43–50
31. Mulligan SJ, MacVicar BA (2004) Calcium transients in astrocyte endfeet cause cerebrovascular constrictions. Nature 431:195–199
32. Oberheim NA, Takano T, Han X et al (2009) Uniquely hominid features of adult human astrocytes. J Neurosci 29:3276–3287
33. Matyash V, Kettenmann H (2010) Heterogeneity in astrocyte morphology and physiology. Brain Res Rev 63:2–10
34. Zhang Y, Barres BA (2010) Astrocyte heterogeneity: an underappreciated topic in neurobiology. Curr Opin Neurobiol 20:588–594
35. Beckervordersandforth R, Tripathi P, Ninkovic J et al (2010) In vivo fate mapping and expression analysis reveals molecular hallmarks of prospectively isolated adult neural stem cells. Cell Stem Cell 7:744–758
36. Ståhlberg A, Andersson D, Aurelius J et al. (2011) Defining cell populations with single-cell gene expression profiling: correlations and identification of astrocyte subpopulations. Nucleic Acids Res 39:e24
37. Eng LF, Ghirnikar RS, Lee YL (2000) Glial fibrillary acidic protein: GFAP-thirty-one years (1969–2000). Neurochem Res 25:1439–1451
38. Okada S, Nakamura M, Katoh H et al (2006) Conditional ablation of Stat3 or Socs3 discloses a dual role for reactive astrocytes after spinal cord injury. Nat Med 12:829–834
39. Herrmann JE, Imura T, Song B et al (2008) STAT3 is a critical regulator of astrogliosis and scar formation after spinal cord injury. J Neurosci 28:7231–7243
40. Renault-Mihara F, Okada S, Shibata S et al (2008) Spinal cord injury: emerging beneficial role of reactive astrocytes' migration. Int J Biochem Cell Biol 40:1649–1653
41. Voskuhl RR, Peterson RS, Song B et al (2009) Reactive astrocytes form scar-like perivascular barriers to leukocytes during adaptive immune inflammation of the CNS. J Neurosci 29: 11511–11522
42. Silver J, Miller JH (2004) Regeneration beyond the glial scar. Nat Rev Neurosci 5:146–156
43. Delaney CL, Brenner M, Messing A (1996) Conditional ablation of cerebellar astrocytes in postnatal transgenic mice. J Neurosci 16: 6908–6918
44. Bush TG, Puvanachandra N, Horner CH et al (1999) Leukocyte infiltration, neuronal degeneration, and neurite outgrowth after ablation of scar-forming, reactive astrocytes in adult transgenic mice. Neuron 23:297–308
45. Sofroniew MV, Bush TG, Blumauer N et al (1999) Genetically-targeted and conditionally-regulated ablation of astroglial cells in the central, enteric and peripheral nervous systems in adult transgenic mice. Brain Res 835:91–95
46. Faulkner JR, Herrmann JE, Woo MJ et al (2004) Reactive astrocytes protect tissue

and preserve function after spinal cord injury. J Neurosci 24:2143–2155
47. Garcia AD, Doan NB, Imura T et al (2004) GFAP-expressing progenitors are the principal source of constitutive neurogenesis in adult mouse forebrain. Nat Neurosci 7:1233–1241
48. Eliasson C, Sahlgren C, Berthold CH et al (1999) Intermediate filament protein partnership in astrocytes. J Biol Chem 274: 23996–24006
49. Pekny M, Johansson CB, Eliasson C et al (1999) Abnormal reaction to central nervous system injury in mice lacking glial fibrillary acidic protein and vimentin. J Cell Biol 145: 503–514
50. Lu YB, Iandiev I, Hollborn M et al (2011) Reactive glial cells: increased stiffness correlates with increased intermediate filament expression. FASEB J 25:624–631
51. Lepekhin EA, Eliasson C, Berthold CH et al (2001) Intermediate filaments regulate astrocyte motility. J Neurochem 79:617–625
52. Potokar M, Kreft M, Li L et al (2007) Cytoskeleton and vesicle mobility in astrocytes. Traffic 8:12–20
53. Potokar M, Stenovec M, Gabrijel M et al (2010) Intermediate filaments attenuate stimulation-dependent mobility of endosomes/ lysosomes in astrocytes. Glia 58:1208–1219
54. Ding M, Eliasson C, Betsholtz C et al (1998) Altered taurine release following hypotonic stress in astrocytes from mice deficient for GFAP and vimentin. Brain Res Mol Brain Res 62:77–81
55. Cho KS, Yang L, Lu B et al (2005) Re-establishing the regenerative potential of central nervous system axons in postnatal mice. J Cell Sci 118:863–872
56. Menet V, Prieto M, Privat A et al (2003) Axonal plasticity and functional recovery after spinal cord injury in mice deficient in both glial fibrillary acidic protein and vimentin genes. Proc Natl Acad Sci U S A 100: 8999–9004
57. Kinouchi R, Takeda M, Yang L et al (2003) Robust neural integration from retinal transplants in mice deficient in GFAP and vimentin. Nat Neurosci 6:863–868
58. Widestrand A, Faijerson J, Wilhelmsson U et al (2007) Increased neurogenesis and astrogenesis from neural progenitor cells grafted in the hippocampus of GFAP−/− Vim−/− mice. Stem Cells 25:2619–2627
59. Davies SJ, Goucher DR, Doller C et al (1999) Robust regeneration of adult sensory axons in degenerating white matter of the adult rat spinal cord. J Neurosci 19:5810–5822
60. Goldshmit Y, Galea MP, Wise G et al (2004) Axonal regeneration and lack of astrocytic gliosis in EphA4-deficient mice. J Neurosci 24:10064–10073
61. Busch SA, Silver J (2007) The role of extracellular matrix in CNS regeneration. Curr Opin Neurobiol 17:120–127
62. Fitch MT, Silver J (2008) CNS injury, glial scars, and inflammation: inhibitory extracellular matrices and regeneration failure. Exp Neurol 209:294–301
63. Alilain WJ, Horn KP, Hu H et al (2011) Functional regeneration of respiratory pathways after spinal cord injury. Nature 475: 196–200
64. Myer DJ, Gurkoff GG, Lee SM et al (2006) Essential protective roles of reactive astrocytes in traumatic brain injury. Brain 129:2761–2772
65. Kimelberg HK, Nedergaard M (2010) Functions of astrocytes and their potential as therapeutic targets. Neurotherapeutics 7: 338–353
66. Parpura V, Heneka MT, Montana V et al (2012) Glial cells in (patho)physiology. J Neurochem 121:4–27
67. Pekna M, Pekny M, Nilsson M (2012) Modulation of neural plasticity as a basis for stroke rehabilitation. Stroke, in press.
68. Lovatt D, Sonnewald U, Waagepetersen HS et al (2007) The transcriptome and metabolic gene signature of protoplasmic astrocytes in the adult murine cortex. J Neurosci 27:12255–12266
69. Cahoy JD, Emery B, Kaushal A et al (2008) A transcriptome database for astrocytes, neurons, and oligodendrocytes: a new resource for understanding brain development and function. J Neurosci 28:264–278
70. Lipton SA, Gu Z, Nakamura T (2007) Inflammatory mediators leading to protein misfolding and uncompetitive/fast off-rate drug therapy for neurodegenerative disorders. Int Rev Neurobiol 82:1–27
71. Nilsson M, Pekny M (2007) Enriched environment and astrocytes in central nervous system regeneration. J Rehabil Med 39:345–352
72. Deller T, Frotscher M (1997) Lesion-induced plasticity of central neurons: sprouting of single fibres in the rat hippocampus after unilateral entorhinal cortex lesion. Prog Neurobiol 53:687–727
73. Turner DA, Buhl EH, Hailer NP et al (1998) Morphological features of the entorhinal-hippocampal connection. Prog Neurobiol 55:537–562
74. Deller T, Del Turco D, Rappert A et al (2007) Structural reorganization of the dentate gyrus

following entorhinal denervation: species differences between rat and mouse. Prog Brain Res 163:501–528

75. Steward O, Kelley MS, Torre ER (1993) The process of reinnervation in the dentate gyrus of adult rats: temporal relationship between changes in the levels of glial fibrillary acidic protein (GFAP) and GFAP mRNA in reactive astrocytes. Exp Neurol 124:167–183
76. Steward O, Torre ER, Phillips LL et al (1990) The process of reinnervation in the dentate gyrus of adult rats: time course of increases in mRNA for glial fibrillary acidic protein. J Neurosci 10:2373–2384
77. Rose G, Lynch G, Cotman CW (1976) Hypertrophy and redistribution of astrocytes in the deafferented dentate gyrus. Brain Res Bull 1:87–92
78. Krohn K, Laping NJ, Morgan TE et al (1995) Expression of vimentin increases in the hippocampus and cerebral cortex after entorhinal cortex lesioning and in response to transforming growth factor beta 1. J Neuroimmunol 56:53–63
79. Jing R, Wilhelmsson U, Goodwill W et al (2007) Synemin is expressed in reactive astrocytes in neurotrauma and interacts differentially with vimentin and GFAP intermediate filament networks. J Cell Sci 120:1267–1277
80. Jensen MB, Finsen B, Zimmer J (1997) Morphological and immunophenotypic microglial changes in the denervated fascia dentata of adult rats: correlation with blood–brain barrier damage and astroglial reactions. Exp Neurol 143:103–116
81. Dehn D, Burbach GJ, Schafer R et al (2006) NG2 upregulation in the denervated rat fascia dentata following unilateral entorhinal cortex lesion. Glia 53:491–500
82. Deller T, Haas CA, Frotscher M (2000) Reorganization of the rat fascia dentata after a unilateral entorhinal cortex lesion. Role of the extracellular matrix. Ann NY Acad Sci 911:207–220
83. Owens T, Babcock AA, Millward JM et al (2005) Cytokine and chemokine inter-regulation in the inflamed or injured CNS. Brain Res Brain Res Rev 48:178–184
84. Bechmann I, Nitsch R (1997) Astrocytes and microglial cells incorporate degenerating fibers following entorhinal lesion: a light, confocal, and electron microscopical study using a phagocytosis-dependent labeling technique. Glia 20:145–154
85. Vuksic M, Del Turco D, Vlachos A et al (2011) Unilateral entorhinal denervation leads to long-lasting dendritic alterations of mouse hippocampal granule cells. Exp Neurol 230:176–185, Epub 2011 Apr 22
86. Steward O, Cotman CW, Lynch GS (1974) Growth of a new fiber projection in the brain of adult rats: re-innervation of the dentate gyrus by the contralateral entorhinal cortex following ipsilateral entorhinal lesions. Exp Brain Res 20:45–66
87. Guthrie KM, Nguyen T, Gall CM (1995) Insulin-like growth factor-1 mRNA is increased in deafferented hippocampus: spatiotemporal correspondence of a trophic event with axon sprouting. J Comp Neurol 352: 147–160
88. Eijkenboom M, Blokland A, van der Staay FJ (2000) Modelling cognitive dysfunctions with bilateral injections of ibotenic acid into the rat entorhinal cortex. Neuroscience 101:27–39
89. Kadish I, Van Groen T (2003) Differences in lesion-induced hippocampal plasticity between mice and rats. Neuroscience 116:499–509
90. Kelley MS, Steward O (1996) The process of reinnervation in the dentate gyrus of adult rats: physiological events at the time of the lesion and during the early postlesion period. Exp Neurol 139:73–82
91. Gwag BJ, Sessler F, Kimmerer K et al (1994) Neurotrophic factor mRNA expression in dentate gyrus is increased following angular bundle transection. Brain Res 647:23–29
92. Deller T, Frotscher M, Nitsch R (1995) Morphological evidence for the sprouting of inhibitory commissural fibers in response to the lesion of the excitatory entorhinal input to the rat dentate gyrus. J Neurosci 15:6868–6878
93. Jensen MB, Hegelund IV, Poulsen FR et al (1999) Microglial reactivity correlates to the density and the myelination of the anterogradely degenerating axons and terminals following perforant path denervation of the mouse fascia dentata. Neuroscience 93:507–518
94. Del Turco D, Woods AG, Gebhardt C et al (2003) Comparison of commissural sprouting in the mouse and rat fascia dentata after entorhinal cortex lesion. Hippocampus 13:685–699
95. Sihlbom C, Wilhelmsson U, Li L et al (2007) 14-3-3 Expression in denervated hippocampus after entorhinal cortex lesion assessed by culture-derived isotope tags in quantitative proteomics. J Proteome Res 6:3491–3500
96. Rosenblum WI, El-Sabban F (1977) Effects of combined parenchymal and vascular injury on platelet aggregation in pial arterioles of living mice: evidence for release of aggregate-inhibiting materials. Stroke 8:691–693

97. Watson BD, Dietrich WD, Busto R et al (1985) Induction of reproducible brain infarction by photochemically initiated thrombosis. Ann Neurol 17:497–504
98. Dietrich WD, Watson BD, Busto R et al (1987) Photochemically induced cerebral infarction. I. Early microvascular alterations. Acta Neuropathol 72:315–325
99. Boquillon M, Boquillon JP, Bralet J (1992) Photochemically induced, graded cerebral infarction in the mouse by laser irradiation evolution of brain edema. J Pharmacol Toxicol Methods 27:1–6
100. Grome JJ, Gojowczyk G, Hofmann W et al (1988) Quantitation of photochemically induced focal cerebral ischemia in the rat. J Cereb Blood Flow Metab 8:89–95
101. Demougeot C, Bertrand N, Prigent-Tessier A et al (2003) Reversible loss of N-acetylaspartate in rats subjected to long-term focal cerebral ischemia. J Cereb Blood Flow Metab 23:482–489
102. Fujioka M, Taoka T, Matsuo Y et al (1999) Novel brain ischemic change on MRI. Delayed ischemic hyperintensity on T1-weighted images and selective neuronal death in the caudoputamen of rats after brief focal ischemia. Stroke 30:1043–1046
103. Yao H, Sugimori H, Fukuda K et al (2003) Photothrombotic middle cerebral artery occlusion and reperfusion laser system in spontaneously hypertensive rats. Stroke 34:2716–2721
104. Petito CK, Morgello S, Felix JC et al (1990) The two patterns of reactive astrocytosis in postischemic rat brain. J Cereb Blood Flow Metab 10:850–859
105. Brenner M (1994) Structure and transcriptional regulation of the GFAP gene. Brain Pathol 4:245–257
106. Schroeter M, Schiene K, Kraemer M et al (1995) Astroglial responses in photochemically induced focal ischemia of the rat cortex. Exp Brain Res 106:1–6
107. Schroeter M, Franke C, Stoll G et al (2001) Dynamic changes of magnetic resonance imaging abnormalities in relation to inflammation and glial responses after photothrombotic cerebral infarction in the rat brain. Acta Neuropathol 101:114–122
108. Nowicka D, Rogozinska K, Aleksy M et al (2008) Spatiotemporal dynamics of astroglial and microglial responses after photothrombotic stroke in the rat brain. Acta Neurobiol Exp 68:155–168
109. Snape MF, Baldwin HA, Cross AJ et al (1993) The effects of chlormethiazole and nimodipine on cortical infarct area after focal cerebral ischaemia in the rat. Neuroscience 53:837–844
110. Bailey SJ, Wood NI, Samson NA et al (1995) Failure of isradipine to reduce infarct size in mouse, gerbil, and rat models of cerebral ischemia. Stroke 26:2177–2183
111. De Ryck M, Keersmaekers R, Duytschaever H et al (1996) Lubeluzole protects sensorimotor function and reduces infarct size in a photochemical stroke model in rats. J Pharmacol Exp Ther 279:748–758
112. Chang YY, Fujimura M, Morita-Fujimura Y et al (1999) Neuroprotective effects of an antioxidant in cortical cerebral ischemia: prevention of early reduction of the apurinic/apyrimidinic endonuclease DNA repair enzyme. Neurosci Lett 277:61–64
113. Kharlamov A, Guidotti A, Costa E et al (1993) Semisynthetic sphingolipids prevent protein kinase C translocation and neuronal damage in the perifocal area following a photochemically induced thrombotic brain cortical lesion. J Neurosci 13:2483–2494
114. Wood NI, Sopesen BV, Roberts JC et al (1996) Motor dysfunction in a photothrombotic focal ischaemia model. Behav Brain Res 78:113–120
115. Shanina EV, Schallert T, Witte OW et al (2006) Behavioral recovery from unilateral photothrombotic infarcts of the forelimb sensorimotor cortex in rats: role of the contralateral cortex. Neuroscience 139:1495–1506
116. Sugimori H, Yao H, Ooboshi H et al (2004) Krypton laser-induced photothrombotic distal middle cerebral artery occlusion without craniectomy in mice. Brain Res Brain Res Protoc 13:189–196
117. Hu X, Wester P, Brannstrom T et al (2001) Progressive and reproducible focal cortical ischemia with or without late spontaneous reperfusion generated by a ring-shaped, laser-driven photothrombotic lesion in rats. Brain Res Brain Res Protoc 7:76–85
118. Schroeter M, Jander S, Huitinga I et al (1997) Phagocytic response in photochemically induced infarction of rat cerebral cortex. The role of resident microglia. Stroke 28:382–386
119. Schroeter M, Jander S, Witte OW et al (1994) Local immune responses in the rat cerebral cortex after middle cerebral artery occlusion. J Neuroimmunol 55:195–203
120. Jander S, Kraemer M, Schroeter M et al (1995) Lymphocytic infiltration and expression of intercellular adhesion molecule-1 in photochemically induced ischemia of the rat cortex. J Cereb Blood Flow Metab 15:42–51
121. Madinier A, Bertrand N, Mossiat C et al (2008) Microglial involvement in neuroplastic

changes following focal brain ischemia in rats. PLoS One 4:e8101

122. Buchkremer-Ratzmann I, Witte OW (1997) Pharmacological reduction of electrophysiological diaschisis after photothrombotic ischemia in rat neocortex. Eur J Pharmacol 320:103–109

123. Elliot KA, Rosenfeld M (1958) Anaerobic glycolysis in brain slices after deprivation of oxygen and glucose. Can J Biochem Physiol 36:721–730

124. Goldberg WJ, Kadingo RM, Barrett JN (1986) Effects of ischemia-like conditions on cultured neurons: protection by low Na+, low Ca2+ solutions. J Neurosci 6:3144–3151

125. Dugan LL, Kim-Han JS (2004) Astrocyte mitochondria in in vitro models of ischemia. J Bioenerg Biomembr 36:317–321

126. Xu L, Sapolsky RM, Giffard RG (2001) Differential sensitivity of murine astrocytes and neurons from different brain regions to injury. Exp Neurol 169:416–424

127. Almeida A, Delgado-Esteban M, Bolanos JP et al (2002) Oxygen and glucose deprivation induces mitochondrial dysfunction and oxidative stress in neurones but not in astrocytes in primary culture. J Neurochem 81:207–217

128. Ouyang YB, Xu LJ, Sun YJ et al (2006) Overexpression of inducible heat shock protein 70 and its mutants in astrocytes is associated with maintenance of mitochondrial physiology during glucose deprivation stress. Cell Stress Chaperones 11:180–186

129. Badawi Y, Ramamoorthy P, Shi H (2012) Hypoxia inducible factor 1 protects hypoxic astrocytes against glutamate toxicity. ASN Neuro 4:231–241

130. Vangeison G, Carr D, Federoff HJ et al (2008) The good, the bad, and the cell type-specific roles of hypoxia inducible factor-1 alpha in neurons and astrocytes. J Neurosci 28:1988–1993

131. Bondarenko A, Chesler M (2001) Calcium dependence of rapid astrocyte death induced by transient hypoxia, acidosis, and extracellular ion shifts. Glia 34:143–149

132. Maloney-Wilensky E, Gracias V, Itkin A et al (2009) Brain tissue oxygen and outcome after severe traumatic brain injury: a systematic review. Crit Care Med 37:2057–2063

Part V

Oligodendroglia and Myelin

Chapter 14

Microtubules and Associated Proteins in Oligodendrocytes, the Myelin Forming Cells of the Central Nervous System

Christiane Richter-Landsberg

Abstract

During the process of myelination, oligodendrocytes spirally enwrap neuronal axons and form multilamellar myelin sheets. Oligodendrocytes in culture are characterized by their numerous membranous extensions containing an elaborate network of microtubules, they are devoid of intermediate filaments, but microfilaments are prominent. To support cell morphology and for the establishment and maintenance of the myelin membrane, an intact, spatially organized cytoskeleton with dynamic properties is essential. A variety of microtubule-associated proteins (MAPs), including MAP2, MAP1B, and tau, regulating growth and stability of microtubules, are present in oligodendrocytes, and also the myelin-specific proteins MBP and CNP associate with microtubules. Furthermore, heat shock proteins interact with the cytoskeleton and in response to cellular stress may be recruited to the cytoskeleton. Glial dysfunction and glial pathology accompany a number of neurodegenerative diseases. During disease progression, proteins with stabilizing or destabilizing properties are recruited to the cytoskeleton and alter microtubule properties. A characteristic hallmark of many diseases is abnormal deposits of cytoskeletal proteins and the accumulation of cellular proteins, such as tau and/or α-synuclein. The identification of microtubule-associated proteins during health and disease will provide insights into our overall understanding of the regulation and mechanistic aspects of microtubule organization and functions. The method described in this chapter provides a rapid and simple way to identify proteins which bind to or are detached from microtubules under normal or pathological situations in living cells.

Key words: Myelin, Tubulin, Microtubule-associated proteins, Tau, Glial inclusion body

1. Introduction

Oligodendrocytes, the myelin forming cells in the central nervous system, are characterized by their numerous membranous extensions, which spirally wrap around neuronal axons and form multilamellar myelin sheaths. They derive from small proliferative precursor cells, migrating out of the germinal zones in the brain and spinal cord, a process which is guided by various regulated signals including growth factors, chemotropic molecules, and

Rolf Dermietzel (ed.), *The Cytoskeleton: Imaging, Isolation, and Interaction*, Neuromethods, vol. 79,
DOI 10.1007/978-1-62703-266-7_14, © Springer Science+Business Media, LLC 2013

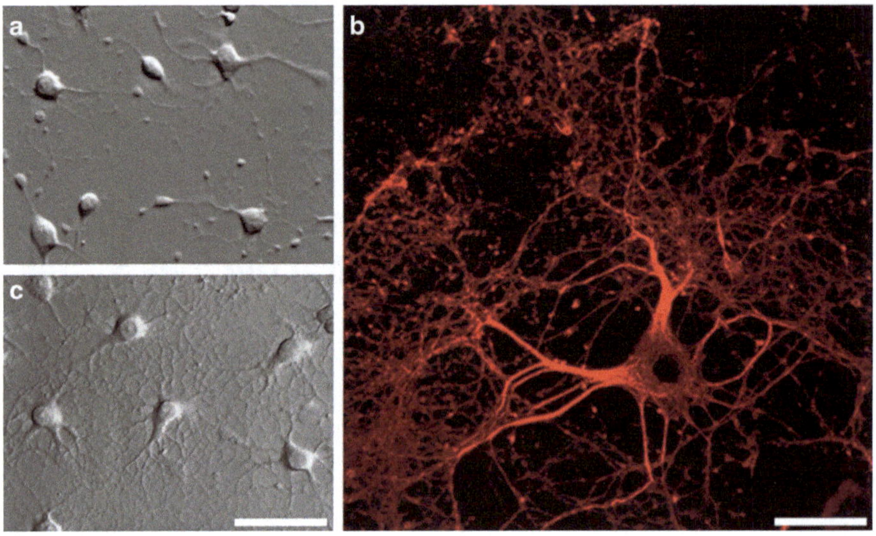

Fig. 1. Oligodendrocyte morphology and microtubule distribution. In (**a** and **c**) Hoffman modulation contrast images of oligodendrocytes derived from the brains of newborn rats were cultured for 2 h, representing immature precursor cells (**a**) or for 6 days, representing mature cells (**c**). Bar in (**a**, **c**) represents 50 μm. During culture maturation cells extend numerous cellular processes. In (**b**) cells after 6 days in culture were fixed with methanol and indirect immunofluorescence was carried out with antibodies against α-tubulin. The picture was taken with a laser scanning confocal microscope (Confocal system Leica TCS NT). Bar in (**b**) represents 25 μm. Note the elaborate network of microtubules extending throughout the cell body and the cellular processes.

contact-mediated mechanisms involving cell surface molecules and different proteins of the extracellular matrix (1–3). The proliferation and differentiation of the early progenitor cells into oligodendrocytes have been studied mainly in cell culture systems, however, many of the data are representative for the development in vivo. The differentiation of precursor cells to myelinating oligodendrocytes is spatially and temporally regulated, and marked by the sequential appearance of myelin-specific proteins. Myelination is a complex tightly regulated process. During the last years it has become evident that neurons and myelinating glia cells communicate with and influence each other through the exchange of signals (1, 4, 5). Oligodendrocytes select axons with a diameter above 0.2 μm and wrap multiple layers of plasma membrane around multiple axons. This process requires the synthesis, sorting and trafficking of large amounts of protein and mRNA in a relatively short time period (2, 3).

To support cell morphology and the establishment and maintenance of the metabolically active myelin membrane, an intact, spatially organized cytoskeleton with dynamic properties is essential (3, 6). Similar to developmental studies, data related to MT function have mainly been derived from in vitro studies. Oligodendrocytes, like nerve cells, display a complex architecture and have an elaborate network of microtubules (MT) (Fig. 1). They are devoid of an intermediate filament system, but microfilaments are prominent (7). The formation, extension, and growth of the oligodendrocyte processes depend on the formation and stability of the MTs. MT-disrupting agents, such as nocodazole, or MT-stabilizing agents, such as taxol, disturb the cell morphology and inhibit transport processes.

MTs provide the tracks for organelle trafficking and the intracellular translocation of myelin gene products including MBP mRNA, which then is translated near the myelin (8, 9). MTs in oligodendrocytes occur in distinct subpopulations, i.e., labile MTs are present in the fine processes and branches, as opposed to relatively stable MTs in the main branches and highly stable MTs in the cell soma (10). This indicates that to enable the early phase of myelin formation a more dynamic MT array may be established locally, while less dynamic MTs may provide structural support and mediate intracellular transport processes at later stages.

1.1. Microtubule-Associated Proteins in Oligodendrocytes

Microtubule-associated proteins (MAPs) regulate the growth and stability of MTs, they bind to the walls of MTs by means of tubulin binding domains which are located in the carboxyl terminal region of the protein. During repeated cycles of assembly and disassembly MAPs copurify with MTs (11). Furthermore, MAPs and proteins recruited to MTs for example during cellular stress or pathological situations, can be identified by isolating MTs from living cells after taxol stabilization (see Methods 2.1.1). In oligodendrocytes a variety of MAPs have been identified, including the high molecular weight family, namely MAP2c, MAP1B, and MAP4, and all six isoforms of the low molecular weight family of the MAP tau (12). Tau proteins are present in immature and mature oligodendrocytes. During in vitro differentiation tau heterogeneity is increased. A decrease in tau phosphorylation is observable, and in mature cells an increase of tau proteins with four MT binding domains (4R-tau) occurs (12). This promotes binding of tau to the MTs leading to an enhanced MT stability which is required during later stages of development to enable the maintenance of the cellular architecture (13).

As mentioned above, oligodendrocytes extend several processes which myelinate distinct axonal segments on different axons. The myelin membrane is well characterized (see also Chapter 15 this volume), it contains a high percentage of lipids and only a small number of major proteins, such as the proteolipid protein (PLP) and its alternatively spliced isoform DM20, the myelin basic proteins (MBP), and 2′,3′-cyclic nucleotide 3′-phosphohydrolase (CNP) which is downregulated in compact myelin. CNP is prominently expressed at the cytoplasmic leaflets of the surface membranes of myelinating oligodendrocytes, while excluded from the compact myelin. Its association with filamentous actin and tubulin has been described (14, 15). CNP binds preferentially to tubulin dimers, as compared to MTs, and promotes MT assembly, process formation and extension and thus has been suggested to function as an MT assembly myelin protein (16). MBP is the one specific myelin protein which is absolutely required for the formation of intact myelin. It is expressed in differentiated cultured oligodendrocytes throughout the flat membranous processes even in the absence of neurons. In compact myelin it is associated with the inner cytoplasmic membrane leaflets, the major dense line. MBP has been reported to function as a microtubule stabilizing protein in differentiated oligodendrocytes (17) and displays MT polymerizing activity, which is regulated by its phosphorylation status (18). Generally, MBP phosphorylation seems to be essential for its localization. Phosphorylated MBP is more prominent in the cell body, while dephosphorylated MBP is abundant in the sheets (18). Early work revealed that after detergent extraction, MBP is closely associated with MTs (14) and that MT disruption with colchicine promoted the disorganization of MBP distribution in cultured oligodendrocytes (15), and is recovered together with tubulin in detergent-insoluble extracts prepared from myelin (19). A recent study demonstrates that MBP can link MTs and actin filaments in vitro, and in vivo is colocalized with both filamentous systems (20). Since MBP is involved in plasma membrane reorganization during myelin formation and interacts with the cytoskeleton, it might play a role in signal transduction and may serve as a scaffold protein that binds other proteins to the cytoskeleton and tether MTs to the cytoplasmic surface of the oligodendrocyte membrane (20, 21). Taken together, the myelin-specific proteins MBP and CNP display MT binding activity and are involved in morphogenetic events, leading to the formation of myelin membranes most likely by interaction with the cytoskeleton.

1.2. Cytoskeletal Disturbances and Glial Pathology in Neurodegenerative Diseases

Filamentous protein inclusions commonly observed in nerve cells and glia are characteristic hallmarks of a variety of neurodegenerative diseases (22, 23). Besides a disease-specific cellular protein, inclusion bodies often contain cytoskeletal proteins, ubiquitin and a variety of heat shock proteins (HSPs) which not only function as molecular chaperones, but also may alter MT polymerization and stability and display properties of MAPs (24–26). In particular members of the HSP70 family were initially characterized as MT binding proteins attaching to polymerized tubulin, and the small HSPs interact with all cytoskeletal elements during health and disease. In response to cellular stress, induced HSPs may be recruited to the cytoskeleton and exert protective effects (26).

1.2.1. Deposits of Cytoskeletal Proteins in Nerve Cells and Glia

Abnormal deposits of cytoskeletal proteins in neurodegenerative diseases include intracellular inclusions of neuronal intermediate filaments or the MAP tau ((27), and chapter 9 this volume). In oligodendrocytes which do not contain an intermediate filament system, the aggregation of tau is a major pathological feature. The accumulation of misfolded proteins in aggregates is likely meant as an attempt to remove otherwise toxic material from the cytoplasm, a process initially aimed as a rescue mechanism and at later stages of the diseases the growing aggregates lead to cell death and neurodegeneration. Not only nerve cells but also astrocytes and oligodendrocytes play a role in the pathogenesis of neurodegenerative diseases. In tauopathies the MAP tau is the major cellular constituent of the lesions, which are most prominent in Alzheimer's disease (AD) and also found in frontotemporal dementias (FTD), such as Pick's disease (PiD), progressive supranuclear palsy (PSP), and corticobasal degeneration (CBD) (23). In AD fibrillary tau accumulations preferentially are formed in neurons, while tau-positive glial inclusions are consistent features in the brains of patients with FTDs. In FTDP-17, tau pathology is present in neurons and glia, while tau-positive glial inclusions, termed coiled bodies, originating in oligodendrocytes are specifically prominent in PSP and CBD (6, 23). Parkinson's disease (PD), dementia with Lewy bodies (DLB) and Multiple system atrophy (MSA) are neurodegenerative diseases collectively called α-synucleinopathies, since they are characterized by the presence of inclusions containing α-synuclein (23). Glial dysfunction and glial pathology accompany these diseases (28).

1.2.2. Glial Cytoplasmic Protein Inclusions

Glial cytoplasmic inclusions (GCIs) originating in oligodendrocytes are the pathological hallmark of MSA, a specific adult onset degenerative disease of the nervous system which has been classified as a primary oligodendrogliopathy (29). α-Synuclein is the major building block of GCIs, representing round or crescent shaped inclusions which are also positive for ubiquitin. GCIs in MSA are positive for α- and β-tubulin, stain intensely with antibodies against ubiquitin, αB-crystallin, and variable reports indicate that they are tau-positive (30).

αB-Crystallin and HSP25/27 belong to the family of small HSPs, which interact with all three cytoskeletal elements (31). HSP25 is specifically prominent in astrocytes. In Rosenthal fibers which represent the characteristic astrocytic inclusions in patients with Alexander's disease, HSP27, αB-crystallin, and ubiquitin coassemble with GFAP (31). As we have demonstrated, proteasomal inhibition by MG-132 induces upregulation of the small HSPs in astrocytes and oligodendrocytes. Immunoblot analysis of MTs isolated from astrocytes or oligodendrocytes after stabilization with taxol (see Methods below), depicts that a proportion of the stress-induced small HSPs bind to the MTs (31–33). In this respect, αB-crystallin is highly inducible, interacts with MTs upon stress, and is almost completely recruited to the MTs. This indicates that it assists in maintaining MT integrity during cellular stress.

Furthermore, proteasomal inhibition in cultured oligodendrocytes causes the formation of inclusion bodies which positively stain with thioflavin S, a histochemical dye specifically binding to crossed β-pleated sheet structures and used to identify fibrillary protein aggregates (32). At the same time, as demonstrated by the sequential extraction of cell lysates using buffers with increasing solubilization capacity, the solubility of tau was decreased and ubiquitinated proteins and αB-crystallin were prominently present in the least soluble fraction.

1.2.3. The Oligodendroglia-Specific Tubulin Polymerization Promoting Protein TPPP/p25α

Another oligodendrocyte-specific protein interacting with MTs has been identified in GCIs, namely p25α or the tubulin polymerization promoting protein (TPPP), which is present in myelin sheaths and most likely linked to myelination. It is considered an MT binding protein with the capability to stabilize the MT network and to promote MT bundling (34). P25α is a phosphoprotein, phosphorylation blocks its MT binding activity and ability to assemble MTs (35, 36). During oligodendrocyte differentiation it is upregulated, accumulates during proteasomal stress (37) and is associated with neuronal and glial inclusions in PD and MSA and other synucleinopathies (38). P25α is a potent inducer of α-synuclein aggregation. Coexpression of both proteins in the oligodendroglial cell line OLN-93 (39) caused α-synuclein aggregation and its phosphorylation at Ser-129, and the rapid disorganization of the MT network and retraction of MTs from the cell soma to the perinuclear region (40). Interestingly, also α-synuclein has been described as an MT-associated protein with MT-polymerizing activity in vitro (41), and has structural and functional similarities to the α-crystallin domain of small HSPs (42). Hence all three proteins, i.e. tau, α-synuclein, and p25α, display MT binding activity, are regulated by phosphorylation and significantly contribute to protein aggregate formation in neurodegenerative disorders.

2. Material and Methods

2.1. Microtubule Binding Assay to Identify Proteins Bound, Detached or Recruited to Microtubules After Stress Situations

In control or healthy cells approximately 50% of the tubulin is present as unpolymerized, soluble globular α- and β-dimer and 50% as polymerized MTs. Soluble tubulin and polymerized MTs can be separated and fractionated from living cells at 37°C by using an MT-stabilizing reassembly buffer (RAB) containing Taxol, GTP, and Triton-X-100 to permeabilize the cells (see below). This procedure enables the rapid isolation of polymerized MTs and the characterization of their bound proteins during control growth conditions and after stress or pathological situations, which might lead to the recruitment of, e.g., heat shock proteins and ubiquitin or the detachment of proteins which are normally associated with the MTs (Fig. 2). This procedure is also easily applicable when only limited amounts of cell material is available which generally is the case working with primary cultures of oligodendrocytes.

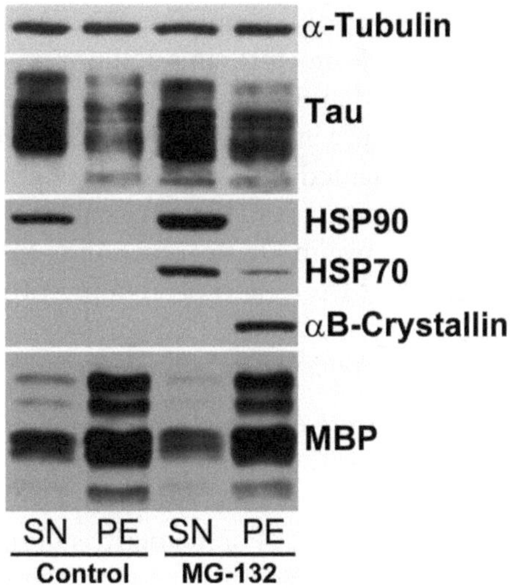

Fig. 2. Microtubule binding assay carried out with oligodendrocytes cultured under control conditions or after proteasomal stress induced by the treatment with MG-132. Oligodendrocytes (7 div) were either untreated (Control) or treated with MG-132 (1 μM, 24 h). Microtubules were isolated as described in the method section, and the supernatant (SN), representing the fraction of soluble cytoplasmic proteins not binding to the microtubles, and the pellet (PE) representing polymerized microtubules and their associated proteins were subjected to immunoblot procedure using antibodies against the proteins indicated on the right. Note that under control conditions the majority of the tau proteins is detectable in the supernatant (SN), MBP is mainly in the pellet (MT fraction) and HSP90 only in the soluble non bound fraction (SN). After proteasomal inhibition (MG-132), heat shock proteins (HSP) are induced and αB-crystallin is recruited to the microtubules and only found in the pellet, while HSP90 and HSP70 are mainly in the supernatant.

The experiments described below were carried out with cultured oligodendrocytes derived from the brains of newborn rats (12, 32). Towards this, cerebral hemispheres were freed of the meninges and mechanically disrupted using a Pasteur pipette. Single cell suspensions were transferred to culture flasks (1 brain/75 cm^2) and kept for 6–8 days in DMEM supplemented with 10% heat-inactivated fetal calf serum (FCS), 50 U/ml penicillin, and 50 μg/ml streptomycin in an incubator at 37°C at 10% CO_2. Oligodendrocytes were prepared from the flasks after 6–8 days when the cellular monolayer was completely covering the surface of the flask. Firstly, cellular debris and microglia growing loosely on top of the mixed glia monolayer were removed by mechanical shaking of the culture flasks for 2 h at 180 rpm using a gyratory shaker. The medium was discarded and the cellular monolayer was washed with PBS (phosphate-buffered saline). Thereafter, fresh growth medium was added, equilibrated in the CO_2 incubator (30 min), and the phase dark cells, representing O-2A progenitors, growing on the astrocytic cell layer were separated by vigorous shaking (for 16 h at 220 rpm) and taken off. This procedure was first described by McCarthy and DeVellis (43). Precursor cells were replated on poly-L-lysine-coated culture dishes (3×10^6 cells/10 cm dish) and kept in serum-free DMEM to which insulin (5 µg/ml), transferrin (5 µg/ml), and sodium selenite (25 ng/ml) (Roche Diagnostics, Mannheim, Germany) was added. Twice per week half of the culture medium was exchanged. After 1 week, these cultures contained differentiated oligodendrocytes with numerous cellular processes (Fig. 1), staining positive for galactocerebroside (GC), myelin basic protein (MBP), myelin-associated glycoprotein (MAG), and proteolipid protein (PLP) antibodies.

2.1.1. Isolation of Microtubules and Soluble Tubulin from Oligodendrocytes

Materials

Preparation of the solutions

Stock solutions for the lysis buffer: 10× RAB (final concentrations are given in square brackets)

Tris [1 M]; $MgSO_4$ [5 mM]; EGTA [10 mM]; NaCl [20 mM], pH 7.0; Phenylmethylsulfonylfluorid (PMSF) (100 mM) in methanol; Boehringer Complete Mini 1 tablet in 1.5 ml aqua bidest (Roche Diagnostics, Mannheim, Germany); taxol (5 mM) (Sigma, St. Louis, USA) in ethanol; GTP (100 µM) in aqua bidest; Triton X-100 10% in PBS.

Working solution of the lysis buffer: 1× RAB

Prepare from the stock solutions to yield the final concentration as given in above:

100 µl RAB (10×), 10 µl PMSF (100 mM), 40 µl GTP (100 µM), 4 µl Taxol (5 mM), 150 µl Boehringer Complete, 10 µl Triton X-100 (10%), add aqua bidest to a final volume of 1 ml.

Sample buffer: 5× Sample buffer (5× SB) (final concentrations are given in square brackets)

5 ml Glycerin (50%); 0.5 ml β-Mercaptoethanol (0.5%); 1 g SDS (10%); 3.125 ml Tris (2 M, pH 6.8) (625 mM); 100 µl EDTA (0.5 M, pH 8.0) (5 mM); 200 µl Bromophenol blue (2.5%) (0.05%); add Aqua bidest to a final volume of 10 ml, pH 6.8.

2.1.2. Method

Cells are grown on poly-L-lysine-coated culture dishes (3×10^6 cells/10 cm dish) for 5 days as described above. To isolate MTs, all solutions and glass ware need to be prewarmed to 37°C to avoid MT depolymerization. Cells are washed with PBS twice and scraped off in 150 µl/10 cm dish RAB pH 7.0 (final concentration: 100 mM Tris, 0.5 mM $MgSO_4$, 1.0 mM EGTA, 2.0 mM Dithiothreitol, 750 mM NaCl; 0.1% Triton X-100, 20 µM Taxol, 2 mM GTP, supplemented with protease inhibitor mixture (Complete; Roche Diagnostics, Mannheim, Germany) and PMSF, see above). The cell suspension is then homogenized in a prewarmed, small glass homogenizer (Dounce homogenizer) with 15 strokes up and down. Immediately thereafter, the suspension is transferred into Beckmann centrifuge tubes (1.5 ml) and centrifuged at 50,000 g for 20 min at 37°C (it is important to use special, durable tubes, since ordinary Eppendorf tubes will break under these conditions). The supernatant (SN) containing the soluble, nonbound fraction is removed into a new tube, and the pellet (P), representing the cytoskeletal fraction containing polymerized microtubules, is stored on ice. The protein concentration in the supernatant is determined. Then the supernatants (SN) of all samples are adjusted to the sample with the lowest protein concentration by adding 1× RAB. The volume of all SN samples is determined and the pellet fractions (P) are taken up in half or a third of the volume of the corresponding SN fraction. This depends on the amount of tubulin and the number of cells one started with and needs to be assessed individually. We try to have equal amounts of tubulin in the pellet and supernatant fraction in the control cells, which makes it easier to compare the relative amounts of binding proteins. 5× Sample buffer (5× SB) is added to all samples and equal amounts are loaded on SDS gels and analyzed by SDS polyacrylamide gel electrophoresis followed by immunoblot procedure (Fig. 2). We load approximately 10 μg of protein each of the supernatant, and 10–20 μl of the corresponding pellet fraction. The volume of the pellet fractions to be loaded varies, depending on the experimental conditions, and need to be adjusted accordingly.

The ratio of unbound proteins (SN) and proteins bound to the MTs (P) is determined by comparing the amount of the proteins in the two fractions in relation to tubulin, and can be quantified by densitometric scanning and calculated as percent bound and unbound in relation to the tubulin immunoreactivity.

3. Perspectives

The oligodendroglia cytoskeleton plays a major role in oligodendrocyte differentiation and myelination (3). A variety of microtubule binding proteins have been described in oligodendrocytes, these include major MAPs, such as MAP1B and tau, and major myelin proteins, such as MBP and CNP. Furthermore, deposits of cytoskeletal proteins in association with, e.g., heat shock proteins are observed in neurodegenerative disorders. During disease progression, proteins with stabilizing or destabilizing properties are recruited to the cytoskeleton and alter MT properties. The procedure described above provides a rapid and simple way to identify proteins which are associated with MTs under normal or pathological situations in living cells. This procedure can also give insight into MT stability as it reflects the status quo of the cells at the time of isolation, e.g., the amount or mass of polymerized and unpolymerized tubulin during specific experimental conditions. However, one needs to keep in mind that taxol, which is often used for in vitro assembly of microtubules and in vitro MT binding assays, may slightly alter tubulin assembly and MT binding properties (11). In this respect, for control experiments it is recommended to use a taxol-free buffer supplemented with glycerol. Taken together, studies characterizing MT-associated proteins in oligodendrocytes and assessing their binding activities during health and disease, will provide a deeper understanding of the mechanisms of cytoskeletal regulation and its role in myelination and demyelination processes.

Acknowledgments

This work was supported by a grant of the Deutsche Forschungsgemeinschaft, Germany. I thank Dr. Olaf Goldbaum for help with the graphic art.

References

1. Bradl M, Lassmann H (2010) Oligodendrocytes biology and pathology. Acta Neuropathol 119(1):37–53
2. Baron W, Hoekstra D (2009) On the biogenesis of myelin membranes: sorting, trafficking and cell polarity. FEBS Lett 584(9):1760–1770
3. Bauer NG, Richter-Landsberg C, Ffrench-Constant C (2009) Role of the oligodendroglial cytoskeleton in differentiation and myelination. Glia 57(16):1691–1705
4. Taveggia C, Feltri ML, Wrabetz L (2010) Signals to promote myelin formation and repair. Nat Rev Neurol 6(5):276–287
5. Simons M, Trajkovic K (2006) Neuron-glia communication in the control of oligodendrocyte function and myelin biogenesis. J Cell Sci 119(Pt 21):4381–4389
6. Richter-Landsberg C (2008) The cytoskeleton in oligodendrocytes. Microtubule dynamics in health and disease. J Mol Neurosci 35(1):55–63
7. Song J, Goetz BD, Baas PW, Duncan ID (2001) Cytoskeletal reorganization during the formation of oligodendrocyte processes and branches. Mol Cell Neurosci 17(4):624–636
8. Carson JH, Worboys K, Ainger K, Barbarese E (1997) Translocation of myelin basic protein mRNA in oligodendrocytes requires microtubules and kinesin. Cell Motil Cytoskeleton 38(4):318–328
9. Smith R (2004) Moving molecules: mRNA trafficking in Mammalian oligodendrocytes and neurons. Neuroscientist 10(6):495–500
10. Lunn KF, Baas PW, Duncan ID (1997) Microtubule organization and stability in the oligodendrocyte. J Neurosci 17(13):4921–4932
11. Kiris E, Ventimiglia D, Feinstein SC (2010) Quantitative analysis of MAP-mediated regulation of microtubule dynamic instability in vitro focus on Tau. Methods Cell Biol 95:481–503
12. Gorath M, Stahnke T, Mronga T, Goldbaum O, Richter-Landsberg C (2001) Developmental changes of tau protein and mRNA in cultured rat brain oligodendrocytes. Glia 36(1):89–101
13. Buee L, Bussiere T, Buee-Scherrer V, Delacourte A, Hof PR (2000) Tau protein isoforms, phosphorylation and role in neurodegenerative disorders. Brain Res Brain Res Rev 33(1):95–130
14. Wilson R, Brophy PJ (1989) Role for the oligodendrocyte cytoskeleton in myelination. J Neurosci Res 22(4):439–448
15. Dyer CA, Benjamins JA (1989) Organization of oligodendroglial membrane sheets. I: association of myelin basic protein and 2′,3′-cyclic nucleotide 3′-phosphohydrolase with cytoskeleton. J Neurosci Res 24(2):201–211
16. Lee J, Gravel M, Zhang R, Thibault P, Braun PE (2005) Process outgrowth in oligodendrocytes is mediated by CNP, a novel microtubule assembly myelin protein. J Cell Biol 170(4): 661–673
17. Galiano MR, Andrieux A, Deloulme JC, Bosc C, Schweitzer A, Job D, Hallak ME (2006) Myelin basic protein functions as a microtubule stabilizing protein in differentiated oligodendrocytes. J Neurosci Res 84(3):534–541
18. Boggs JM (2006) Myelin basic protein: a multifunctional protein. Cell Mol Life Sci 63: 1945–1961
19. Gillespie CS, Wilson R, Davidson A, Brophy PJ (1989) Characterization of a cytoskeletal matrix associated with myelin from rat brain. Biochem J 260(3):689–696
20. Boggs JM, Rangaraj G, Heng YM, Liu Y, Harauz G (2011) Myelin basic protein binds microtubules to a membrane surface and to actin filaments in vitro: effect of phosphorylation and deimination. Biochim Biophys Acta 1808(3):761–773

21. Fitzner D, Schneider A, Kippert A, Mobius W, Willig KI, Hell SW, Bunt G, Gaus K, Simons M (2006) Myelin basic protein-dependent plasma membrane reorganization in the formation of myelin. EMBO J 25(21):5037–5048
22. Dohm CP, Kermer P, Bahr M (2008) Aggregopathy in neurodegenerative diseases: mechanisms and therapeutic implication. Neurodegener Dis 5(6):321–338
23. Jellinger KA (2009) Recent advances in our understanding of neurodegeneration. J Neural Transm 116(9):1111–1162
24. Liang P, MacRae TH (1997) Molecular chaperones and the cytoskeleton. J Cell Sci 110(Pt 13):1431–1440
25. Richter-Landsberg C (2007) Heat shock proteins: expression and functional roles in nerve cells and glia. In: Richter-Landsberg C (ed) Heat shock proteins in neural cells. Springer/Landes Bioscience, New York/Austin, TX, pp 1–12
26. Richter-Landsberg C, Goldbaum O (2003) Stress proteins in neural cells: functional roles in health and disease. Cell Mol Life Sci 60(2):337–349
27. Cairns NJ, Lee VM, Trojanowski JQ (2004) The cytoskeleton in neurodegenerative diseases. J Pathol 204(4):438–449
28. Fellner L, Jellinger KA, Wenning GK, Stefanova N (2011) Glial dysfunction in the pathogenesis of alpha-synucleinopathies: emerging concepts. Acta Neuropathol 121(6):675–693
29. Wenning GK, Stefanova N, Jellinger KA, Poewe W, Schlossmacher MG (2008) Multiple system atrophy: a primary oligodendrogliopathy. Ann Neurol 64(3):239–246
30. Jellinger KA, Lantos PL (2010) Papp-Lantos inclusions and the pathogenesis of multiple system atrophy: an update. Acta Neuropathol 119(6):657–667
31. Richter-Landsberg C, Goldbaum O (2007) Small heat shock proteins and the cytoskeleton. In: Richter-Landsberg C (ed) Heat shock proteins in neural cells. Springer/Landes Bioscience, New York/Austin, TX, pp 13–24
32. Goldbaum O, Richter-Landsberg C (2004) Proteolytic stress causes heat shock protein induction, tau ubiquitination, and the recruitment of ubiquitin to tau-positive aggregates in oligodendrocytes in culture. J Neurosci 24(25):5748–5757
33. Bauer NG, Richter-Landsberg C (2006) The dynamic instability of microtubules is required for aggresome formation in oligodendroglial cells after proteolytic stress. J Mol Neurosci 29(2):153–168
34. Vincze O, Tokesi N, Olah J, Hlavanda E, Zotter A, Horvath I, Lehotzky A, Tirian L, Medzihradszky KF, Kovacs J et al (2006) Tubulin polymerization promoting proteins (TPPPs): members of a new family with distinct structures and functions. Biochemistry 45(46):13818–13826
35. Acevedo K, Li R, Soo P, Suryadinata R, Sarcevic B, Valova VA, Graham ME, Robinson PJ, Bernard O (2007) The phosphorylation of p25/TPPP by LIM kinase 1 inhibits its ability to assemble microtubules. Exp Cell Res 313(20):4091–4106
36. Hlavanda E, Klement E, Kokai E, Kovacs J, Vincze O, Tokesi N, Orosz F, Medzihradszky KF, Dombradi V, Ovadi J (2007) Phosphorylation blocks the activity of tubulin polymerization-promoting protein (TPPP): identification of sites targeted by different kinases. J Biol Chem 282(40):29531–29539
37. Goldbaum O, Jensen PH, Richter-Landsberg C (2008) The expression of tubulin polymerization promoting protein TPPP/p25alpha is developmentally regulated in cultured rat brain oligodendrocytes and affected by proteolytic stress. Glia 56(16):1736–1746
38. Kovacs GG, Gelpi E, Lehotzky A, Hoftberger R, Erdei A, Budka H, Ovadi J (2007) The brain-specific protein TPPP/p25 in pathological protein deposits of neurodegenerative diseases. Acta Neuropathol 113(2):153–161
39. Richter-Landsberg C, Heinrich M (1996) OLN-93: a new permanent oligodendroglia cell line derived from primary rat brain glial cultures. J Neurosci Res 45(2):161–173
40. Kragh CL, Lund LB, Febbraro F, Hansen HD, Gai WP, El-Agnaf O, Richter-Landsberg C, Jensen PH (2009) {alpha}-synuclein aggregation and ser-129 phosphorylation-dependent cell death in oligodendroglial cells. J Biol Chem 284(15):10211–10222
41. Alim MA, Ma QL, Takeda K, Aizawa T, Matsubara M, Nakamura M, Asada A, Saito T, Kaji H, Yoshii M et al (2004) Demonstration of a role for alpha-synuclein as a functional microtubule-associated protein. J Alzheimers Dis 6(4):435–442, discussion 443–9
42. Kim TD, Choi E, Rhim H, Paik SR, Yang CH (2004) Alpha-synuclein has structural and functional similarities to small heat shock proteins. Biochem Biophys Res Commun 324(4):1352–1359
43. McCarthy KD, De Vellis J (1980) Preparation of separate astroglial and oligodendroglial cell cultures from rat cerebral tissue. J Cell Biol 85:890–902

Chapter 15

Myelin Proteome Analysis: Methods and Implications for the Myelin Cytoskeleton

Olaf Jahn, Stefan Tenzer, Nicole Bartsch, Julia Patzig, and Hauke B. Werner

Abstract

Myelin, the multilayered membrane that enwraps and insulates neuronal axons for fast signal propagation, is a plasma membrane specialization of oligodendrocytes and Schwann cells in the central and peripheral nervous system, respectively. Here we provide our lab protocols for the purification of myelin from mouse brains and for gel-based and gel-free proteomic applications, as well as a brief discussion with respect to our current knowledge of the myelin cytoskeleton.

Key words: Schwann cell, Oligodendrocyte, Myelin, Proteome, Proteomics, Sciatic nerve, Spinal cord, Charcot-Marie-Tooth disease, Neuropathy, Leukodystrophy

1. Introduction

Myelin, the multilayered membrane that enwraps and insulates neuronal axons for fast signal propagation, is a plasma membrane specialization of myelinating glial cells in the central and peripheral nervous system of vertebrates (1). Myelin also supports axonal survival upon immunological or genetic challenges (2). Myelin can be considerably enriched from homogenized nervous tissue by sucrose gradient centrifugation as the low-density membranous material that accumulates as the interphase between 0.32 and 0.85 M sucrose. This comparatively simple method was introduced by William Norton and Shirley Poduslo in the early 1970s (3). Since then, only moderate changes have been applied, such as the use of a Polytron instead of a Dounce homogenizer (4), the addition of a second density gradient centrifugation step that reduces axonal contaminations (5), and the inclusion of protease inhibitors in all solutions. The method has proven to robustly

Rolf Dermietzel (ed.), *The Cytoskeleton: Imaging, Isolation, and Interaction*, Neuromethods, vol. 79, DOI 10.1007/978-1-62703-266-7_15, © Springer Science+Business Media, LLC 2013 achieve a high yield of pure myelin across various laboratories. The application of proteomic techniques to the myelin-enriched fraction is still relatively novel and has been applied to CNS myelin purified from mice (6–9), rats (10, 11), and humans (12, 13), and to mouse PNS myelin (14). Myelin proteome analysis has been recently reviewed with respect to technical aspects (15), a meta-analysis is available (16), and the application to myelin-related diseases has been discussed (17). Here we provide our lab protocols for the purification of myelin from mouse brains and for gel-based and gel-free proteomic applications, as well as a brief discussion of our current knowledge of the myelin cytoskeleton.

2. Materials and Protocols

2.1. Protocol 1: Purification of a Myelin-Enriched Fraction from the CNS

Materials (Centrifuge and Tubes)

1. Beckmann XL-70 Ultracentrifuge and SW-41 Ti swing-out rotor (or equivalent).
2. Six corresponding adaptors.
3. Beckman Centrifuge Tubes, Ultra-clear, 14 × 89 mm, order no: 344059. The maximum working volume for these tubes is 12 mL. Twenty-four tubes are needed for one experiment.

Preparations

1. For 0.85 M sucrose dissolve 29.09 g sucrose in 100 mL distilled water. For 0.32 M sucrose take 37.65 mL of 0.85 M sucrose and fill the volume up to 100 mL with distilled water. Sucrose solutions should be sterile-filtered and can be kept at 4°C for 1–2 days.
2. Pre-cool centrifuge and centrifuge tubes to 4°C.
3. Pre-cool all glassware, solutions, and distilled water to 4°C.
4. Before homogenization, supplement the sucrose with protease inhibitor (e.g. Roche complete).

Procedure (optional):

Mice or rats may be transcardiacally perfused with physiological salt solution to clear capillaries and thus diminish contaminations of the myelin sample with proteins derived from blood.

Procedure:

Step 1: Brain Homogenate Preparation

- Dissect brain and place in a centrifuge tube containing 0.32 M sucrose. Use 6 mL sucrose for one half brain or one spinal cord. Brains should be dissected quickly and immediately transferred into sucrose solution with protease inhibitors. If it is necessary to freeze the brains, freezing should be quick

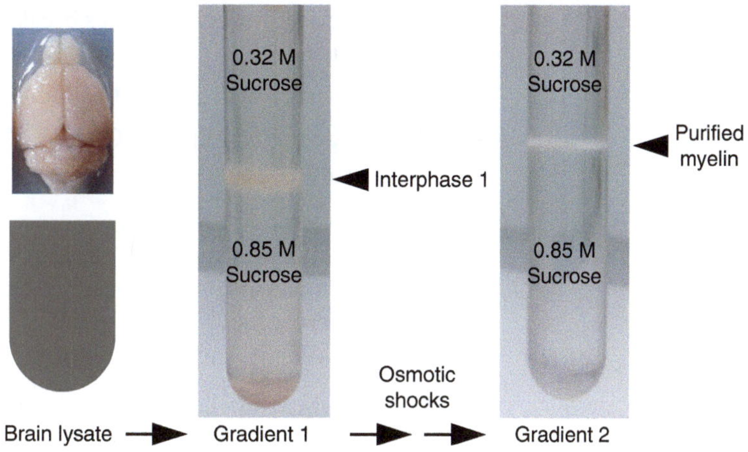

Fig. 1. During sucrose gradient centrifugation, the myelin-enriched fraction accumulates as the interphase between 0.32 and 0.85 M sucrose. For details see protocol 1.

(dry ice) and thawing should be in sucrose solution with protease inhibitors.

- Homogenize tissue using Homogenizer Kinematica AG Polytron PT3000 (or equivalent). Work on ice. Clean the homogenizer after each sample.
- 400 μL brain homogenate may be kept (−20°C) for later use.

Step 2: First Sucrose Gradient

- Fill the centrifugation tubes with 6 mL 0.85 M sucrose.
- Layer homogenate (in 0.32 M sucrose) carefully over the 0.85 M sucrose.
- Tare the tubes with 0.32 M sucrose and make certain that all six tubes have exactly the same weight (maximum difference 0.01 g).
- Centrifuge for 30 min at 4°C at 75,000 × *g* (equals 24,600 rpm when using the SW41Ti).

Step 3: Washing

- Collect the band in the 0.32 M/0.85 M sucrose interphase with a Pasteur pipette and place it into a new, clean tube. As myelin has the lowest density of any nervous system membrane it accumulates at the 0.32 M/0.85 M sucrose interphase, visible as a fluffy white cloud (see Fig. 1). Mitochondria, nuclei, synaptosomes, microsomes, etc. are largely pelleted, but a small proportion is trapped in the myelin layer as contaminants.
- Fill the tubes with cold distilled water. It is important to fill the tubes to the top before each centrifugation step. Leave some space above the solution to prevent spilling, but not too much

free space as it may cause tubes breaking and spoiled samples in the adaptor.

- Tare the tubes with distilled water.
- Centrifuge for 15 min at 4°C at 75,000 × *g* (equals 24,600 rpm when using the SW41Ti).

Step 4: First Osmotic Shock

Osmotic shock in hypotonic solution (e.g., water) and subsequent centrifugation at low speed is required to diminish cytoplasm and microsomes.

Carefully discard the supernatant.

- Resuspend each pellet in 1 mL distilled water.
- Fill the tubes with distilled water, and incubate for 10 min on ice.
- Meanwhile tare the tubes with distilled water.
- Centrifuge for 15 min at 4°C at 12,000 × *g* (9,800 rpm when using the SW41Ti).

Step 5: Second Osmotic Shock

- Carefully discard the supernatant, remove water completely.
- Resuspend each pellet in 1 mL distilled water.
- Fill the tubes with distilled water and incubate for 10 min on ice.
- Meanwhile tare the tubes with distilled water.
- Centrifuge for 15 min at 4°C at 12,000 × *g* (9,800 rpm when using the SW41Ti).

Step 6: Second Sucrose Gradient

- Carefully discard the supernatant, remove water completely.
- Resuspend each pellet in 1 mL of 0.32 M sucrose, then fill the volume to 6 mL.
- Fill each centrifugation tube with 6 mL 0.85 M sucrose solution.
- Layer 6 mL of myelin fraction (in 0.32 M sucrose) carefully over 0.85 M sucrose.
- Tare the tubes with 0.32 M sucrose.
- Centrifuge them for 30 min at 4°C at 75,000 × *g* (equals 24,600 rpm when using the SW41Ti).

Step 7: Washing

- Collect the band at the 0.32 M/0.85 M sucrose interphase (see Fig. 1) with a Pasteur pipette and place it into a new, clean tube.
- Fill the tubes with cold distilled water.

- Tare the tubes with distilled water.
- Centrifuge for 15 min at 4°C at 75.000 × g (equals 24.600 rpm when using the SW41Ti).
- Carefully discard the supernatant, remove water completely.
- Dissolve the pellet in 200 μL TBS (supplemented with protease inhibitors).

2.1.1. Notes

From one half mouse brain, a yield of approximately 600 μg purified myelin can be expected. For quantification, a Lowry assay (e.g., Biorad) or the 2-D Quant kit (GE Healthcare) are recommended. The addition of a third density gradient centrifugation was suggested (18) but has not been commonly applied. We recommend that if myelin from different conditions is to be directly compared, e.g., mutant mice with altered myelin protein or lipid composition and their littermate controls, the samples should be prepared in parallel, i.e., centrifugation is done in the same rotor run.

Quality control has often been performed by enzymatic assays testing for the specific activities of 2′,3′-cyclic nucleotide phosphodiesterase (CNP) (18, 19) or carbonic anhydrase (CA) (20) in purified myelin compared to homogenates. The enzymatic activity of CNP in purified myelin is up to tenfold higher than in brain homogenate. Metabolic assays are not optimal to determine myelin purity, as CNP and CA define only a subcompartment of myelin, and enzyme activities may fade during preparation or storage. Therefore quality control is preferentially performed by immunoblot, in which the enrichment of various myelin marker proteins with the myelin purification can be tested when the same amount (protein equivalent) of myelin and homogenate is loaded side by side (Fig. 2). Frequently used markers include proteolipid protein (PLP), myelin basic protein (MBP), myelin and lymphocyte protein (MAL), and CNP. The abundance of myelin proteins in purified CNS myelin is 20-50 fold higher compared to brain homogenate (Fig. 2). Using appropriate antibodies, the reduction of contaminants (e.g., neuronal, blood, and nuclear proteins, mitochondria) can be assessed. Transmission electron microscopy can additionally be used for morphological analysis of the myelin-enriched fraction but requires experience to determine the structures of various organelles and membranes.

Modifications to the myelin-enrichment protocol include the preparation of low, medium, and high-density subfractions of previously purified myelin (18, 21–23), the simultaneous preparation of myelin and rough microsomes using a "floating-up" protocol (24), the extraction of peripheral membrane proteins from myelin (25), the purification of myelin from peripheral nerves (4, 14, 18), and the preparation of a fraction enriched for the axon-myelin contact sites at the nodes of Ranvier, termed axogliosomes. The latter subfraction has also been subjected to proteome analysis (11, 12, 26). For these applications we refer the reader to the cited papers.

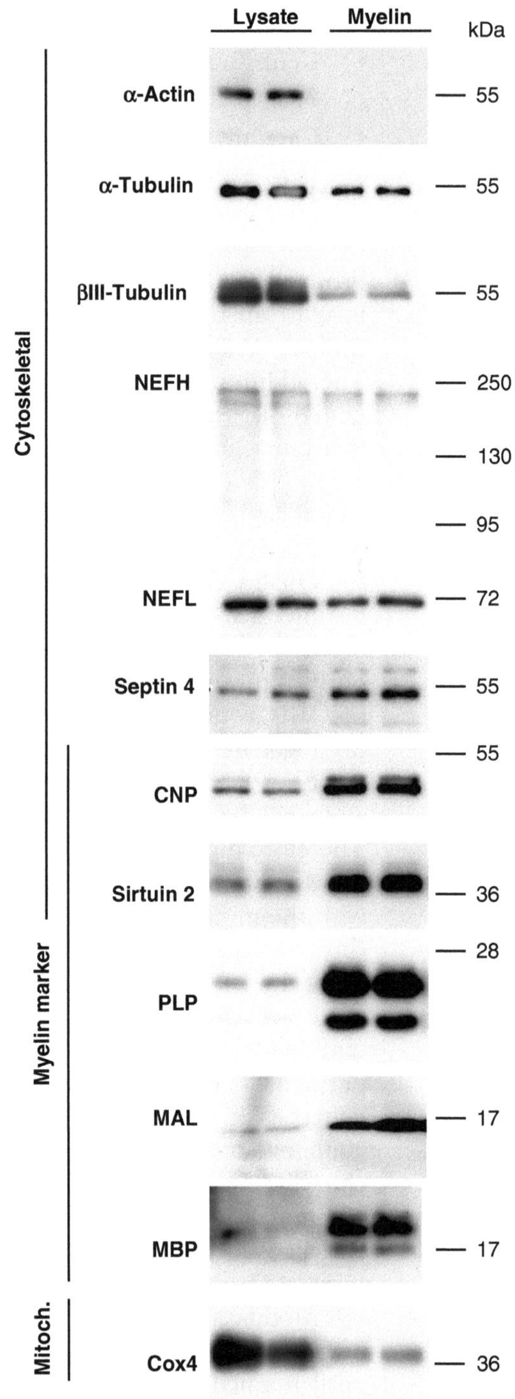

Fig. 2. Detection of cytoskeletal and myelin marker proteins in the myelin-enriched fraction by immunoblot. Equal amounts of brain homogenate and of the myelin-enriched fraction purified from mouse brain were loaded and detected with antibodies specific for the indicated proteins. Note that myelin marker proteins are enriched as a result of myelin purification, while contaminants such as mitochondria are diminished.

2.2. Gel-Based Proteome Analysis

Myelin is characterized by its high lipid content (70–75% of dry weight), the related presence of very hydrophobic transmembrane proteins, and the exceptionally high abundance of PLP, a hydrophobic tetraspan, and MBP, a highly basic protein. Together, PLP and MBP account for approx. 25% of the total CNS myelin proteins (15), which implements a challenge for proteomic analysis by conventional two-dimensional gel-based approaches involving isoelectric focusing (2D-IEF/SDS-PAGE) considering their major shortcomings such as the under-representation of membrane proteins, problems with displaying basic and hydrophobic proteins, and a limited dynamic range (see also Chapter 5 this volume). Nevertheless, when the amount of available material allows, 2D-IEF/SDS-PAGE is an attractive approach for routine profiling of intact proteins from complex mixtures and still has the highest resolving power in this respect. It is thus not surprising that first insights into the myelin proteome came from two-dimensional protein maps generated by 2D-IEF/SDS-PAGE (6). This was not possible without incremental improvements in 2D-gel technology, of which the use of the zwitter-ionic detergent amidosulfobetaine-14 (ASB-14) instead of the common 3-((3-cholamidopropyl)-dimethylammonio)-1-propanesulfonate (CHAPS) (27) was probably the most important one. Although with ASB-14 a slightly decreased resolution in IEF and some horizontal smearing may be observed, a substantially increased solubilization of myelin proteins outbalances this effect (6, 8, 15). The improved accessibility of the myelin proteome by 2D-IEF/SDS-PAGE also paved the way for differential analyses on the basis of the 2D-differential fluorescence intensity gel electrophoresis (2D-DIGE) technology (see (28) for protocol). When applied to myelin purified from a mouse model for the hypomyelinating leukodystrophy Pelizaeus-Merzbacher Disease, 2D-DIGE revealed the NAD^+-dependent deacetylase sirtuin-2 as a novel myelin protein that may mediate the function of PLP in maintaining axonal integrity (8).

2.2.1. Protocol 2: 2D-IEF/SDS-PAGE

For large-scale 2D-IEF/SDS-PAGE (18 × 20 cm, 1 mm gel thickness), ~600 μg myelin is first delipidated and precipitated with methanol/chloroform according to (29):

- Take 150 μL myelin preparation (typical protein concentration ~4 μg/μL).
- Add 600 μL methanol (vortex), 300 μL chloroform (vortex), 450 μL water (vortex).
- Centrifuge for 1 min at 9,000 × *g* (phase separation, proteins are in the interphase).
- Discard upper phase.
- Add 450 μL methanol (vortex).

- Centrifuge for 5 min at 16,100 × *g* (proteins precipitate and are pelleted).
- Discard supernatant.
- Dry pellet (carefully use a stream of nitrogen, do not dry completely).
- Solubilize in 50 μL lysis buffer (mixing (not vortexing), ultrasonic bath, and/or freeze/thaw cycles may help to resolubilize the pellet; prevent from heating above 37°C (protein carbamylation)).
- Centrifuge for 2 min at 10,000 × *g*, take supernatant.
- Determine protein concentration (e.g., 2D-Quant Kit, GE Healthcare; required if exact gel load is an issue as for 2D-DIGE).

The sample is then filled up to 360 μL with rehydration buffer and centrifuged (2 min, 10,000 × *g*). The supernatant is applied to immobilized pH gradient (IPG) strips (18 cm, nonlinear pH 3–11) through rehydration loading and IEF is performed for about 46 kV h (50 V/12 h, 300 V/1 h, 1,000 V/1 h, 1,000–3,000 V/0.5 h, 3,000 V/3 h, 3,000–8,000 V/0.5 h, 8,000 V/4 h). After completion of IEF, the proteins are reduced in SDS equilibration buffer containing 10 mg/mL DTT (2× 5 min), and subsequently alkylated in SDS equilibration buffer containing 25 mg/mL iodoacetamide (1× 10 min). The IPG strip (together with a dry filter paper containing molecular weight marker) is transferred onto 20 × 20 cm self-casted 8–16% acrylamide gradient gels (allow to polymerize at least over night, 4°C) and embedded in agarose (avoid using pre-stained marker for 2D-DIGE as they may give strong fluorescence signals). Second dimension separation is performed under denaturing conditions at room temperature (6–7 mA per gel, ca. 20 h).

Lysis buffer: 7 M urea, 2 M thiourea, 2% ASB-14, 1% ampholytes, 0.6% DTT (for 2D-DIGE, the sample is first solubilized in 50 μL labeling buffer (30 mM Tris–HCl pH 9.0, 7 M urea, 2 M thiourea, 2% ASB-14) and 50 μL lysis buffer are added after labeling).

Rehydration buffer: 7 M urea, 2 M thiourea, 2% ASB-14, 0.5% ampholytes, 0.3% DTT.

SDS equilibration buffer: 50 mM Tris–HCl pH 8.8, 6 M urea, 30% glycerol, 2% SDS, 0.002% Bromophenol blue.

Running buffer: 190 mM glycine, 25 mM Tris, 0.1% SDS.

2.2.2. Protocol 3: 2D-16-BAC/SDS-PAGE

Despite the use of ASB-14, 2D-IEF/SDS-PAGE fails to appropriately display abundant transmembrane myelin marker proteins such as PLP, myelin-associated glycoprotein (MAG), myelin oligodendrocyte glycoprotein (MOG), or oligodendrocyte-specific protein (OSP/claudin-11). Thus, alternative 2D gel systems are required to deepen the proteome coverage while retaining the benefits of displaying intact proteins. In the most common version of these systems, the charge-dependent separation in the first dimension (i.e., the IEF) is replaced by a size-dependent separation in the presence of the cationic detergent 16-benzyldimethyl-*n*-hexadecylammonium chloride (16-BAC) (16-BAC sample buffer: 22.5% urea, 5% 16-BAC, 5% glycerol, 37.5 mM DTT, 0.025% Pyronin Y) (30). Due to the similar separation principle in both dimensions, proteins are typically dispersed along a diagonal rather than distributed over the entire gel area. Accordingly, these gel systems have a lower resolution compared to 2D-IEF/SDS-PAGE, but can resolve highly basic and even membrane-spanning proteins (31). Application of 2D-16-BAC/SDS-PAGE to mouse CNS myelin not only resulted in the identification of novel myelin-associated proteins, but also readily enabled displaying of the transmembrane myelin marker proteins PLP, MAG, MOG, and OSP (8, 32, 33).

For large-scale 2D-16-BAC/SDS-PAGE (20 x 20 cm, 1 mm Gel thickness) ~100 μg myelin is resolubilized in 2 mM EGTA (pH 7.0) and centrifuged (10 min, 16.000 × g, 4°C). Optionally, soluble and membrane-associated proteins can be depleted by subjecting ~400 μg myelin to six consecutive washing/centrifugation cycles (2× 50 mM Tris, 150 mM NaCl, 1× 1 M KCl, 1× 100 mM Na_2CO_3, 2× 2 mM EGTA). The final pellet is solubilized in 100 μL 16-BAC sample buffer and separated on self-casted 7.5% acrylamide 16-BAC gels with 4% acrylamide stacking gel (30). Casting of the acidic 16-BAC separation gel is sophisticated: (1) use 0.75 mm gel thickness to facilitate transfer to the second dimension, (2) dissolve 16-BAC by warming in a water bath, (3) prepare solutions of ascorbic acid, $FeSO_4$, and H_2O_2 freshly as otherwise polymerization may fail, (4) overlay gel with potassium phosphate buffer (75 mM, pH 2.1), (5) allow to polymerize at least over night, 4°C. After the first dimension, gels are fixed in isopropanol/acetic acid/H_2O (3.5/1/5.5), stained with Coomassie Brilliant Blue (CBB) R250, and whole lanes of interest are excised (fixation/staining is optional, but in our hands increased the precision of lane-cutting). Gel strips are equilibrated in 100 mM Tris–HCl pH 8.8 (2× 15 min), incubated with 3× Laemmli sample buffer (pH 8.8, 2× 5 min), and transferred onto 20 × 20 cm self-casted 8–16% acrylamide gradient gels (allow to polymerize at least over night, 4°C) and embedded in agarose. Second dimension separation is performed as above.

2.2.3. Protocol 4: 1D-SDS-PAGE

Although 1D-SDS-PAGE has a much lower resolution than 2D gel systems, it is a powerful technique to prepare "problematic" proteins (e.g., transmembrane proteins, proteins with extreme isoelectric point) for mass spectrometric protein identification. The lower resolution at the protein level is compensated by chromatographic separation of the proteolytic peptides prior to mass spectrometry (MS), an approach commonly referred to as GeLC-MS/MS. For systematic protein identification, 25 μg myelin is separated on precast NuPAGE Bis-Tris 4–12% gradient gels (8×8 cm, 1 mm gel thickness) according to the manufacturer's protocols (Invitrogen). By using the MOPS running buffer system, resolution can be improved for the molecular weight range above the dominant PLP band (~25 kDa), while MBP and other small proteins are separated more efficiently when using the MES buffer system. Due to the high lipid content of myelin, protein loads >10 μg per lane may require delipidation prior to 1D-SDS-PAGE (see protocol 2), at least when optimal separation in the low molecular weight range (<20 kDa) is required.

2.2.4. Protocol 5: Gel Staining and Protein Identification

For gel staining, we recommend colloidal Coomassie staining according to Neuhoff (34). Fix the gel with 40% (v/v) ethanol, 10% (v/v) acetic acid for at least 60 min. Wash the gel with water for 2× 10 min (gel should be completely covered). Stain the gel in dye working solution at least over night. Transfer the gel into new box with 1% (v/v) acetic acid. Wash the gel with 1% acetic acid until all Coomassie particles are removed. In our hands, the modified protocol given below allows the detection of 5–10 ng protein per 1D gel band.

Dye stock solution: 0.1% (w/v) CBB *G250*, 2% (w/v) ortho-phosphoric acid, 10% (w/v) ammonium sulfate (prepare the stock solution at least 24 h before staining, do not filter).

Dye working solution: 80% (v/v) dye stock solution, 20% (v/v) methanol (prepare the dye working solution immediately before use by adding one part of methanol to four parts of the resuspended dye stock solution while stirring).

2.2.5. Notes: Gel Staining and Protein Identification

All steps are performed on a shaking table at room temperature. Use only glass containers or plastic boxes specified for gel staining. Low quality plastic containers may release plasticizers that interfere with mass spectrometric protein identification. Gels can be documented with a conventional flatbed scanner or with a near-infrared fluorescence imager, the latter allowing for direct quantification (35). If mass spectrometric protein identification is desired, it is recommended to excise gel plugs/bands immediately. Gel plugs/bands can be stored in appropriate vials/microplates at –20°C until further processing. For mass spectrometric protein identification, 2D-IEF/SDS-PAGE is most commonly interfaced to MALDI-MS, preferably through an automated platform in the 96-well format (36). To account for the lower resolution of 2D-16-BAC/SDS-PAGE and particularly 1D-SDS-PAGE (resulting in more complex peptide mixtures after in-gel digestion), these gel systems are usually interfaced to nano-flow LC-MS/MS setups, i.e., the proteolytic peptides are separated by reversed-phase HPLC before they enter the mass spectrometer. Gel pieces are either excised as cylindric plus of 1.5 mm diameter from 2D gels, or entire lanes from 1D gels are sliced into 24 pieces by a self-constructed punching device. In both cases, uniformity of the gel pieces is crucial when further processing by automated in-gel digestion procedures is desired.

2.3. Gel-Free Proteome Analysis

In the recent years, gel-free techniques have emerged in myelin proteome analysis to overcome the limitations of gel-based methods, in particular those of 2D-IEF/SDS-PAGE (see above). These LC-MS/MS-based techniques are commonly referred to as shotgun approaches (see also Chapters 5 and 7 this volume), as separation at the protein level is omitted. Instead, purified myelin is proteolytically digested (at the expense of information related to protein integrity such as protein size and charge) and separation takes place at the level of proteolytic peptides before interfacing with MS. When considering the tremendous complexity of such peptide mixtures, this scenario obviously poses a challenge for chromatography and mass spectrometry. However, due to the recent technical advancements, modern shotgun approaches appear to be particularly attractive for tackling the myelin proteome with its several orders of magnitude of protein abundance and its many highly basic, hydrophobic, and transmembrane proteins. Advantages of these techniques include that they (1) have a high dynamic range, (2) are not biased toward certain protein classes, and (3) require only low amounts of total protein (up to 100-fold less than 2D-IEF/SDS-PAGE). Finally, LC-MS/MS-based approaches enable the label-free quantification of proteins from complex mixtures, which allowed for the systematic reassignment of protein abundance in CNS myelin (15, 16).

To take advantage of latest nanoscale high-resolution separation techniques (e.g., UPLC) for LC-MS/MS analyses, careful sample preparation is a crucial step. In particular, direct-injection setups without trapping columns, which feature increased peak capacities and thereby lower detection limits due to reduced analyte interference essentially require particle-free sample preparation. Additionally, the aim of achieving optimal chromatographic resolution and sensitivity requires substances interfering with chromatographic separation or electrospray ionization to be removed or at least minimized during sample preparation. Therefore, sample preparation methods should be carefully chosen and specifically tailored to the specific properties of the intended sample. For analysis of small (<30 μg) sample amounts, various approaches have been described, including in-solution digestion after protein precipitation and filter-aided sample preparation (FASP) (37), which uses an ultrafiltration device for tryptic digestions.

2.3.1. Protocol 6: Myelin Sample Preparation for LC-MS/MS

The unique composition of myelin implies special requirements for sample preparation as to solubilization of transmembrane proteins and lipid removal. In our hands, incubation of purified myelin in lysis buffer for 15 min in an ice-cooled ultrasonic bath resulted in complete solubilization of myelin proteins. Afterwards, most lipids can easily be removed by centrifugation (13,000 × *g*, 4°C, 15 min). Subsequently, solubilized proteins are precipitated before tryptic in-solution digestion. This step further reduces lipid content of the sample. For small sample amounts, optimal protein recoveries can be obtained using a commercially available kit (Calbiochem ProteoExtract Protein Precipitation Kit). Of note, we highly recommend to slightly modify the manufacturer's protocol by increasing the washing steps of the precipitate to three to ensure complete lipid removal. Like in the methanol/chloroform precipitation protocol mentioned above, extra care should be taken to avoid over-drying of the resulting protein pellet that can result in poor solubilization in subsequent steps.

Lysis buffer: 7 M urea, 2 M thiourea, 10 mM Tris pH 8.0, 5 mM DTT.

2.3.2. Protocol 7: In-Solution Digestion and Identification of Myelin Proteins

Special detergents are required to obtain optimal proteome coverage, ensuring full solubilization and unfolding of the proteins prior to in-solution tryptic digestion. For this purpose, the protein pellet (approx. 30 μg total myelin protein) is resuspended with 0.2% RapiGest detergent (Waters), sonicated for 10 min followed by incubation for 15 min at 80°C in a thermomixer. Subsequently, proteins are reduced with 8 mM DTT for 30 min at 56°C, followed by the addition of 15 mM iodoacetamide to block cysteine residues by carbamidomethylation for 45 min at RT in the dark. Excess iodoacetamide is then quenched by adding DTT to a final concentration of 16 mM followed by the addition of 0.5 μg sequencing grade trypsin (Trypsin Gold, Promega). After overnight incubation at 37°C, RapiGest detergent is removed by lowering the pH of the samples to pH 2 using 100 mM HCl. This causes hydrolysis and precipitation of the detergent, which can then be completely removed from the sample by centrifugation (13,000 × *g*, 30 min, 4°C). This centrifugation step also eliminates any particles; however, extra care should be taken not to stir the pellet when transferring the supernatant (i.e., the tryptic peptides) to a new vial.

In our studies, the samples are then analyzed by nanoUPLC coupled to a QTOF mass spectrometer operated in data-independent alternate scanning mode (MS^E), allowing for identification and quantification of over 400 proteins in a single experiment (14, 15, 38).

2.4. The Myelin Cytoskeleton

The compact compartment of myelin is apparently lacking cytoskeletal elements, as cytoplasm is squeezed out during compaction. Any discussion of the myelin cytoskeleton thus focuses on the non-compact compartments of myelin, namely adaxonal (i.e., inner tongue), paranodal, incisural, and abaxonal (i.e., outer tongue) myelin, as well as on the developmental stages of myelination before compaction has been completed. A large number of novel myelin-associated proteins have been identified by proteome analysis (14–16), and individual validation and functional analysis have now become available for some of them. This section will briefly highlight some aspects of the myelin cytoskeleton with regard to myelin proteome analysis. For a detailed review of the oligodendroglial cytoskeleton see (39) and Chapter 14 this volume, and aspects particular to Schwann cells have been covered by (40).

2.4.1. Actin

The actin cytoskeleton is required for normal developmental myelination, which involves that highly motile glial cell processes are extended. This was shown by analyses of mouse mutants lacking particular actin filament-regulating proteins. Importantly, mice lacking WAVE1 (Wiskott-Aldrich syndrome protein (WASP) family member 1, also termed WASF1) were hypomyelinated in the CNS (41), and mice lacking N-WASP (neural Wiskott-Aldrich syndrome protein, also termed WASL) specifically from Schwann cells were hypomyelinated in the PNS (42, 43). Different from actively myelinating oligodendrocytes, the processes of mature oligodendrocytes that comprise a compact sheath are much less motile but rather stationary in the long term. Indeed, filamentous actin is not very abundant in mature oligodendrocytes (44), and, by immunoblot, α-actin cannot be detected in purified CNS myelin without overexposure (see Fig. 2). Consequently, WAVE1 and N-WASP were not detected by MS in mature CNS myelin (15). However, myelin is not devoid of actin-associated proteins. For example, the expression of the actin-binding protein ermin (also termed juxtanodin) in oligodendrocytes increases at a late stage of myelination, and ermin is localized to noncompact mature myelin (45, 46) and has also been detected in mature CNS myelin by LC-MS^E (Table 1). Ermin may be involved in cytoskeletal rearrangements during myelin wrapping, compaction, or plasticity.

2.4.2. Microtubules

Microtubules are the main cytoskeletal class in oligodendrocytes (47–51) and in Schwann cells (52, 53), including mature myelin, in which microtubules can be recognized by electron microscopy as localized to the non-compacted compartments. The normal development and function of myelin strongly relies on the microtubule-based trafficking of transport vesicles (54–56), myelin proteins and mRNA (57, 58) encoding the essential (59, 60) myelin protein MBP to the sites of active myelination in the glial cell periphery (61, 62), where its translation is tightly controlled (63, 64). This view has recently been supported by analyses of zebrafish lines harboring mutations affecting the kinesin motor protein Kif1b (65) or tubulin alpha 8-like 3a (66). Among other defects, mutant zebrafish are hypomyelinated connected to a failure to efficiently traffic and specifically localize mbp-mRNA in the glial processes.

Table 1
Relative abundance of myelin proteins

Myelin protein		%	Cytoskeleton-associated protein in myelin		%
Proteolipid protein	PLP	15.4	2′,3′-Cyclic nucleotide phosphodiesterase	CNP	3.7
Myelin basic protein	MBP	6.8	Sirtuin 2	SIRT2	0.8
Myelin oligodendrocyte glycoprotein	MOG	1.2	Neurofilament light chain	NEFL	0.4
Myelin-associated glycoprotein	MAG	0.9	Tubulin polymerization promoting protein	TPPP	0.2
Oligodendrocyte-specific protein	OSP	0.7	Septin 4	SEPT4	0.08
Carbonic anhydrase 2	CA2	0.1	Ezrin	EZR	0.08
			Radixin	RDX	0.06
			Moesin	MSN	0.06
			Cofilin 1	CFL1	0.05
			Ermin	ERMN	0.02

The relative abundance of selected myelin proteins and cytoskeleton-associated proteins in the myelin-enriched fraction purified from mouse brains was determined by LC-MS^E and is given in percent. Classical myelin proteins constitute less than 35% of the total myelin protein, as the increased number of myelin-associated proteins newly identified with proteomic methods goes along with a lower relative abundance of any individual protein compared with previous estimates based on band intensity after 1D-SDS-PAGE and protein staining

The NAD^+-dependent deacetylase sirtuin 2 (SIRT2) has been newly identified by proteome analysis as an abundant myelin constituent (8), and, in the CNS, is mainly expressed by oligodendrocytes (67, 68). SIRT2 has been suggested to deacetylate α-tubulin at least in vitro (67, 69), which may affect microtubule stability and related cellular functions (70) though functional analysis in vivo has not been reported yet. Myelin proteins with suggested enzymatic functions to posttranslationally modify tubulin and microtubules are among the most abundant myelin proteins (Table 1), including SIRT2, CNP (71, 72), and TPPP (Tubulin polymerization promoting protein, also termed p25alpha) (73, 74). We expect that the systematic exploitation of the myelin proteome datasets regarding proteins associated with or modifying microtubules will provide further insight into the myelin cytoskeleton.

2.4.3. Intermediate Filaments

Although Schwann cells are devoid of hemidesmosomes (40), Schwann cells and the non-compact compartment of peripheral myelin comprises intermediate filament proteins such as neurofilament light chain (NEFL) (14, 75). NEFL mutations in humans lead to a subtype of Charcot-Marie-Tooth disease with neurological features that suggest both axonal and Schwann cell pathology (76, 77). Intermediate filaments are commonly assumed to be lacking in oligodendrocytes (39), at least in culture. Interestingly, NEFL was readily identified in purified CNS myelin by mass spectrometry (Table 1) and immunoblot (Fig. 2), raising the possibility that NEFL, and possibly other intermediate filament proteins, are indeed constituents of both PNS and CNS myelin, in which they may contribute to structural stability.

2.4.4. Septins

Unexpectedly, myelin proteome analysis has led to the identification of several members of a fourth class of filament-forming proteins, termed the septins (8, 13, 78, 79). Septins are not only abundant CNS myelin constituents, a subset of them are actually enriched in CNS myelin compared to brain homogenate, different from actin, tubulin, and intermediate filaments (see Fig. 2; Table 1). The function of the septin cytoskeleton in myelin is currently under investigation and is expected to yield new insight into the relevance of cytoskeletal filaments for the structure and function of myelin.

3. Concluding Remark

Proteome analysis has provided a valuable resource (16) for the systematic molecular and genetic analysis of myelin with respect to functions in myelin development, stability, plasticity, and glia-axonal interactions in the context of white matter physiology and pathophysiology. While the techniques presented here have proven a valuable aspect of the methodological spectrum applied in current myelin research, the exploitation of their capabilities is still in an early phase.

Acknowledgments

We thank N. Schaeren-Wiemers and K.-I. Nagata for antibody gifts, D. Hesse, T. Liepold, and M. Uecker for proof-reading, and K.-A. Nave for continuous support. Our work is supported by the Bundesministerium für Bildung und Forschung (BMBF DLR-Leukonet), the European Commission (EU-Health FP7 LeukoTreat), and the Deutsche Forschungsgemeinschaft (DFG-SFB490, Z3).

References

1. Sherman DL, Brophy PJ (2005) Mechanisms of axon ensheathment and myelin growth. Nat Rev Neurosci 6:683–690
2. Nave KA (2010) Myelination and support of axonal integrity by glia. Nature 468:244–252
3. Norton WT, Poduslo SE (1973) Myelination in rat brain: method of myelin isolation. J Neurochem 21:749–757
4. Wiggins RC, Fuller GN (1981) Analysis of distribution of rat sciatic nerve protein among soluble, insoluble, and myelin subfraction. Neurochem Res 6:719–727
5. Haley JE, Samuels FG, Ledeen RW (1981) Study of myelin purity in relation to axonal contaminants. Cell Mol Neurobiol 1:175–187
6. Taylor CM, Marta CB, Claycomb RJ, Han DK, Rasband MN, Coetzee T, Pfeiffer SE (2004) Proteomic mapping provides powerful insights into functional myelin biology. Proc Natl Acad Sci U S A 101:4643–4648
7. Vanrobaeys F, Van Coster R, Dhondt G, Devreese B, Van Beeumen J (2005) Profiling of myelin proteins by 2D-gel electrophoresis and multidimensional liquid chromatography coupled to MALDI TOF-TOF mass spectrometry. J Proteome Res 4:2283–2293
8. Werner HB, Kuhlmann K, Shen S, Uecker M, Schardt A, Dimova K, Orfaniotou F, Dhaunchak A, Brinkmann BG, Mobius W, Guarente L, Casaccia-Bonnefil P, Jahn O, Nave KA (2007) Proteolipid protein is required for transport of sirtuin 2 into CNS myelin. J Neurosci 27: 7717–7730
9. Manrique-Hoyos N, Jurgens T, Gronborg M, Kreutzfeldt M, Schedensack M, Kuhlmann T, Schrick C, Bruck W, Urlaub H, Simons M, Merkler D (2012) Late motor decline after accomplished remyelination: impact for progressive multiple sclerosis. Ann Neurol 71: 227–244
10. Baer AS, Syed YA, Kang SU, Mitteregger D, Vig R, Ffrench-Constant C, Franklin RJ, Altmann F, Lubec G, Kotter MR (2009) Myelin-mediated inhibition of oligodendrocyte precursor differentiation can be overcome by pharmacological modulation of Fyn-RhoA and protein kinase C signalling. Brain 132:465–481
11. Roth AD, Ivanova A, Colman DR (2006) New observations on the compact myelin proteome. Neuron Glia Biol 2:15–21
12. Dhaunchak AS, Huang JK, De Faria Junior O, Roth AD, Pedraza L, Antel JP, Bar-Or A, Colman DR (2010) A proteome map of axoglial specializations isolated and purified from human central nervous system. Glia 58: 1949–1960
13. Ishii A, Dutta R, Wark GM, Hwang SI, Han DK, Trapp BD, Pfeiffer SE, Bansal R (2009) Human myelin proteome and comparative analysis with mouse myelin. Proc Natl Acad Sci U S A 106:14605–14610
14. Patzig J, Jahn O, Tenzer S, Wichert SP, de Monasterio-Schrader P, Rosfa S, Kuharev J, Yan K, Bormuth I, Bremer J, Aguzzi A, Orfaniotou F, Hesse D, Schwab MH, Mobius W, Nave KA, Werner HB (2011) Quantitative and integrative proteome analysis of peripheral nerve myelin identifies novel myelin proteins and candidate neuropathy loci. J Neurosci 31:16369–16386
15. Jahn O, Tenzer S, Werner HB (2009) Myelin proteomics: molecular anatomy of an insulating sheath. Mol Neurobiol 40:55–72
16. de Monasterio-Schrader P, Jahn O, Tenzer S, Wichert SP, Patzig J, Werner HB (2012) Systematic approaches to central nervous system myelin. Cell Mol Life Sci 69(17): 2879–2894
17. Werner HB, Jahn O (2010) Myelin matters: proteomic insights into white matter disorders. Expert Rev Proteomics 7:159–164
18. Larocca JN, Norton WT (2007) Isolation of myelin. Curr Protoc Cell Biol. Chapter 3: Unit3.25
19. Prohaska JR, Clark DA, Wells WW (1973) Improved rapidity and precision in the determination of brain 2′,3′-cyclic nucleotide 3′-phosphohydrolase. Anal Biochem 56: 275–282
20. Cammer W, Fredman T, Rose AL, Norton WT (1976) Brain carbonic anhydrase: activity in isolated myelin and the effect of hexachlorophene. J Neurochem 27:165–171
21. Cammer W, Bieler L, Fredman T, Norton WT (1977) Quantitation of myelin carbonic anhydrase-development and subfractionation of rat brain myelin and comparison with myelin from other species. Brain Res 138:17–28
22. Sheads LD, Eby MJ, Sampugna J (1977) Myelin subfractions isolated from mouse brain. Studies of normal mice during development, quaking mutants, and three brain regions. J Neurobiol 8:67–89
23. Menon K, Rasband MN, Taylor CM, Brophy P, Bansal R, Pfeiffer SE (2003) The myelin-axolemmal complex: biochemical dissection and the role of galactosphingolipids. J Neurochem 87:995–1009
24. Colman DR, Kreibich G, Frey AB, Sabatini DD (1982) Synthesis and incorporation of myelin polypeptides into CNS myelin. J Cell Biol 95:598–608

25. Caroni P, Schwab ME (1988) Two membrane protein fractions from rat central myelin with inhibitory properties for neurite growth and fibroblast spreading. J Cell Biol 106:1281–1288
26. Huang JK, Phillips GR, Roth AD, Pedraza L, Shan W, Belkaid W, Mi S, Fex-Svenningsen A, Florens L, Yates JR 3rd, Colman DR (2005) Glial membranes at the node of Ranvier prevent neurite outgrowth. Science 310:1813–1817
27. Taylor CM, Pfeiffer SE (2003) Enhanced resolution of glycosylphosphatidylinositol-anchored and transmembrane proteins from the lipid-rich myelin membrane by two-dimensional gel electrophoresis. Proteomics 3:1303–1312
28. Viswanathan S, Unlu M, Minden JS (2006) Two-dimensional difference gel electrophoresis. Nat Protoc 1:1351–1358
29. Wessel D, Flugge UI (1984) A method for the quantitative recovery of protein in dilute solution in the presence of detergents and lipids. Anal Biochem 138:141–143
30. Hartinger J, Stenius K, Hogemann D, Jahn R (1996) 16-BAC/SDS-PAGE: a two-dimensional gel electrophoresis system suitable for the separation of integral membrane proteins. Anal Biochem 240:126–133
31. Braun RJ, Kinkl N, Beer M, Ueffing M (2007) Two-dimensional electrophoresis of membrane proteins. Anal Bioanal Chem 389:1033–1045
32. Yamaguchi Y, Miyagi Y, Baba H (2008) Two-dimensional electrophoresis with cationic detergents: a powerful tool for the proteomic analysis of myelin proteins. Part 2: analytical aspects. J Neurosci Res 86:766–775
33. Yamaguchi Y, Miyagi Y, Baba H (2008) Two-dimensional electrophoresis with cationic detergents, a powerful tool for the proteomic analysis of myelin proteins. Part 1: technical aspects of electrophoresis. J Neurosci Res 86:755–765
34. Neuhoff V, Arold N, Taube D, Ehrhardt W (1988) Improved staining of proteins in polyacrylamide gels including isoelectric focusing gels with clear background at nanogram sensitivity using Coomassie Brilliant Blue G-250 and R-250. Electrophoresis 9:255–262
35. Luo S, Wehr NB, Levine RL (2006) Quantitation of protein on gels and blots by infrared fluorescence of Coomassie blue and Fast Green. Anal Biochem 350:233–238
36. Jahn O, Hesse D, Reinelt M, Kratzin HD (2006) Technical innovations for the automated identification of gel-separated proteins by MALDI-TOF mass spectrometry. Anal Bioanal Chem 386:92–103
37. Wisniewski JR, Zougman A, Nagaraj N, Mann M (2009) Universal sample preparation method for proteome analysis. Nat Methods 6: 359–362
38. Kramer-Albers EM, Bretz N, Tenzer S, Winterstein C, Mobius W, Berger H, Nave KA, Schild H, Trotter J (2007) Oligodendrocytes secrete exosomes containing major myelin and stress-protective proteins: Trophic support for axons? Proteomics Clin Appl 1:1446–1461
39. Bauer NG, Richter-Landsberg C, Ffrench-Constant C (2009) Role of the oligodendroglial cytoskeleton in differentiation and myelination. Glia 57:1691–1705
40. Scherer SS, Arroyo EJ (2002) Recent progress on the molecular organization of myelinated axons. J Peripher Nerv Syst 7:1–12
41. Kim HJ, DiBernardo AB, Sloane JA, Rasband MN, Solomon D, Kosaras B, Kwak SP, Vartanian TK (2006) WAVE1 is required for oligodendrocyte morphogenesis and normal CNS myelination. J Neurosci 26:5849–5859
42. Jin F, Dong B, Georgiou J, Jiang Q, Zhang J, Bharioke A, Qiu F, Lommel S, Feltri ML, Wrabetz L, Roder JC, Eyer J, Chen X, Peterson AC, Siminovitch KA (2011) N-WASp is required for Schwann cell cytoskeletal dynamics, normal myelin gene expression and peripheral nerve myelination. Development 138:1329–1337
43. Novak N, Bar V, Sabanay H, Frechter S, Jaegle M, Snapper SB, Meijer D, Peles E (2011) N-WASP is required for membrane wrapping and myelination by Schwann cells. J Cell Biol 192:243–250
44. Capani F, Ellisman MH, Martone ME (2001) Filamentous actin is concentrated in specific subpopulations of neuronal and glial structures in rat central nervous system. Brain Res 923:1–11
45. Brockschnieder D, Sabanay H, Riethmacher D, Peles E (2006) Ermin, a myelinating oligodendrocyte-specific protein that regulates cell morphology. J Neurosci 26:757–762
46. Wang T, Jia L, Lv B, Liu B, Wang W, Wang F, Yang G, Bu X, Yao L, Zhang B (2011) Human Ermin (hErmin), a new oligodendrocyte-specific cytoskeletal protein related to epileptic seizure. Brain Res 1367:77–84
47. Lunn KF, Baas PW, Duncan ID (1997) Microtubule organization and stability in the oligodendrocyte. J Neurosci 17:4921–4932
48. Terada N, Kidd GJ, Kinter M, Bjartmar C, Moran-Jones K, Trapp BD (2005) Beta IV tubulin is selectively expressed by oligodendrocytes in the central nervous system. Glia 50: 212–222
49. Wilson R, Brophy PJ (1989) Role for the oligodendrocyte cytoskeleton in myelination. J Neurosci Res 22:439–448

50. Wu C, Chang A, Smith MC, Won R, Yin X, Staugaitis SM, Agamanolis D, Kidd GJ, Miller RH, Trapp BD (2009) Beta4 tubulin identifies a primitive cell source for oligodendrocytes in the mammalian brain. J Neurosci 29:7649–7657
51. Hildebrand C, Remahl S, Persson H, Bjartmar C (1993) Myelinated nerve fibres in the CNS. Prog Neurobiol 40:319–384
52. Kidd GJ, Andrews SB, Trapp BD (1994) Organization of microtubules in myelinating Schwann cells. J Neurocytol 23:801–810
53. Perkins GA, Sosinsky GE, Ghassemzadeh S, Perez A, Jones Y, Ellisman MH (2008) Electron tomographic analysis of cytoskeletal cross-bridges in the paranodal region of the node of Ranvier in peripheral nerves. J Struct Biol 161:469–480
54. Feldmann A, Amphornrat J, Schonherr M, Winterstein C, Mobius W, Ruhwedel T, Danglot L, Nave KA, Galli T, Bruns D, Trotter J, Kramer-Albers EM (2011) Transport of the major myelin proteolipid protein is directed by VAMP3 and VAMP7. J Neurosci 31:5659–5672
55. Kippert A, Trajkovic K, Rajendran L, Ries J, Simons M (2007) Rho regulates membrane transport in the endocytic pathway to control plasma membrane specialization in oligodendroglial cells. J Neurosci 27:3560–3570
56. Larocca JN, Rodriguez-Gabin AG (2002) Myelin biogenesis: vesicle transport in oligodendrocytes. Neurochem Res 27:1313–1329
57. Kursula P, Lehto VP, Heape AM (2001) The small myelin-associated glycoprotein binds to tubulin and microtubules. Brain Res Mol Brain Res 87:22–30
58. O'Connor LT, Goetz BD, Couve E, Song J, Duncan ID (2000) Intracellular distribution of myelin protein gene products is altered in oligodendrocytes of the taiep rat. Mol Cell Neurosci 16:396–407
59. Popko B, Puckett C, Lai E, Shine HD, Readhead C, Takahashi N, Hunt SW 3rd, Sidman RL, Hood L (1987) Myelin deficient mice: expression of myelin basic protein and generation of mice with varying levels of myelin. Cell 48:713–721
60. Readhead C, Popko B, Takahashi N, Shine HD, Saavedra RA, Sidman RL, Hood L (1987) Expression of a myelin basic protein gene in transgenic shiverer mice: correction of the dysmyelinating phenotype. Cell 48:703–712
61. Breitschopf H, Suchanek G, Gould RM, Colman DR, Lassmann H (1992) In situ hybridization with digoxigenin-labeled probes: sensitive and reliable detection method applied to myelinating rat brain. Acta Neuropathol 84:581–587
62. Brosamle C, Halpern ME (2002) Characterization of myelination in the developing zebrafish. Glia 39:47–57
63. Laursen LS, Chan CW, Ffrench-Constant C (2011) Translation of myelin basic protein mRNA in oligodendrocytes is regulated by integrin activation and hnRNP-K. J Cell Biol 192:797–811
64. White R, Gonsior C, Kramer-Albers EM, Stohr N, Huttelmaier S, Trotter J (2008) Activation of oligodendroglial Fyn kinase enhances translation of mRNAs transported in hnRNP A2-dependent RNA granules. J Cell Biol 181:579–586
65. Lyons DA, Naylor SG, Scholze A, Talbot WS (2009) Kif1b is essential for mRNA localization in oligodendrocytes and development of myelinated axons. Nat Genet 41:854–858
66. Larson TA, Gordon TN, Lau HE, Parichy DM (2010) Defective adult oligodendrocyte and Schwann cell development, pigment pattern, and craniofacial morphology in puma mutant zebrafish having an alpha tubulin mutation. Dev Biol 346:296–309
67. Li W, Zhang B, Tang J, Cao Q, Wu Y, Wu C, Guo J, Ling EA, Liang F (2007) Sirtuin 2, a mammalian homolog of yeast silent information regulator-2 longevity regulator, is an oligodendroglial protein that decelerates cell differentiation through deacetylating alpha-tubulin. J Neurosci 27:2606–2616
68. Southwood CM, Peppi M, Dryden S, Tainsky MA, Gow A (2007) Microtubule deacetylases, SirT2 and HDAC6, in the nervous system. Neurochem Res 32:187–195
69. North BJ, Marshall BL, Borra MT, Denu JM, Verdin E (2003) The human Sir2 ortholog, SIRT2, is an NAD+-dependent tubulin deacetylase. Mol Cell 11:437–444
70. Perdiz D, Mackeh R, Pous C, Baillet A (2011) The ins and outs of tubulin acetylation: more than just a post-translational modification? Cell Signal 23:763–771
71. Bifulco M, Laezza C, Stingo S, Wolff J (2002) 2′,3′-Cyclic nucleotide 3′-phosphodiesterase: a membrane-bound, microtubule-associated protein and membrane anchor for tubulin. Proc Natl Acad Sci U S A 99:1807–1812
72. Lee J, Gravel M, Zhang R, Thibault P, Braun PE (2005) Process outgrowth in oligodendrocytes is mediated by CNP, a novel microtubule assembly myelin protein. J Cell Biol 170: 661–673
73. Hoftberger R, Fink S, Aboul-Enein F, Botond G, Olah J, Berki T, Ovadi J, Lassmann H, Budka H, Kovacs GG (2010) Tubulin polymerization promoting protein (TPPP/p25) as a

marker for oligodendroglial changes in multiple sclerosis. Glia 58:1847–1857

74. Skjoerringe T, Lundvig DM, Jensen PH, Moos T (2006) P25alpha/Tubulin polymerization promoting protein expression by myelinating oligodendrocytes of the developing rat brain. J Neurochem 99:333–342
75. Previtali SC, Zerega B, Sherman DL, Brophy PJ, Dina G, King RH, Salih MM, Feltri L, Quattrini A, Ravazzolo R, Wrabetz L, Monaco AP, Bolino A (2003) Myotubularin-related 2 protein phosphatase and neurofilament light chain protein, both mutated in CMT neuropathies, interact in peripheral nerve. Hum Mol Genet 12:1713–1723
76. Nicholson G, Myers S (2006) Intermediate forms of Charcot-Marie-Tooth neuropathy: a review. Neuromolecular Med 8:123–130
77. Vance JM (2000) The many faces of Charcot-Marie-Tooth disease. Arch Neurol 57: 638–640
78. Ogawa Y, Rasband MN (2009) Proteomic analysis of optic nerve lipid rafts reveals new paranodal proteins. J Neurosci Res 87: 3502–3510
79. Buser AM, Erne B, Werner HB, Nave KA, Schaeren-Wiemers N (2009) The septin cytoskeleton in myelinating glia. Mol Cell Neurosci 40(2):156–166

INDEX

Rolf Dermietzel (ed.), *The Cytoskeleton: Imaging, Isolation, and Interaction*, Neuromethods, vol. 79, DOI 10.1007/978-1-62703-266-7, © Springer Science+Business Media, LLC 2013

MIX
Papier aus verantwortungsvollen Quellen
Paper from responsible sources
FSC® C105338

If you have any concerns about our products,
you can contact us on
ProductSafety@springernature.com

In case Publisher is established outside the EU,
the EU authorized representative is:
Springer Nature Customer Service Center GmbH
Europaplatz 3, 69115 Heidelberg, Germany

Printed by Libri Plureos GmbH
in Hamburg, Germany